NATIVE PLANTS OF THE SOUTHEAST

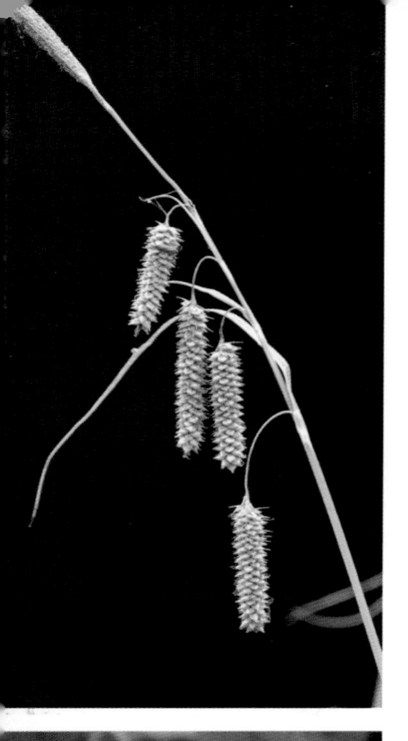

NATIVE PLANTS OF THE SOUTHEAST

A Comprehensive Guide to the Best 460 Species for the Garden

Larry Mellichamp

photographs by Will Stuart

Timber Press
Portland | London

To the North Carolina Native Plant Society, the second oldest native plant society in North America, which is committed to promoting education, appreciation, and preservation of our native flora, and which has provided generous financial support for the University of North Carolina at Charlotte Botanical Gardens public projects with native plants, and whose members have provided countless hours of camaraderie and information exchange at meetings and field trips;

and

To all native plant societies in the Southeast that seek to provide educational outreach to all people in support of understanding and enjoying native plants.

Published in 2014 by Timber Press, Inc.

The Haseltine Building
133 S.W. Second Avenue, Suite 450
Portland, Oregon 97204-3527
timberpress.com

6a Lonsdale Road
London NW6 6RD
timberpress.co.uk

Printed in China
Book design by Laken Wright
Layout and composition by Christi Payne

Library of Congress Cataloging-in-Publication Data

Mellichamp, Larry.
Native plants of the Southeast: a comprehensive guide to the best 460 species for the garden/Larry Mellichamp; photographs by Will Stuart.—1st ed.
p. cm.
Includes bibliographical references and index.
ISBN 978-1-60469-323-2
1. Native plants for cultivation—Southern States. 2. Native plant gardening—Southern States. I. Stuart, Will. II. Title.
SB439.24.S66M45 2014
635.0975—dc23 2013014425

A catalog record for this book is also available from the British Library.

Contents

Preface

How did I come to write a book about gardening with native plants in the Southeast?

When I was about ten years old something happened that changed my life, but I did not know it at the time, only in retrospect. One summer day my Great Aunt Annie let me pick a flower from a special patch in her garden that she said was just for me. What a shock, what a thrill. I did not really know what it meant. I remember picking the flower; I even remember what it was—a "wild" petunia, according to my aunt.

You don't realize what had just happened.

My aunt had a meticulous garden: a garden so well laid out in beds, so precisely managed through the year, so off-limits that no child (rarely an adult) would have a measurable half-life of time exploring the beds, so critical that only she could weed and water, so spotless that the bees didn't dare land on a petal for fear of leaving a foot print. No one could step off the path, touch anything, much less pick a flower. My heavens! I did not even know back then that flowers could be picked. My aunt had said that it was time I got to pick a flower.

Why did she let me do it? I don't know. Perhaps it was like a special coming-of-age ritual, an early taste of an initiation into the inner sanctum of adulthood. She knew that I liked coming to her house, but the best I could do was help mow the lawn and water the ancient mistletoe cactus on the porch (which I still grow a piece of). Did she have a scheme, did she set a trap, and did she know what would happen?

My aunt taught fourth grade in a small South Carolina town her whole career. I guess she knew kids. She let me pick that flower so I would become a botanist someday. That was it. I know it. (If not, it's a good explanation anyway.) And the trick worked, though it took another decade for the plan to be fulfilled.

This story, I think, points to a problem. City folks then, as now, did not have a personal relationship with plants and flowers, especially wildflowers and native plants. People hardly even noticed them, much less thought that there was any reason to do anything with them. And, if a plant was thought special, most folks still did not know how to get to know it personally.

My parents, for example, never knew a thing about a wild plant or the secrets of nature. They planted multiflora rose as a living screen because they saw an ad in a magazine, and they lined up privet as a hedge on our property line. They did not know anything important existed beyond the weedy buffer zone that separated the mowed lawn from the wilderness harboring snakes and poison ivy. I did not know any better than they did at that young age, though as a Boy Scout I was able to get out into the woods and get my first glimpse of the secrets of nature.

Back then, native animals had wild places to live and did not venture into yards to browse. Homeowners rarely saw a deer, and if they did, it was a kick. Canadian geese were rare birds in the Carolinas, and it was a marvel to see them at a local wildlife refuge. Such encounters were the limited connections most city folk had with the natural world.

Life has changed greatly since that time. Knowledge, if not experience, is more accessible because of television and the Internet. More people are simply aware of nature, the plights and pleasures of the wild kingdom. There is more told about animals to be sure, but plants get their time to shine—as well.

From the vantage point of my own little narrow view of the world, I think the new era started with Rachael Carson and her book *Silent Spring*, followed by Earth Day in 1970. Soon came the Endangered Species Act (1973), an increase in books about wildflowers and birds, more-frequent gardening shows on television, and the rise to prominence of botanical gardens and nature centers. Before long people were talking about global warming and climate change. Little by little, awareness increased regarding the importance of the environment and the role of other organisms that share the planet. Perhaps the NASA space program helped stimulate our interest in the blue planet as described from space, that it is finite and we need to take better care of it.

When I picked that flower in my aunt's garden, the structure of DNA was hardly known, the details about the web of life were not common knowledge, and I had been taught very little about nature lore. Picking that flower did not come too soon to prepare me for the events to come.

I went on to graduate school at the University of Michigan (finishing in 1976) and became a botanist. I loved teaching, was very fortunate to become indoctrinated into the realm of field botany, and soon discovered it was to become my passion. No one talked too much about growing native plants at that time. However, I met Fred Case in Saginaw, Michigan, in 1971. He studied native plants and was an avid wildflower grower, and he opened the world of native plant cultivation to me. In 1983, the first conference on landscaping with native plants was held in Cullowhee, North Carolina, and this was the beginning of a completely new chapter in the popular movement for everyday people to become more knowledgeable about the role of nature in their lives. The dogma was to plant more natives and try to break the long-time practice of the overuse of (invasive) exotic plants in home and roadside landscapes.

Those early founders of what came to be known as the Cullowhee Conference were certainly visionaries, and I thank them every day for what they started. Gardening became for me a useful distraction from the daily grind, as well as a creative and challenging endeavor to see what I could acquire and grow.

I was fortunate to see two amazing Southeastern wildflower gardens, which lead to my early realization that this was an interesting preoccupation. One was at the mountain home of Tom and Bruce Shinn in Leicester, North Carolina, and the other was the renowned native garden of Emily Allen in Winston-Salem, North Carolina. The encouragement of these gardens and gardeners was profound in my life.

No one influenced me more in understanding the role of plants in the lives of people than Ritchie Bell, botanist, environmental activist, and first director of the North Carolina Botanical Garden in Chapel Hill. Likewise, Ken Moore, who was assistant director at that time, constantly harped on the importance of using native plants. Jim Matthews, my undergraduate mentor at the University of North Carolina at Charlotte, took me out on countless field trips to collect scientific specimens and study plants all across the region. What a thrill. These trips provided a solid foundation in the principles of plant identification.

When I returned to UNC Charlotte as director of the modest botanical gardens begun in 1966 by my other influential professor, Herbert Hechenbleikner, I realized that I had to devote a portion of the Gardens' effort and resources to growing and promoting native plants. Much of the advice I present here is based on intimate involvement with propagating and growing many of the species in the University's seven-acre native plant garden, The Van Landingham Glen.

One of the greatest successes of UNC Charlotte Botanical Gardens has been holding annual plant sales that specialize in providing hard-to-find locally grown native wildflowers, ferns, shrubs, and trees. Out of this experience and in response to a rising interest from local gardeners, the Gardens began offering a series of courses leading to a Certificate in Native Plant Studies. The response has been encouraging, and today the University is embarking on the development a new garden, one of the first ever at a public garden to demonstrate specifically the use of natives in the home landscape. By creating a relatable, interesting, and attractive garden that exclusively utilizes native plants, we hope to bridge a mental and physical gap between home landscapes and the natural landscapes that we have forgotten we are a part of. We hope the new garden will become a display that inspires and stimulates gardeners with a new awareness of the region's native plants.

This, then, is a brief version of how I came to write a book about native plants of the Southeast. From the beginning of this project, one of my goals was to complement the species accounts with photographs that capture that splendor. To this end, I enlisted a friend and colleague, Will Stuart, to provide images for this book. Will travelled to the far corners of the Southeast in search of beauty and has brought much of it back for you to see. I am proud that he was willing to become a part of this extensive book project, and I am thrilled that he captured so many birds and insects utilizing the plants.

Will's photographic journey began in the 1970s while he was teaching biology in upstate New York. To illustrate the richness and diversity of a mature maple-beech forest, he sought to capture the beauty of emerging spring ephemerals on 35-mm slides, which he shared with his students. Forty years later, the woodlot where he captured many of his early images is now another suburban housing tract.

Will relocated to North Carolina in 1997 and subsequently traveled to scores of natural areas and gardens throughout the Southeast, always looking for another native plant species to add to his life list. He went digital in 2004 and his first purchase was a Canon macro lens. Many of his photographs allow you an up-close, "bug's eye" view of a leaf, berry, or blossom. In recent years, his interests have expanded to include the birds and butterflies that depend upon the health of native plant communities. According to Will, knowing where to go and when to go are the best-kept secrets for successfully photographing native plants.

When asked about a favorite moment while working on this book, Will recalled an early July day on the Blue Ridge Parkway. He had staked out a location near Mount Pisgah where a colony of *Lilium superbum* was blossoming near trees festooned with *Aristolochia* vines. As the morning sunshine warmed the air, scores of pipevine swallowtails began to flock to the lilies, often two or three on a blossom. Will's challenge was to find a butterfly that was not coated with a mass of yellow pollen. And he found just the right photo.

In sum, the book that you hold in your hands is the result of my strong desire to demonstrate the beauty and diversity of native plants of southeastern United States. More than that, it is a guide to selecting and utilizing those plants in the home landscape, a trend that began, or accelerated, in the 1970s with the rise of the native plant movement. In this book, I want to entice the reader to take a closer look at native plants by presenting outstanding photos of selected species so painstakingly captured by my friend Will Stuart. My plan has been to give you enough information about the plants to help you decide how to use them; and if you already have native plants on your property, to be able to identify them and appreciate their special value. I encourage you to reconsider how you use a plant you already know, or to try to grow a new species that strikes your fancy. Get to know our Southeast native plants personally.

Introduction

This book is about Southeast native plants and how to select and use them in the home landscape. Most gardeners are well aware of the wealth of plant life to be found in the region, famous for its magnolias, azaleas, and blueberries, among others. What may come as a surprise, however, is how many Southeast native plants have become mainstream members of the gardening world, even if the plants can't always be found at the local garden center. At worst, the average gardener may not recognize the plants they are buying as American natives. At best, the plants are known and sought out because they are indicative of the region and because they tolerate the extremes of climate, namely, the cold winters of the mountains and the prolonged summer heat and humidity in the Piedmont and coastal plain. Gardeners will continue to celebrate these natives, use them in the best landscape situations, and enjoy them as much as any plant, without concern for their origins.

The region covered in this book is at once traditional, and then again arbitrary. The Southeast certainly includes Virginia, North Carolina, Tennessee, South Carolina, Georgia, Alabama, northern Florida, Mississippi, and that little bit of the toe of Louisiana east of the Mississippi River. It is arbitrary in that plant distributions are not based on state boundaries. The Mississippi River is a good western boundary, but some species cross over and others don't. Because of this arbitrariness, I was able to choose to include some species that have just barely crept into the region, such as ostrich fern in the extreme northern counties of Virginia, but had to exclude some great species because they do not cross the Mississippi River natively, such as Arkansas bluestar and giant coneflower.

Of course, I still can't include all the good native plants that I have met in 45 years of botanizing and gardening in the Carolinas and immediate surroundings. I apologize up front for leaving out your favorite species because I didn't think it is a good candidate in general for the Southeast, or because I have not met and grown it, or just because it was lower on the list before we ran out of room. I am especially aware that much of Florida has been given short shrift because so many of its unique plants come from very specialized habitats nestled down in warm Zone 9 and are simply unfamiliar to me. Actually, I include several wonderful plants that originate in Florida, such as Florida torreya and scarlet sage.

This book will help you make choices about using common and lesser-known natives in your home landscape. It is my goal to not only extoll the virtues of well-known plants, but also to introduce you to uncommon natives that should become more widely tried and tested. Some day they may become mainstream themselves.

What Is a Native Plant?

There is unnecessary controversy over the definition of a native plant. Simply put, a species is native to the Southeast if it was growing in the region before European settlement. The plant might have come by forces of nature or by the activities of animals and humans, but it lives and reproduces on its own and seems to fit in with the other native flora and fauna. Furthermore, a native plant is one that cannot be known to have come from an exotic place. For example, both oxeye daisy and Queen Anne's lace are thoroughly

Cardinal flower (*Lobelia cardinalis*) is one of many Southeast natives celebrated by gardeners everywhere.

naturalized in the Southeast, but we know that they are natives of Europe.

A problem with this definition may be that we cannot always know precisely what was here because no accurate lists exist from the early centuries. Certain species native to a limited region of North America, such as Osage-orange from Arkansas, Oklahoma, and Texas and catalpa from south-central Mississippi east just into Georgia, have been spread by early humans and have naturalized far beyond their original ranges.

Wind and migrating birds continually carry seeds about, and occasional plant establishments occur in what seem to be isolated regions. For example, Wright's cliff brake fern found a niche in North Carolina from its native desert home in the American Southwest, presumably by spores carried on the wind, and the scarlet hibiscus grows among other natives in a population in south-central North Carolina, far removed from its main Gulf Coast range. How the hibiscus got to North Carolina is a mystery, although it is widely cultivated in modern times. Do these species count as natives? I think they have to. They may seem oddly out of place, and we don't know when and how they got here, but it is reasonable for them to be here. There are many other examples of unusual distributions that are the result of climate changes and plant migrations over the past fifteen thousand years since the end of Pleistocene glaciation.

Furthermore, many useful medicinal and edible wild plants have been spread by modern humans and domesticated animals, making it sometimes difficult to know what is truly native in a region. Good cases there include coneflowers and goldenseal. Are they native in the Carolinas, or are they escaped and naturalized from their long history of being grown as medicinals? Since they are clearly native in adjacent regions, I believe we can count them as Southeast natives without worry.

It does not matter much to me whether a plant can be proven truly native to a specific region, as long as it came from somewhere reasonably close and fits in with the flora without being an obnoxious weed. In considering a suspect species, we would waste more time trying to prove its region of nativity (though this could be interesting detective work for some) and not enjoy it for what it brings. The fact is, by now, humans have disturbed so much native habitat and spread so many plants around in various activities including gardening, agriculture, military maneuvers, and simple family travel vacations, that it would not be surprising to find almost anything almost anywhere in an almost-suitable habitat. The environment would select against a truly unfit species from becoming established even it were adventive for a while. It is sad that certain native species, such as peppervine and groundsel tree, have infiltrated regions of North America where they have become weedy pests. This is a consequence of the dynamic nature of our sprawling society, though the vast majority of our worst weeds are introductions from other continents and hemispheres.

Why Grow Natives?

Much has been written about the benefits of growing native plants. Three advantages commonly mentioned for growing natives is that they are better adapted than exotics, they are unlikely to escape and become nuisances, and they provide food for native wildlife, while exotics do not. Let's look at these benefits more closely.

1. **Natives are better adapted than exotics.** If this were so, then we would be growing all natives around our homes. The popular landscape exotics are clearly better adapted because they have been selected for the tough conditions of the home site for many decades. They usually are grown quickly, planted crudely, and walked away from—and they have to survive. While they prefer to be treated better, they can be grown with minimal watering, fertilizing, spraying, and other maintenance.

   Just because a plant is native does not mean it can be easily established in cultivation. Some natives are very difficult to establish, even though they may live in your neighborhood. They might have very exacting requirements. Ebony spleenwort and running-cedar are examples of natives that rarely transplant well and are difficult to propagate.

   Other natives are easily grown, but may require extra light or watering to keep them performing the way you expect. Some common roadside wildflowers like orange-flowered butterfly-weed are not so easily established in a flower bed, particularly if you overwater to enhance the other kinds of perennials you may also grow. It is imperative that natives be placed where they perform best in order for you to appreciate their performance.

   In contrast to the hard-to-grow natives, some very rare natives have turned out to be easily grown in cultivation. Georgia aster, pale purple coneflower, and white wild indigo are examples of species rare in the wild, but good-doers in cultivation. The point is that it can work both ways. Some species are easier to establish than others, and you cannot say that a native is automatically better adapted outside its immediate region than an exotic, or vice versa (that an exotic is better).

A mockingbird samples the bright red fruits of winterberry (*Ilex verticillata*).

2. **Natives are unlikely to escape and become a nuisance.** This may be true in a natural situation, as in naturescaping, but is not necessarily true in all cases. In managed woodland where native trees and shrubs are grown and the forest is thinned to enhance the "good" specimens, seedlings of dogwoods, maples, oaks, and many other native species will become more abundant in the cleared space, to the point of requiring constant removal if you do not want them to fill up the forest with saplings. That is what they have evolved to do, and in a truly native setting they have to compete to get a foothold. You give them even a little advantage, and they will take it. However, they certainly are not going to leave the woodland and go become a pest in your sunny flower garden. In this way, they are different from invasive exotic plants.

   Many other exotic species do not spread at all. I rarely see natural reproduction from exotic junipers, magnolias, and others. I have seen only one *bona fide* example of kudzu spreading from seeds (it certainly takes over where it has been planted, however). I am not trying to defend the exotics. The list of exotics in cultivation that spread is much longer than the list of those that do not, but you can't say that natives are innocent in this regard. It could be that our definition of what is troublesome is the real issue. Some folks would not mind if dogwoods did take over the woods.
3. **Natives provide food for native birds and insects, while exotics do not.** There is no question that native animals rely much more on native plants than on exotics. The larvae of many butterflies and beetles, for example, will eat only native species. In this regard, there has been a decline in the abundance of certain native insects because their habitats and food plants have been lost to development.

   Douglas Tallamy's book *Bringing Nature Home* is a startling revelation of the damage we have done to natural plant-animals associations through disturbance and overuse of exotics in the home landscape. Tallamy enumerates many examples of plant species that native insects will eat, or will not eat, and shows that natives are overwhelmingly preferred. His book has created something of a mini firestorm, stimulating deep commitment in many folks to do something about the overplanting of exotics and try to reverse the decline in the diversity of native birds and insects. These creatures are intricately linked in the web of life, as are we all, and any disruption in those interactions causes the whole system to weaken.

The remedy, according to Tallamy, is to plant more natives, create islands of natural vegetation in a sea of urban uniformity, and let native birds and insects eat a little bit of our property. I have absolutely no problem with doing that.

The plight of birds and insects is not totally dire, however. Much as birds may relish eating the berries of natives like Eastern red cedar and American holly, they will eat nonnatives to some extent, as well. We have all witnessed a flock of robins or cedar waxwings converge on a nonnative Burford holly in February or March and eat every berry in a matter of minutes. Sometimes I wish the birds would leave the berries alone on native hollies a little longer for us to enjoy.

I also find that many common songbirds are quite content to eat at bird feeders (those are nonnative grains) and hummingbirds fight over the white-sugar water in our "fast-food" hummingbird feeders. We may not be doing these birds any favors. It would be better to incorporate more actual plants in our landscapes that are attractive to songbirds and hummingbirds and allow the birds to eat naturally. Cardinal flower, Eastern columbine, and Indian-pink are enticing for humans, and they look great in our gardens as well. To appeal to the seed eaters, plant coneflowers, sunflowers, and black-eyed Susans. I realize that we can't replace all of our window-level bird feeders with plants, but we can try to be as natural as possible.

So, then, why should we plant natives if they don't have all the benefits we have been led to believe they have? They *are* more difficult to find at nurseries. They often cost more than mass-produced exotics. Some native perennials take longer to grow and often don't look very nice all season, especially if they have been eaten up by native insects.

Well, the real reason is simply that they are beautiful plants with attractive flowers, foliage, and fruits. They are local plants with seasonal appeal, like local-grown foods. They are regional plants that convey a sense of place. And they are American plants that make us proud of our heritage.

Many underutilized natives are beautiful and interesting, and they should be grown more, but people don't know about them. They are not part of mainstream merchandising, partly because there has been no market demand for them in the past and they are more difficult to acquire and grow in large numbers. Nurseries have been content to grow what they have always grown, those species and cultivars that do best and make them the most money. Those nurseries that do grow unusual exotics spend a lot of time and money touting their virtues. Fewer nurseries point out the virtues of natives. Why don't more nurseries promote natives? In part, it is the old attitude that the more different, the more exotic (from far away) something is, the more wonderful and desirable it becomes. This rationale has been used to sell many products in the past.

We conveniently forget that the bad comes with the good. If it turns out that some natives don't work so well, we still must embark more strongly on a trajectory to understand and utilize our native plants if we want to see them rise to a higher stature or level of acceptance and desirability. Instead of saying things like, "Oh, that's just a wild plant from the woods," we should be saying more things like, "And yes, of course it's beautiful, it's one of our natives."

Part of the solution, then, is to seek out nurseries that carry native plants and buy from them. As more natives are sold, more will be grown. The price should come down, and new species and cultivars will be brought into cultivation for testing.

One of the reasons I grow natives in my landscape is because they remind me of their natural habitats with the associated plants, birds, mammals, and insects that also are part of a certain ecosystem. Native plants are part of a little community, all interacting. Never forget that a species in cultivation has a home somewhere in nature where it is adapted to a particular set of environmental conditions. The only reason we can grow it in our gardens is that it is adaptable enough to accommodate the less-than-perfect conditions that we provide. And for this ability we are eternally grateful. To have the beauty of a flower in our own back yard is a gift where nature shares her citizens, and we should be appreciative of the gesture.

What Habitats Are Found in the Southeast?

The Southeast is a very diverse region for plant habitats. It can be divided into physiographic provinces based on topography, soil, and bedrock type. These are the regions that I will use when giving the natural ranges of species.

The **Coastal Plain Province** is fairly uniform in its flatness, with well-drained sandy-loam soils stretching along

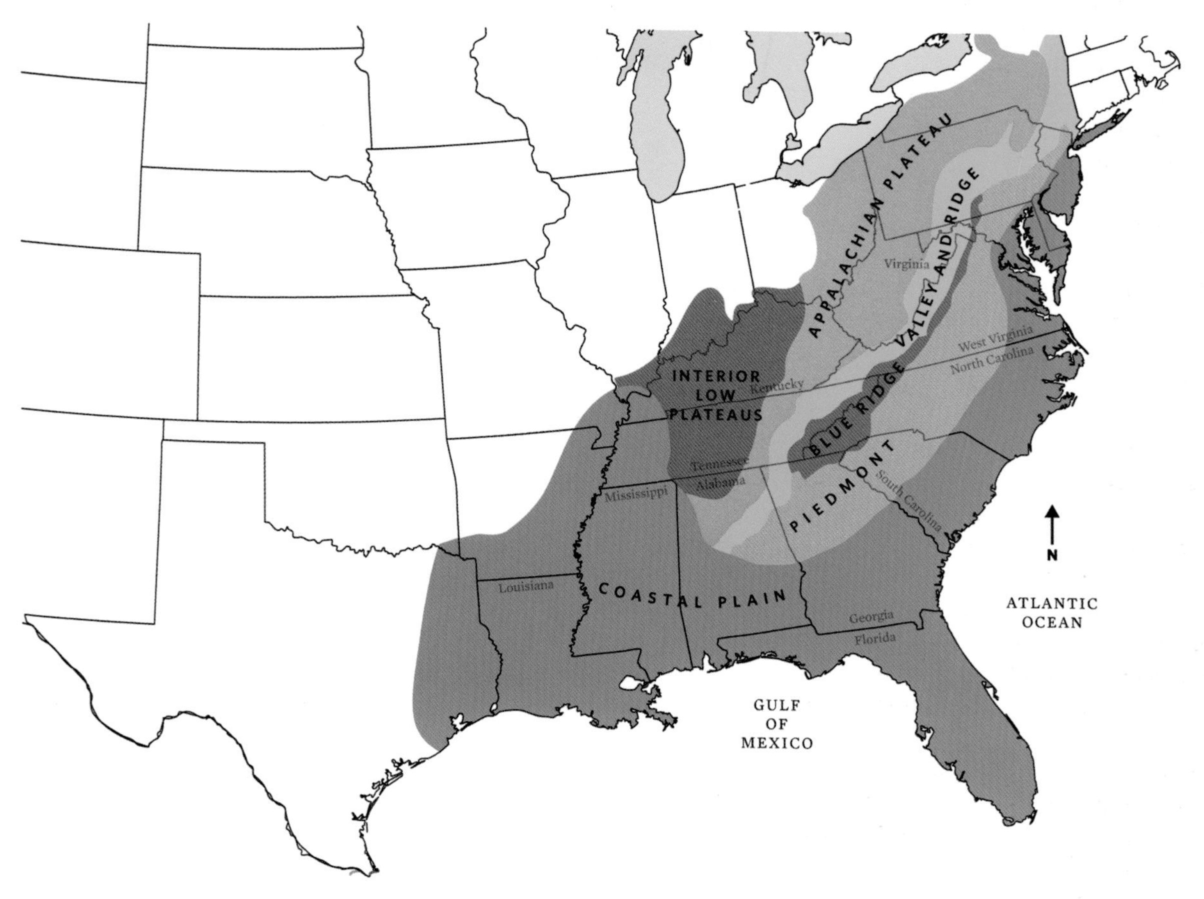

Physiographic regions of the Southeast.

the coastline and up the Mississippi corridor. The Sandhills Region of this province is a narrow belt just inland where deep white sand accumulated along an ancient shoreline.

The **Piedmont Province**, literally "foot of the mountains," is a wide region of rolling hills and small leftover remnants of worn-down mountains. The soils are more clay based.

The **Blue Ridge Mountains** are ancient formations of mostly granitic rocks, eroded and rounded over 250 million years. They have never been glaciated nor flooded. Changes in elevation and north-to-south slope exposures provide many diverse habitats for plants in a cool, moist climate.

The regions in Tennessee and Northern Alabama are underlain by limestone rocks, and the **Valley and Ridge Province**, the **Cumberland Plateau Province**, and the **Interior Low Plateaus Province** provide diverse habitats much like the Piedmont, but with rich soils of a less acidic nature.

There are, of course, gaps in distributions, and plants can occur in far-flung (disjunct) locations without obvious reason. These are the interesting cases that botanists like to study.

A major factor that helps explain plant distributions in the Southeast is the effect of Pleistocene glaciation. The glacial period ended some ten to twelve thousand years ago, and although glaciers never came into the Southeast, the climate was much colder then, and most species migrated southward to seek habitats that were more equitable. Many species remained in the mountains and simply survived, while others went extinct and only their relatives survive today in China and Japan.

Within the physiographic provinces of the Southeast are recognizable habitats. We might at first glance simply refer to them as fields, forests, swamps, and marshes, and that would be a reasonable generality. We can break them down a little further by designating more specific habitat types based on dominant trees (or lack thereof) reflecting degrees of moisture and soil type. Two native plant guides that specifically give more details of habitats and their characteristic species are *Wildflowers and Plant Communities of the Southern Appalachian Mountains and Piedmont* by Timothy Spira and *A Guide to the Wildflowers of South Carolina* by Richard Porcher and Douglas Rayner.

Along the southernmost Blue Ridge Parkway is an extensive area known as Graveyard Fields. Devoid of forest, it offers a bountiful array of shrubs and wildflowers, with especially fiery fall colors.

MOUNTAIN COMMUNITIES

At the highest elevations in the Blue Ridge Mountains, above 4500 feet, is **spruce-fir** forest. Here red spruce (*Picea rubens*) and Fraser fir (*Abies fraseri*) dominate the canopy. They grow mixed with yellow birch, mountain ash, smooth serviceberry, red-fruited elderberry, Catawba rhododendron, and Allegheny blackberry. Herbaceous plants such as southern lady fern, pink-flowered turtleheads, Canada mayflower, and painted trillium grow in the layer under the tree and shrub canopy.

Above the spruce-fir forest level, at 6000 feet, near the summit of Mount Mitchell and the high peaks of Roan Mountain, for example, the forest is pure Fraser fir with these other species in open areas. The climate is harsh, but the views are spectacular. The soil is always moist, acidic, and organic. Few, if any, of these high mountain species can thrive south of the cooler portions of Zone 7 in gardens. When visiting these places, note the spectacular displays of red berries on mountain holly, red-fruited elderberry, and especially mountain ash.

TOP Spruce and fir species dominate the Blue Ridge Mountain forests.
BOTTOM Rhododendrons along a mountain stream.

At these high elevations are two habitats of great interest in the southern mountains: grassy balds and heath balds. It is not known with certainty how they formed: they may be holdovers from glacial times when alpine meadows formed here. They may also have been heavily grazed as early herders of sheep, cattle, and other livestock moved their flocks according to the changing seasons. **Grassy balds** are treeless areas dominated by grasses and sedges, some herbaceous wildflowers, and a few scattered patches of trees and shrubs, especially flame, swamp, and sweet azaleas. **Heath balds** are dominated by Catawba rhododendron, blueberries, mountain fetterbush, and other members of the heath family. The soils are thin, moist, and very acidic.

Below 4500 feet, the woodlands are dominated by northern red oak, along with yellow birch, striped maple, chestnut oak, and occasional stump sprouts of the American chestnut. These woods can be rich with shrubs and wildflowers, especially members of the heath family, alternate-leaf dogwood, and witchhazel, along with ferns such as southern lady, hay scented, New York, and shining clubmoss; and certain wildflowers like wake-robin, white snakeroot, leather flower, wild yam, bigleaf aster, whorled loosestrife, and bee balm. Here again, while many of these species grow at lower elevations, most plants from these higher realms need cool summers to thrive in the lowlands.

Heartleaf foamflower (*Tiarella cordifolia*) in Piedmont woods.

Pitcherplant (*Sarracenia* spp.) bog community.

At 1500–3000 feet are the richest **mixed hardwood forests** and **cove forests**, especially well developed on north-facing slopes where small streams cut a wide V-shaped cleft in the mountain. Here the unmatched rich assemblage of trees encompasses white ash, basswood, sweet birch, yellow buckeye, black cherry, Canada hemlock, Fraser magnolia, sugar maple, tulip-poplar, white oak, red oak, white pine, silverbell, and sourwood. Hickory and beech occur in some spots that are drier and more acidic. Shrubs in this community include mountain holly, wild hydrangea, striped maple, pawpaw, sweet shrub, and spicebush. The climate is moist and moderate, and the soil is rich and organic. Ferns are lush, consisting of broad-beech, southern brittle, interrupted, wood-ferns, rattlesnake, silvery glade, and northern maidenhair fern. The forest floor is carpeted with favorite ephemeral wildflowers like bloodroot, Dutchman's-britches, foamflower, wild geranium, dwarf crested iris, yellow lady's-slipper orchid, phlox, trilliums, windflower, and many, many more.

In the mountains, the pines (like pitch, table mountain, and scrub), Carolina hemlock, and dwarf juniper are found on drier slopes, ridges, and rock outcrops, especially where fire has burned. White pine and Canada hemlock inhabit the moister coves and slopes because they can reproduce in the shade.

Throughout the mountains are exposed rock outcrops where certain plants grow in crevices and on thin soils. Where the soil is a little thicker, a shrubby meadow will develop, full of pink-shell azaleas, blueberries, huckleberries, Carolina rhododendrons, red chokeberry, red maple, wild-raisin, serviceberry, and St.-John's-worts. These often turn spectacular fall colors promoted by the highly acidic soils and full-sun exposure. The species from this lower elevation are much easier to cultivate in Zone 7 and the cooler parts of Zone 8.

PIEDMONT COMMUNITIES

The elevation of the Piedmont (or Valley and Ridge and Interior Low Plateaus provinces in Tennessee) slowly declines from 1500 feet to the coastal plain, and **mixed forests** comprise oaks, hickories, and pines. These sites are usually drier than those in the mountain communities, with less organic soils, and fewer species.

Coral-bean (*Erythrina herbacea*) with black vented oriole.

Virtually all the forests in the Piedmont have developed since extensive cotton farming was abandoned in the 1930s. The forests are dominated by red, white, black, chestnut, post, and southern red oaks, along with white ash, basswood, black gum, several species of hickory, persimmon, red maple, southern sugar maple, sassafras, and tulip-poplar. Growing among the hardwoods are shortleaf, scrub, and some loblolly pines. Red cedars persist, though do not reproduce, from the earlier days when they came into an old field within ten years of farming abandonment and dominated for the next twenty years or so. Wildflowers can be common in a lowland forest or floodplain, but are sparse on drier slopes. Dense populations of Virginia spring beauty, bloodroot, wild geranium, Wherry's foamflower, green-and-gold, dwarf iris, and Christmas fern are readily found in the region, but are not nearly as lush as in the moist mountain forests. The famous American chestnut once dominated many of these forests, as in the mountains, but has become virtually extinct as large trees due to the chestnut blight (beginning in 1904) and remains only as frequently encountered stump sprouts that are kept from developing into trees by the persistent fungus.

COASTAL COMMUNITIES

Moving towards the coastal plain and Atlantic Ocean, you pass through a narrow stretch called the **Sandhills**. In this region, the soils are dominated by deep, white sand, remnants of ancient coastline sand dunes. The heat from the sun reflecting off the sand creates a harsh and bright environment; and the well-draining sandy soils offer little in the way of moisture-retention during drought. The plants of this region must be adapted to severely hot, dry conditions. Fire is often a factor in determining the mosaic of vegetation across the landscape.

In places where some organic matter has accumulated and a bit more moisture exists, such as in swamps near creeks and rivers and on rocky bluffs above rivers at the Fall Line (the distinct break between the coastal plain and the Piedmont), the slopes are home to well-developed hardwood forests, as in the Piedmont. Otherwise, the landscape is dominated by loblolly and longleaf pine. Ironically, these same pines can also grow well in swamplands and wet saturated soils along with bald cypress and pond cypress in the wettest portions of this region.

Rich assemblages of shrubs and wildflowers occur in the sandhills, in both open, very sandy soils and more shady sites in pine forests. These plants need a specialized site in cultivation to thrive, namely, sandy and very well drained soil. A rock garden or raised bed can help create such a site in areas where it is not natural. In northern Florida, however, where sandy and well-drained soil is typical, many interesting plants are found, such as goat's-rue, tree blueberry, prickly-pear cactus, Barbara's-buttons, dwarf hawthorns, wild plums, sundial lupine, staggerbush, wax-myrtle, red sage, and yucca.

The outer coastal plain presents a spectrum of habitats of great interest and diversity. Sunny, open meadows commonly referred to as **pitcherplant bogs** begin to appear. These low-nutrient sites have sandy-peaty soils that stay permanently wet and are burned by fire every five years or so. They may be open grassy meadows, or they can be populated by scattered longleaf pines. They may be variously referred to as savannas or pine flatwoods, depending on how thick the pines are. Long leaf pine is the main pine, and is adapted to periodic fire because the very long needles (up to 18 inches) allow the quick fires to burn over their tops and preserve the growing tip hidden down in the base. You will see stands of various aged saplings along roadsides and the edges of the wetter boggy ground.

Pitcherplant bogs have been determined to be the most diverse habitats in the temperate zone, with up to 40 species of vascular plants (not counting mosses and algae) per square meter. This high diversity is primarily due to the harshness of the full-sun environment, lack of nutrients, and periodic fires that keep any one species from dominating any one site. The chapter on bog plants enumerates many of the more desirable species that will thrive in a wet sandy-peaty situation. You do not have to burn your bog garden, but you must keep each species under control, as some will aggressively colonize a cultivated bog garden.

The pitcherplant bog habitat may blend sharply or gradually into the **pocosin** (swamp-on-a-hill) habitat in many places. These wetlands are found in Atlantic coastal regions where a slightly elevated wet depression is filled with thick woody vegetation. Very few herbaceous plants live there. A well-formed pocosin can be so dense that you would need a boardwalk cut through the vegetation to be able to easily traverse it. Some twenty species of woody plants may be typically found in a pocosin, many of them evergreen, including loblolly, longleaf, and pond pines, Atlantic white cedar, red bay, loblolly-bay, shining fetterbush, honey-cups, titi, wax-myrtle, inkberry holly, gallberry holly, pond cypress, service berry, sweet-bay magnolia, and laurel-leaf catbrier, one of the most vicious of spiny vines.

The sites with the taller and denser vegetation can be termed **high pocosins**. On slightly drier ground moving away from the high pocosin you can transition into a *short* (or shrubby) **pocosin** lacking dense tall vegetation, where many of the most wonderful ornamental shrubs are found. These include dwarf fothergilla, coastal sweet-pepperbush, Virginia-willow, inkberry holly, myrtle-leaved holly, honey-cups, shining fetterbush, dwarf wax-myrtle, coastal azalea, dwarf serviceberry, creeping blueberry and southern sheep-kill laurel.

The short pocosin can transition into a sunnier and wetter pitcherplant bog, or a drier pine flatwoods. The habitats of the world-famous Venus flytrap and pixie-moss are in these moist-to-wet transition zones (or ecozones) between sunnier and shadier, wetter and drier habitats. These very desirable plants are usually found abundantly only in perfect habitats that are limited in extent. The coastal region, more so than most areas, is a mosaic of habitats depending on soil type, moisture, nutrition, and fire frequency. The diversity of plants is astonishing, and the display of flowers and colorful fruits is amazing.

Closer to the coastline, the landscape is dominated by wetlands, both freshwater from rivers and natural lakes called bay lakes, and brackish water from tidal activity right on the Intracoastal waterway, small creeks, and river mouths. A **swamp** is one type of wetland, usually with standing water, sometimes 2–3 or more feet deep after heavy rains, dominated by trees, shrubs, and woody vines. Swamps in the Southeast are home to shrubby buttonbush, Virginia-willow, climbing hydrangeas, and smilax vines. Bald cypress can dominate a swamp, but may be joined by red maple, pumpkin ash, water hickory, water tupelo, swamp black gum, swamp chestnut oak, water oak, overcup oak, and sycamore.

The second type of wetland, a **marsh**, is a wet habitat with no trees, dominated by herbaceous plants. Freshwater marshes can be diverse habitats for nonwoody species (see the chapter on aquatics). Here, standing water provides a habitat where plants never have to worry about drying out, and they compete vigorously for the lush nutrients brought in by runoff from the land and floodwaters from upstream. **Brackish marshes** are salty to some degree, influenced by ocean water and tides, and are much less diverse, even to the point of being dominated by one or two species of grass or grasslike plants covering many acres as you get closer to the ocean. The nutrient-rich water supports sea and bird

TOP Managed burn in a pocosin.
BOTTOM Bald cypress (*Taxodium distichum*) swamp.

Coastal sand dunes with sea oats (*Uniola paniculata*).

life, but not many flowering plants are found there, and these would not be easy to grow in cultivation. One exception is seashore mallow, an excellent herbaceous plant for the garden.

The coastal woody plant community closest to the Atlantic Ocean, the **maritime forest**, overlooks the brackish marshes and the ocean and is dominated by live oak. Other tree species can grow there, such as red bay, loblolly pine, red buckeye, wild olive, or devilwood and yaupon holly. There are not many wildflowers in the shady maritime forest due to the evergreen nature of the live oak canopy and the very sandy well-drained soil. One of the most famous shrubs from this habitat is yaupon holly. It has small, not-so-spiny leaves, beautiful red berries, and is very adaptable in cultivation. The maritime forest accepts the full brunt of hurricane and storm winds, and must be resilient. When the forest is disturbed to build houses, it usually creates an opening in the wind-swept canopy triggering the beginning of decline.

This forest forms on very old, stable sand dunes along the shore, with open **dunes** toward the ocean where the sand shifts. Here it is stabilized by sea oats and other sand-binding grasses. Between the ridges of sand in the open sunlight, but somewhat protected from the wind, are blanket-flower, yuccas, prickly-pear cacti, and a most remarkable coastal species, coral-bean. The striking red-flowered perennial coral-bean is unusual in that it can be strictly herbaceous in the north of its range (North Carolina) and become woodier and even treelike as you go south into Florida and west to Texas. I have treated it in the chapter on shrubs. It requires almost pure sand to thrive and can bloom all summer on elongating flower stalks.

How Do Natives Get into Cultivation?

Native plants grow all around us. We see them along roadsides, where they may form conspicuous yellow stands of goldenrods in the fall best viewed as impressions of color, or on walks and hikes, where we can perform intimate analysis of the flowers of a tiny spring beauty or peer into the throat of a conspicuous turtlehead. We can see them at local botanical gardens or private homes and marvel at their beauty and diversity. At some point, we have a desire to have some of them in our gardens, for various reasons.

Acquiring a wild plant can be as simple as digging up a root-ball from a ditchbank, lifting a Christmas fern from roadside woodland, or getting a division from a friend who has something we want in their flower bed. Some folks would think nothing of doing this as their fancy strikes them, whether they take from private or public lands, and whether the plant is common or rare, easy or hard to grow. So, let me say right off, it is not proper to dig a wild plant without the permission of the landowner. That could be difficult to obtain if you are out in a remote hunt-club "wilderness," unless members were given blanket permission to collect seeds prior to setting out on the excursion.

Other folks would say it is improper to dig any plant from the wild or collect its seeds, even with permission. They would wait for the plant in question to naturally disperse itself into their yard and find a suitable home. However, if it were that easy, those plants would be there already.

Between these two extremes is a happy medium whereby wild plants can come into cultivation and be available to everyone. Let me quickly point out that for every plant in cultivation, native or exotic, someone had to bring that plant into cultivation in some formal way. They first would have to notice the plant and want to grow it, then either dig it up or find a seedling nearby. They would transport the plant in a suitable manner, replant it into a suitable location, water it for several days, and hover over it to monitor its condition much like a stay in an intensive care unit after major surgery. After all, you have greatly disturbed an established plant, and you don't know how much damage you have done to the root system or how easily it will recover in its new situation.

Transplantation can be done with a high degree of success if you are experienced. There are professional plant hunters who go about this, but more often, it is amateur naturalists who have learned by experience to do the right thing when collecting plants. These individuals have developed a keen eye to spot something unusual in the wild, such as a nontypical growth form, a more floriferous specimen, or a leaf with more color markings. Even a nearby seedling may not be the same, nor would you necessarily be able to gather seeds at that moment, and the collector would have to take the exact plant in question.

One of the most common variants we see in nature is the pure white, or albino, form of an otherwise colorful wildflower. These plants stand out easily, and we seem to be attracted to them. There is probably a white variant of every wildflower. Some are in commercial production, such as *Iris cristata* 'Tennessee White'. Other common examples are doubled flowers, those that are conspicuously larger or smaller than typical, and those with extra petals.

In general, it is always better to collect just seeds, even if you have to mark the plant when it is in bloom and come back when the seeds are ripe. With seeds, you do not have to worry about survival of a growing plant during transport. More importantly, in growing several plants from seeds, you would be able to select the plants that survive under cultivation conditions, and these would be a notch better adapted than if you dug young seedlings from the wild. It is analogous to the notion that baby animals born in the zoo, who never knew the wild, make better adjusted zoo citizens than mature animals dragged in off the range and put in a pen. The latter would truly suffer from the traumatic change.

Professional horticulturists and naturalists working for botanical gardens, nature centers, and arboretums worldwide collect seeds with the proper permits. They know what they are doing and their likely prospects for success. They know when to proceed and when not to bother. They have suitable facilities back home for receiving and processing the living material, often accompanied by dried specimens of the plants as vouchers for proper identification. These samples will become scientifically valuable museum specimens and should be preserved in an herbarium (pressed and dried plant collections for scientific study) with collection locality records.

In 2011, I was part of a professional expedition with a joint team organized by the Mount Cuba Center in Hockessin, Delaware, and the Edinburgh (Scotland) Botanical Garden. The team went into the mountains of North Carolina and Georgia and collected seeds and plant parts (bulbs, rhizomes, and cuttings) from more than 200 species. The samples went back to the two institutions for processing, propagation, and evaluation. A certain percentage, hopefully high, will survive to become available for display and study, and likely propagation for future generations of cultivated stock.

Nursery owners also bring unusual plants into cultivation and grow them for testing and propagation. Plants that have been selected and propagated from whatever source can be taken to an appropriate wholesale grower for mass production, and then small plants are sold to retail outlets for public sales.

There is nothing wrong with interested and knowledgeable amateur individuals being able to acquire, grow, breed, select, and propagate native plants at home. They could do all of the same things professionals might do, in some cases even better if they have more time and a keen eye. Nonprofessionals are limited only by their facilities and perhaps by lack of contacts to have their plants brought into general distribution. Many great plants have been developed this way, and the propagator gets to name the selections. I hope to do more of this when I retire.

Albino form of the usually blue-purple dwarf iris (*Iris verna*).

Another way native plants get into cultivation is by chance. Someone spots an unusual plant variant in a private garden or in a nursery or botanical garden where many plants are grown from seeds for evaluation or where plants self-sow in a naturalized population. The unusual plant is propagated and given a cultivar name. Many cultivar names honor the person in whose garden the original plant was spotted. This process of selecting and naming is very important as it brings some of the very best plants available into the public market place.

If you want to grow only wild-type specimens, unchanged from the wild, and this is an excellent way to go, you just need to start with two or three original seed-grown specimens and let them cross-pollinate and start a population of seedlings. As the population grows, seedlings will find their favorite niches in your garden. Someday you might find a special "winner" among the offspring. The special named cultivars will not breed true to their type from seed and must be asexually propagated. Many are self-sterile and will make no seed at all.

To those who say it is improper to collect wild plants, I avow that without such activity, no good plants would be brought into cultivation. This kind of work has been going on for centuries. It is a take-off from the early herbalists who brought in wild plants for medicinal purposes and grew them nearby for ready access. Today we have countless licensed herb collectors who bring in wild plants for use by drug companies and nurseries; bringing in a wild plant to grow for ornamental purposes is no crime, and must be done if we are to have new plants for the garden coming along. However, it must be done in the proper manner so plants are not improperly treated and wasted. The point is to get new material from the wild for propagation. Wild collecting for direct resale is not okay. Even if they were rescued from imminent destruction, they may not be readily adaptable to transplanting and the buyer will lose them and blame you for selling him a bad plant. Wild plants should usually be propagated and established before resale to the public. Of course, if you are collecting rare and endangered species for horticultural purposes, you must have the proper permits from your state plant protection department, and indicate to them your purposes for working with rare plants.

It has always been very popular for native plant societies to sponsor plant rescues. These are local outings where a group will go with shovels and spades and dig desirable plants from a site that is to be developed and destroyed. In this case, permission would have to be granted from the developer, and sometimes a waiver of liability must be signed by each participant so the owner is not held responsible for an accident. The pros of this practice are that you get to save common (or uncommon) plants from sure destruction, and you get some "free" plants for your efforts. The cons are that you are digging plants you may not know much about or even if you really want them. What is worse is that plants dug from the wild, especially at the wrong time of year, may not transplant well and you have wasted your efforts. You have to learn how deep to dig, how much soil to take, how to de-select the undesirables, and to get the target specimens moved into their new homes in a timely manner. This is all part of an educational endeavor undertaken by various organizations and is well worth the loss of a few wild plants that would be gone anyway.

When a new plant is brought into cultivation, it should be given a cultivar name, and propagated asexually to create a batch of identical plants that can be trialed and tested for suitableness in a garden setting. Cultivar is a horticultural term meaning "cultivated variety." It describes plants grown strictly under the watchful eye of humans (not in the wild) and comes with rules about how the name should be formed. Cultivars can come from the wild or be found as a variant in a garden. For a cultivar name to be valid, it must be described and registered with the appropriate institution that keeps records on such cultivars of specific groups of plants. In this way, people can learn the history of a cultivar. For example, the International Carnivorous Plant Society keeps a listing of new cultivars of all carnivorous species in various genera, and publishes them periodically in their quarterly Carnivorous Plant Newsletter. A cultivar name will usually be in the native tongue of the originator (not in Latin as in scientific names) and be set in single quotes in formal writing.

There is a strong trend towards naming more and more cultivars, especially if they are more unusual than the typical "wild type." Cultivars are not necessarily hybrids, though they can be. Cultivars have specific characteristics that make them desirable, and they will usually display that trait year after year in the garden. While a particular cultivar may be derived from a wild plant, it is not necessarily representative of the wild populations of plants. Indeed, a cultivar is usually quite distinct from its typical wild relatives. It is still derived from a native, and may sexually reproduce and produce progeny similar to its wild relatives. In my view, such plants should be allowed in native plant collections.

Established cultivars may be artificially hybridized with each other, or with nonnative species to create new and better garden plants. These may be more problematic to accept if you are a native purist. For example, the new smoke tree cultivar 'Grace' (*Cotinus* 'Grace') is a hybrid between our native *C. obovatus* and the European smoke bush (*C. coggygria*). I recommend it as a more compact smoke tree specimen.

Phlox stolonifera 'Bruce's White' is a good example of a white form of the wild blue creeping phlox, found originally by Bruce Shinn, and named after her.

How Do We Incorporate Natives in Our Landscapes?

There are many ways to utilize natives in the home landscape. I will divide this discussion into two parts: traditional landscaping and naturescaping.

TRADITIONAL LANDSCAPING

Traditional landscaping uses certain widely available plants from mainstream nurseries, planted in the same way no matter what region, to give an effect in keeping with general design principles. Examples of plants used in this way are Japanese and Chinese hollies, junipers, azaleas, privet, cleyera, abelia, and boxwoods. This approach is understandable in that many people do not want to spend much time with their home landscape, checking out the options, making decisions on types of plants, worrying about what will grow and what will be low maintenance. They want the simplest, cheapest solution that will keep them in good standing with the neighbors and the homeowner's association. They may truly believe that what they see in the typical home landscape is the only way to landscape a yard. If the soil is not right for "traditional" plants, the homeowner changes the soil. If problems arise and the plants don't grow well, the homeowner applies chemical fertilizer, pesticides, soil amendments, and extra water in an attempt to "fix" the situation. In short, in traditional landscaping there is an agribusiness chemical solution for every circumstance.

The worst aspect of this scenario is the constant need to prune the plants around the house to keep them from becoming too large. This problem arises because the plants that are generally sold as being well adapted are fast-growing species that produce a salable (and inexpensive) plant in a short time. In the long term, the constant need to prune is bad for the plant and the homeowner. I suggest that you make plant choices such that you never have to prune anything routinely unless you are shearing a hedge or want to create ornamental topiaries or espaliers.

Using the same handful of traditional landscape plants repeatedly leads to a decrease in diversity. All landscapes come to look somewhat the same, and they change little through the seasons. If there were to be a new disease of, say, Burford hollies, our established Southeastern landscapes would lose a tremendous number of plants. Twenty years ago a situation exactly like this occurred in the Southeast with the almost total loss of the widely planted "miracle shrub" known as red-tip photinia due to a fungal disease. People are also recognizing the severe shortcoming of the extremely popular and widespread flowering Bradford pear trees that break apart in ice storms as they age. Of course, this is not to say that there aren't some wonderful nonnative landscape plants. Where would we be in the Southeast without the summer-flowering crepe-myrtles, the winter-blooming camellias, and the seemingly infinite array of gorgeous spring-flowering evergreen azaleas?

In a traditional landscape, the lawn is an indispensable feature. Now, lawns can be beautiful, rewarding, and functional, but the fact is lawns are the most unnatural of settings for plants. No place in nature harbors a massive area of grass that is uniformly grazed to 2 inches high year-around, routinely overwatered and overfertilized, poked full of holes and overseeded twice a year. Even the American prairies and the African savannas were not that uniform. Because they are so unnatural, lawns require much care—establishing, watering, fertilizing, mowing, raking, pest control, repairing. Then, after all that, you can have disfiguring attacks from insects and fungus during hot, humid summer weather. Yet, the lawn is the all-American home feature, the playground for pets and kids, a beautiful expanse to set off the home and grounds, an open space to make you feel proud of your property . . . and a possession (read "obsession") that requires more time, energy, and money than all your other outdoor amenities combined. We may not be able to dispense with the American lawn (what would guys do on weekends), but we can convert some of it, little by little, to a mulched woodland natural area, creative mixed border, interesting shrub collection, or colorful perennial bed. And we can choose to replace conventional plants with native plants.

In the home landscape, the traditional specific functional uses for woody plants are as foundations, specimens, accents, borders, hedges, privacy screening, groundcovers, and shade trees (see box). These uses should be understood so that you can make a sketch of your property and indicate where you want shade, a privacy hedge, a grove of spring flowering shrubs, or an all-around perfect specimen that commands central attention. In making these decisions, you are carrying out a design plan, and it requires knowledge of the plants and their characteristics, environmental requirements, and maintenance needs.

Many simple books will help you work out a design plan and make some choices from the wide array of common plants found at nurseries, garden centers, and big-box stores. Only a few publications highlight the virtues of native plants in a systematic way. No matter what types of plants you are looking for, when you go to the garden center and ask for advice on what to plant, be prepared to address the following questions:

What function do you want the plant to play in your landscape?
What do you want the plant to look like?
Will the plant receive shade or sun in the proposed site?
What is your soil like—clay, sand, wet, dry?
How large can the plant get without pruning? (How much space do you have?)
Do you want to have to prune the plant?
Are you looking for flowers and colorful fruits, or fall color?
How much money do you want to spend? (Will you start with larger or smaller specimens?)

In a way, purchasing a plant is like buying a piece of furniture, an object of art, or even a pet. You want to get the right fit and feel for your interests and temperament.

To utilize natives in traditional landscaping, you would look for a native that fits the traditional use, and evaluate it to see if all of its traits are satisfactory in that capacity. The difference would be that you might have a harder time finding a good native species as they are not as often carried by traditional nurseries, just as you may have more difficulty finding a handmade piece of furniture from a specialized artisan.

NATURESCAPING

Naturescaping differs from traditional landscaping in that it utilizes plants that are native to the site of the immediate region. These plants can be found growing wild in your county or province. For example, to re-create a moist oak-hickory woods community found commonly in the Piedmont region of the Southeast, the broad palette of plants would include half a dozen native azaleas, the same number of viburnums, several deciduous hollies, and chalk

FUNCTIONAL USES OF PLANTS

The examples given here are of traditional exotic species because these forms will be most familiar to most people. Finding a native to substitute is the modern challenge

1. **Foundation plants** are evergreens that fit near the house to give the foundation a context, to hide the ground, to soften the corners, and to border the windows. In common practice, these are often boxwoods, hollies, privets, tea-olives, abelias, junipers, and evergreen azaleas. All of these must be pruned several times a year. Usually this involves shearing them into round meatballs or square boxes. Often overgrown specimens are overpruned to tame them or to delay the need to prune so often, but such drastic pruning usually leads to decline in plant health and appearance. I suggest that you look for native and exotic plants that do not have to be routinely pruned. There are some examples of both available to help you achieve this end; they are slower growing and can cost more initially.

2. **Specimen plants** stand alone in the landscape to show off their superior appearance. Generally, they offer beauty for more than one season and improve with age. Traditional plants used for this purpose are small flowering or foliage trees, such as Japanese maples, deciduous magnolias, flowering dogwoods, cinnamon-bark crepe-myrtles, or weeping cherry.

3. **Accent plants** have one or more better-than-average traits, such as showy flowers or colorful berries, and are generally used in mixed borders or groupings where you want some to stand out from time to time. Traditionally trees and shrubs used for this purpose are camellias, special azaleas, firethorn, viburnum, and hydrangea. For variety, maybe a gardenia or nandina could be thrown into the mix.

4. **Border plants** are small shrubs usually planted along a straight or curving edge of a property to act as a showplace for seasonal interest and a demarcation and even screening of the site. A mixed border has plants of similar size and growth rate, such as azaleas mixed with nandinas, where each plant brings a different feature to the mix.

5. **Hedges** are borders of one species (or cultivar) where the plants are grown close together and pruned regularly to create a living wall for the purpose of demarcation and privacy. They take a lot of time to clip and shape into a living fence and thus are seen less often today, or are poorly maintained. You must prune a hedge wider at the bottom, narrower at the top, so the lower branches receive enough sunlight to remain leafy.

6. **Privacy screenings** are a form of hedge, only taller. They have the same effect as a fence. Ideally, the plants used for privacy screenings are modest sized needle-leaved evergreens that grow 20–30 feet tall and remain narrow, because pruning such walls is virtually impossible. Traditional plants used are Leyland cypress and columnar arborvitae.

7. **Groundcovers** are low-growing plants that occupy space and fill in areas which otherwise would have bare soil, mulch, or lawn. They can be ground-hugging or slightly taller and usually are evergreen. Because they are fast growing and are naturally spreading, they may eventually need to be confined by an impenetrable boundary like a sidewalk, or by mowing or periodic edging. Traditional plants used for this function are monkey-grass, Mondo-grass, periwinkle vine, and English-ivy. Groundcovers for full sun are more difficult to find, with creeping junipers probably the best. I will tell you up front that finding a sun-loving evergreen groundcover among native plants is difficult. In the mountains, you may have more choices with a cooler and wetter climate. In the coastal plain with moderate rainfall, you may be able to support creeping blueberry. In the Piedmont and interior low plateau, you may best be served with creeping phlox and native northeastern shore juniper.

8. **Shade trees** are usually natives, either left on the property when the home was built, or planted early on to provide shade for the future. I cannot think of a single decent shade tree that is not native. Of course, for something meant to live that long (100 years would not be an unusual life-span for a shade tree) it should be chosen and planted with care, so as to not crowd others, be too near the house, or be naturally weak and short-lived. Unfortunately, many trees planted on the cheap do just that. Avoid these natives: black cherry, sweet-gum, black locust, black walnut, black willow, box-elder, cottonwood, hackberry, American elm, silver maple, sycamore, white pine, and water oak; and non-natives chinaberry, Leyland cypress, princess tree, Siberian elm, tree-of-heaven, weeping willow, Lombardy poplar, or hybrid poplar.

Ostrich fern (*Matteuccia struthiopteris*) edges a stone pathway in a naturescaped garden.

maple. Among the numerous small flowering trees would be dogwood, redbud, umbrella magnolia, and silverbell. All are perfect in part shade to part sun and require only moderate watering during the first summer to get them established. Supplemental watering in subsequent years would keep them lush, but they would not need constant attention.

Naturescaped sites are considered informal and have no lawn. Plants are arranged randomly as they might be encountered in a forest, or in groupings according to kinds, blooming season, sun versus shade, along a trail, or other designs. In the best case, you would not water, prune, fertilize, or manipulate the site very much, only thin out existing trees or limbs to give your chosen plants the optimal sunlight. You would even leave dead twigs and spent flowers for other organisms to utilize. Natural mulch would cover the ground, and you might have spring ephemerals or evergreen wildflowers in the ground layer, also with minimal care. Any extra care you provide only enhances their performance.

The benefits to naturescaping are low maintenance, less water and chemical usage (or none, really), and attraction of more native birds and insects to the site, but the site can support only a limited number of such animals naturally. You are creating a more balanced little ecosystem rather than a traditional landscape. Every organism utilizes every part of the site, from the fresh flowers and fruits to the rotting twigs and leaves. You are not overfeeding the birds and squirrels with packaged seeds. On the down side, the site will not look as formal, lush, or pristine as you might have been used to. If that is a problem for you, you would need to decide your degree of acceptance and perhaps aim for a hybrid situation where you utilize many natives in more traditional settings, providing more care and getting more reward, or you keep some traditional exotic plants in prime locations and place natives in surrounding zones further from the house. I like the notion of using the best plant for a situation, whether native or exotic.

CREATING A MEADOW

One word here about meadows, also referred to as prairies. You cannot talk about a true prairie without also talking about periodically burning it. Anything less is just a meadow. There are very few sites in the Southeast where true tall-grass prairies exist, and these are mostly remnants of what may have been prairies when the climate was cooler and drier (during glacial times) and fires burned more randomly and kept the vegetation across the region in more of a mosaic of different stages of redevelopment. It is painstaking work to create a meadow, and you will need luck to have a site with the right soil type and rainfall. You cannot scatter out the contents of "meadow-in-a-can" mixes and expect a flourishing beautiful meadow. That is a myth (but it sells many meadow plant seeds). You would need to plant individual plants or seeds at exactly the right time of year, keep them watered, remove weeds by hand, and follow through until your desirable plants get large enough to live and reproduce on their own. I have seen only three such human-made meadows of natives in North Carolina, and they are wonderful.

What Factors Are Important for Plant Selection and Cultivation?

Gardening requires patience and acceptance of a variety of situations. Not every plant will grow for you like you saw it growing elsewhere, or you may grow something better than anyone else may. That is part of the fun of gardening; it's about the journey and the challenges, not necessarily the conclusion.

Remember your grandmother's guideline (or somebody's grandmother) when planting something new: "The first year it sleeps, the second year it creeps, the third year it leaps." This is pretty much true in the Southeast with its long growing seasons where the winters of anticipation, the springs of overactivity, the summers of interminable heat, and the autumns of reflection make us always wish for a different season and a new year to be upon us.

HARDINESS

The Southeast is a great place for gardening, with its mild winters and gradual seasonal change. The region is somewhat paradoxical as I have defined it, with the high mountains much colder (Zones 6 and 7) than the lowlands (Zones 8 and 9).

Hardiness is defined as the ability of a plant to withstand cold winter temperatures. Cold tolerance varies from plant to plant within a species, based on where it originated and what climate it had to endure. A single species that grows widely from Maine to Florida, such as red maple (*Acer rubrum*), will be differently adapted at one end of its range than at the other. And, just because a plant is sold as *Acer rubrum* does not mean it can live in both extremes equally well. There will be local strains of the species adapted along the continuum of temperatures from one extreme to the other. You can no more take a southern-born specimen up north than you can bring a northern member down south.

In contrast, it is critical to understand that plants from mountain zones in the Southeast are not always heat tolerant. This factor is more of a concern in the region than cold tolerance. We rarely lose native plants to cold (unless they are improperly grown in pots of wet soil during severe cold periods with temperatures below 15°F). During the summer, when night temperatures are above 70°F for long periods (usually 2 months or more), a plant in the Southeast is forced to run its basal metabolism at a higher rate during its night-time rest period, thus using up much of its stored energy that would go into growing and flowering. Instead, the plant runs down and wastes away (or "melts in the heat" as some southerners would say). This situation is paradoxical and counterintuitive, because we have been conditioned by most of our older books and plant information tags to believe that the only limitation we have to worry about is cold hardiness.

Southern plants can be taken up north with far greater success, but northern and high-mountain plants cannot be brought into the hot Southeast in many cases. If you try, you need to keep the plant out of direct, hot afternoon sun and keep it well drained so soil-borne diseases do not weaken it further while it is dealing with the stresses of heat and humidity. You also need to keep it from being stressed by drying out unexpectedly.

The solution to this problem then is to obtain plants from southern sources when possible, or at least get those that were grown in the Southeast for a minimum of one full year to show that they are adapted to the heat. That is the primary rule I live by when obtaining plants, unless I want to experiment and take a chance. About half the plants I've gotten from nurseries in the high mountains (above 2300 feet), the Pacific Northwest, and north of Pennsylvania have stayed alive in my garden. Not bad odds for the situation.

Heat tolerance is the single biggest factor for successful cultivation in the Southeast. You can alter a site by adding shade, modifying the soil, and providing more water, but you can't make it cooler at night. The best you can do is look for microclimates: plant on a hill or north-facing slope in a ravine, look to see where native plants have sought a cooler refuge down by a stream. It's best not to even think of planting Russell hybrid lupines or Pacific giant delphiniums. Zone 8 in Portland, Oregon, may be the same temperature in winter as Zone 8 in Portland, Alabama, but the summers are very different.

The American Horticultural Society has produced a heat-zone map to allow gardeners to begin to understand the effects of heat on plant growth. (See the book *Heat Zone Gardening* by Marc H. Cathey, or the AHS website at www.ahs.org/pdfs/05_heat_map.pdf.) Since many native plants have not been formally rated for heat tolerance, I will not refer to that map in the plant descriptions, although it would be to your advantage to guide yourself to think along those lines. However, the heat zones closely follow the cold zones, and in the plant descriptions I will mention whether a plant can tolerate heat based on the familiar cold hardiness zones. That is, a plant hardy to Zone 4 or 5 (very cold) will not be happy in the heat of our southeastern Zone 8. And when you see a book or catalog indicating a plant is hardy in Zones 4–7, that indicates it will also not be happy in Zones 8 and 9.

Bloodroot (*Sanguinaria canadensis*) thrives in full to part shade under deciduous trees.

BREAKING DORMANCY

There is another concept to consider that will help you understand and grow plants better in the Southeast, namely, that northern and mountain plants do not fare well because the winters might be too warm in the lowlands. All temperate zone plants need a winter cold period to break dormancy. The reason you cannot grow native ferns, wildflowers, or shrubs as houseplants is because they will go dormant and never come back to life without a cold period to break dormancy.

As a rule, plants in the Southeast need 40 days of cold at or below 40°F. Plants in Zone 7 and colder need a longer cold period to break dormancy, say, up to three months. Introducing a northern plant into a southeastern garden will not allow it to meet its minimum cold requirement, and the plant will neither leaf out nor grow properly in spring. The term for exposing a plant to cold temperatures is vernalization. It is one more reason to get your plants from southern nurseries, no more than 2 states away from you.

Vernalization is also necessary for seeds. They require a cold period to break dormancy before they will germinate, and in the plant descriptions in this book, I generally indicate the conditions required for seed germination. When I say, "stratify three months," it means to store the seeds in moist soil for 90 days (more-or-less) such that they spend at least one month below 40°F. (The term stratification, analogous to vernalization, implies placing the seeds between layers—strata—of moist soil, an old gardening technique.) This requirement is typically met by planting seeds outdoors in the fall so that they germinate the following spring. There are exceptions to this rule. Some seeds need no pretreatment, such as columbine and foamflower. Others require hardly any cold, but it helps stimulate more uniform or a greater percentage of germination.

In summary, plants need to be vernalized and seeds need to be stratified to meet dormancy requirements.

LIGHT LEVELS

Now for the most confusing part of gardening: defining the light levels for garden situations. Sun. Shade. Part sun, part shade. Filtered sun, filtered shade. Open shade. What do they all mean? Unless a source defines these terms, it may not be clear what is meant.

The main concerns in the Southeast are to moderate exposure to the hot afternoon sun and to avoid deep shade, where light quality is low from filtering down through all the layers of leaves in the tree canopy. Removing a few unneeded saplings or branches in a woodland garden will do wonders to improve understory plant growth. As a rule, the more sun exposure, the more flowers. No plant likes the full brunt of hot afternoon sun. Some can—and must—tolerate it; others are disfigured or damaged by it. It certainly puts stress on a plant in a garden situation that is not found in the wild. You will do well to select and place your specimens according to their sun needs and tolerances, and this, more than any other single factor, will help determine your gardening success.

I have seen wonderful example of plants in less sun than recommended, and they looked even better than if they had been grown in full sun because their leaves were not as thick and upswept. Many plants are very adaptable and will do as best they can to survive in any conditions given them. While animals can just move to a more hospitable place, plants have to stay and take it, so they adjust the size and angle of their leaves to face the sun or avoid it.

The sequence that I espouse from low to high light is shade, part shade (includes open shade, filtered shade, and high shade), part sun, and sun. These are the terms that I use in the plant descriptions to suggest light levels, but even here there is room for adjustment as lighting is not an exact science. Since the sun moves across the sky, light varies throughout the day in the same spot. Here are my definitions of these lighting conditions:

Sun means full sun all day long. This is good for grasses, meadow plants, aquatic and marsh plants, roadside plants, and rock outcrop plants. Inhabitants of pitcherplant bogs can take full sun because their soil is always moist, though there are often pine trees around. Plants have to be tough to take the full brunt of south-exposure to the full sun all afternoon; they have to be able to tolerate the stress of less water available during hot times. Many plants require full sun; many tolerate it; some suffer from it. Some plants in the Southeast simply go dormant when summer comes and avoid full sun altogether (especially bulbous plants). Others can come up and die down as the rains come and go (like rain-lilies). Usually, keeping the soil moist will allow a plant to tolerate full sun and not develop scorched leaves. Otherwise, afternoon shade is required. Northern plants that like full sun would definitely prefer some afternoon shade in the South.

Part sun means morning sun only, with afternoon protection from the sun. Ideally, plants in part sun still can see the open sky in the afternoon, but direct sun does not warm the leaves for long. A site behind a house (north or east side) gives the plant high-quality unfiltered light all day, as does a site in a garden open to sun but surrounded by tall trees around the periphery so that afternoon shade comes in parts of the garden. In places like these, there is abundant light, but not direct sun all day. Many sun-loving plants can grow well in such sites.

Part shade means the plant is under high shade from tall trees, with few low limbs and no dense understory trees. There are openings in the canopy where much sun can come in, but never for so long that plant leaves heat up. Part shade could be called open shade, filtered shade, or high shade. It is ideal for most woodland plants that never grow in direct sun.

Shade means full shade. Plants in shade hardly get a ray of direct sunlight. Deep shade is found under a dense canopy of low limbs (like under a thick dogwood or beech, or an evergreen hemlock), and is the least desirable place to plant anything, although many plants could survive. Ferns are ideal in deep shade since they do not have to bloom. Some wild plants grow in dense shade, but then they just tolerate it and do not flower well. They are waiting for an opening in the canopy someday.

LIGHT TERMS

- Sun indicates full sun all day long.
- Part sun indicates morning sun only, with protection from the afternoon sun.
- Part shade indicates short periods of direct sunlight but mostly indirect light.
- Shade indicates no direct sunlight.

PLANTING TIME

My last piece of specific gardening advice is to plant in the early fall (late September into October) and water plants once well (maybe twice if October is dry). Do this for all your plants. Fall planting will do more than anything to improve your chances of successful gardening because the plants will have three months of stress-free time to grow roots to get ready for winter, and then be established somewhat going into their first stressful summer.

If you plant in spring, which is traditional up north and very popular in the Southeast (it is a wonderful time to see flowers and be excited about planting), do plan to monitor watering and heat stress very closely that first summer. Water well at lest once a week, making sure the water penetrates the new and old soil and keeps the roots moist without being too wet.

How Do We Obtain and Identify Native Plants?

There are several places to look when it's time to acquire a native plant already in cultivation. Local garden centers have plants that are mass-produced on a national scale. Some natives are among them, such as maidenhair fern, orange butterfly-weed, and various selections of coneflower and phlox. Alternatively, specialty nurseries grow and sell the better of the natives, both via mail-order and on-site. Get to know the staff at these nurseries, establish a relationship with them, and you may be able to encourage them to grow new species that are of interest to you. Most nurseries have web sites, and there are clearing houses to help find local nurseries.

Botanical gardens are another source of information. The North Carolina Botanical Garden has a great web site with a listing of native plant sources at http://ncbg.unc.edu/. Botanical gardens often hold plant sales on their premises. The UNC Charlotte Botanical gardens has three sales a year involving natives (http://gardens.uncc.edu). Check with the closest botanical garden or native plant society in your area.

A good way to get plants is by joining a local or regional plant society. The North Carolina Native Plant Society, for example, the oldest regional plant society in the country, has two annual plant auctions to raise money, and a seed exchange. The Georgia Botanical Society has regular field trips and outings to see native plants. Individuals bring common and rare plants that they have propagated and the fun begins. The North American Rock Garden Society and state and regional perennial plant association have seed exchange programs where many great plants are available.

Some individual growers offer native plants on eBay or other means of advertising (market bulletins, for example). Be careful when buying plants you can't see; they may not be alive or they could be the wrong plant. Unfortunately, some plant sellers buy natives cheaply from local collectors who dig the plants from the wild and sell them as is. This results in millions of plants dying from improper care, and people throwing their money away. Do not buy cheap plants from dealers you do not know.

Eventually you will need to find information about a plant: what it is, how to propagate it, how to grow it, what it's good for, where it grows, and so forth. Here again you can turn to your local native plant society, botanical garden, or nature center. It should have a library with books on various plant topics. The Internet, of course, is a fast and seemingly easy way to track down information, but sometimes you have to wade through unnecessary information and you can find conflicting details, simply because it provides everything on a subject.

Knowing the correct name of a species is the first step in unlocking information about the plant. While common names are good, they can be confusing and misunderstood across different regions. Sometimes two unrelated plants are given the same common name, or a name that is popular in one region is unknown in another. For these reasons and others, Latin species names (a two-part designation consisting of a genus name and specific epithet) are more accurate for precisely identifying a species. That said, Latin plant names can be tedious and they are subject to change, but you have to work with them.

Several web sites are designed to help you identify plants. An excellent example is www.namethatplant.net. You can type in a name or trait of the plant, and the site will help you narrow down the possibilities. It even has recordings so you can hear the Latin name pronounced (with some neat southern accents). The North Carolina Native Plant Society (www.ncwildflower.org) has a database of wildflower photographs that have been helpful for identification. Many other organizations do likewise, so check out the sources in your state or region.

Finally, the ultimate source of inspiration and information comes when you see a plant in its native haunts and learn what conditions it requires. Every state has a native plant society that plans field trips to interesting habitats across its state. Join your society; go on the trips; find out who is most knowledgeable and ask them questions. You will learn a lot and see some great sites that will thrill you. Meet with other folks in your community to practice the art of propagating and growing plants. Volunteer at a botanical garden or nature center. Get involved. The more people who know and appreciate native plants and their habitats, the more plants we can preserve for future generations to enjoy.

[W]e cannot win this battle to save species and environments without forging an emotional bond between ourselves and nature—for we will not fight to save what we do not love (but only appreciate in some abstract sense). So let them all continue—the films, the books, the television programs, the zoos, the little half acre of ecological preserve in any community, the primary school lessons, the museum demonstrations, even (though you will never find me there) the 6:00 a.m. bird walks. Let them continue and expand because we must have visceral contact in order to love. We really must make room for nature in our hearts.

—STEPHEN JAY GOULD

Hiking the heath balds in the Roan Highlands provides native plant enthusiasts with spectacular views of Catawba rhododendron (*Rhododendron catawbiense*) in full bloom.

How to Use This Book

The plant descriptions in this book are divided into nine groups. Each chapter begins with an introduction to the group. The species descriptions follow, arranged in alphabetical order by scientific name and presenting certain vital information in a prescribed format.

NAMES. Plants are listed by their current scientific name. The nomenclatural authority used is an amazing work on the native plants of the Southeast, *Flora of the Mid-Atlantic States*, by Alan S. Weakley of the University of North Carolina Herbarium. We call it "Weakley's Flora." It is continuously updated in a searchable, on-line version and thus is the latest word in plant identification (*www.herbarium.unc.edu/flora.htm*). This is not your everyday source of native plant information, and certainly not horticultural advice, but it has technical identification keys, distribution maps, and nomenclatural notes on the species. In a very few cases I have chosen not to follow Weakley's preferred name for a species, especially when he suggests the new name is controversial. Where an older Latin name might be of use for looking up a plant in the literature, it's included as a synonym.

Sarracenia leucophylla

Next come one or two of the most acceptable common names. These names are far from standardized and, for some people, are unimportant. Don't be afraid to learn a few Latin names and start to make associations; you will be glad you did in the end.

Following the plant's common name(s) is the family name. A family is a closely related group of genera that have something in common, usually floral structure. Botanists use families a great deal. Some of the common plant family names are very familiar to gardeners: rose, magnolia, iris, lily, and aster. All scientific family names end in *-aceae*; Rosaceae is one example. Learning about a family is the first step in unlocking information about a plant, because many members of a family behave alike and have similar floral structure and growth behavior.

HABITAT & RANGE. These are very general descriptions. Habitat is stated when it's clearly definable within a region. What is useful to know is if a plant comes from a more sunny or shady location, wet or dry place, whether rocky or rich soil, whether disturbed or stable areas, its propensity to tolerate flooding or salt spray, and so on. Range is described most often as "mostly throughout the Southeast," meaning a plant has been found in nearly every region. This tells you it's widely adapted. If it's mountains only, you know it likes cooler night temperatures and soil that is more organic. There is not much need to know if, for example, a plant is absent in Tennessee, but the knowledge will save you from looking for it if you live there (or vice versa). If you really want to study distributions, which are fascinating, go to *www.bonap.org/genera-list.html* and browse by genus to find excellent up-to-date maps of distribution by county for every recognized species in North America. You will find several interesting ways to search and view there.

ZONES. This is the range of hardiness zones in which the plant is believed to be able to survive. It does not correlate with the natural range of the plant. In the Southeast, we will be dealing with Zones 6 and 7 in the mountains, and 8 and 9 throughout the bulk of the region. I will use these zones to refer to heat tolerance more so than cold. It is the summer night temperatures that are important, and most high-mountain and northern species can't withstand the prolonged night temperatures above 65°F typical throughout Zones 8 and 9. (See the back of this book for addresses of hardiness maps on the Internet and a list of temperatures associated with each hardiness zone.)

SOIL. I have indicated a general soil type as "moist, well drained." Deviations from this might indicate drier or sandy or extra well drained. If the plant comes from a swamp or wetland, I will indicate it can tolerate seasonal flooding, meaning it can take very wet roots, but not necessarily continually standing water. Don't worry too much about soil in general. If you have plants currently growing well, you should be able to grow most plants in that soil. Avoid pure red clay and pure white sand, unless you know the plant can take it.

Aquilegia canadensis with pipevine swallowtail.

LIGHT. This is a general indication of range of exposure to direct sunlight light. A plant could survive in full shade, but it might grow and bloom better for you in part shade or part sun (open shade). In general, a plant will want as much light as possible, but woodland species appreciate protection from full afternoon sun. This could be accomplished by placing them on the north or east side of a house or large tree, where the plant can "see some sky" but not suffer the full brunt of the hot sun.

DESCRIPTION. This is a brief diagnosis to help you envision the useful features of the plant or to help differentiate it from a similar species. Within each entry, I indicate whether the species is evergreen or deciduous, then its growth habit, whether clumping or creeping, then growth pattern and mature size. Use this as a general guide, and realize that most plants will grow larger as they get older. The width of a tree is generally two-thirds its height. Don't take these sizes too precisely, as different factors (such as nutrition and water availability) affect the ultimate growth rate. Then I describe salient features of the leaves, whether alternate or opposite, and simple or compound, as knowing these are critical first steps for proper identification. Then I indicate whether they turn fall color. The absence of data implies that it's not useful in a given case; for example, there may be no notable fall color if I don't mention it. Then for flowers, I indicate how showy they are, the color, fragrance, size, shape, arrangement on the plant, and peculiarities, and especially blooming dates. I've specified seasons rather than months, since March is late winter in North Carolina, but February might be late winter in Florida. Generally, in the Carolinas, the northern parts of Zone 8, December–February is winter (with daytime freezes possible), March–May is spring; June–August is summer, and September–November is autumn. Last is the fruit, where I give a general description and note whether it's showy or not, and the maturation time for those who wish to attempt to collect seeds. Remember to collect seeds just before they get ripe so you don't miss them; let them ripen in the shade in a dish or paper bag (not plastic) at home. Finally, I may indicate some other notable fact, such as a special bird or insect associate, interesting bark, or growth pattern.

PROPAGATION. Here I give general indications of the conditions necessary to induce germination; usually it's a cold period for most natives. Cutting of stems in summer is usually before flowering begins, and usually with the aid of a rooting hormone. Because of the many variations in practices, I urge you to consult one of several excellent propagation manuals listed in the bibliography by Cullina (2000), Deno (1993), Dirr (2009), and Phillips (1985).

LANDSCAPE USES. Here I give indications of traditional landscape use. A foundation plant is suitable for planting near the house or a building. A specimen refers to an outstanding plant that can be featured in a prominent place. An accent is a better-than-average specimen that can be placed in a mixed border or flower bed. Groupings can be either masses or several specimens together, usually in odd numbers rather than single specimens for a better effect. Ground cover refers to plants that will grow low and spread to form a mass. Hedge plants indicate tall, evergreen species that can be used for screening. Plants for naturalizing in informal woodland are those that can be used as scattered specimens, or as background, or as focal point, or however you might see them fit in the landscape. Plants best left as natives in the woods are those that should not be removed if you are clearing out understory; these would be more desirable to keep on the property than some others would. Of course, you can try a plant in any setting you may desire and see how you like it.

EASE OF CULTIVATION. I have indicated whether easy, moderate, difficult, or very difficult. Use these as general guidelines and follow your best horticultural practices. This book is not a how-to-grow-it manual, but I have given some recommendations where useful. Here is where experience and patience come in. I have had to move some plants three times before they "took hold." Be creative and take some chances, but know your plants and their most important needs. Putting a shade-loving plant in the full sun and trying to keep it alive with water will just compound other problems, and vice-versa. Talk with knowledgeable growers and find out what has worked in your region.

AVAILABILITY. This is very difficult to quantify; all you really need is one source to say a plant is available, but I indicate *rare* if I could not find a listing in the Internet but I know it to be available; *infrequent* if I could find two sources; *frequent* if I could find three or four; and *common* if I easily found five sources and quit looking. Get involved in a network of plant people by way of on-line list-serves, chatrooms, state and local native plant societies, master gardener groups, and local botanical institutions and find out who has what and how to get it.

NOTES. This is where I highlight the best features of the species and indicate any concerns as well as any suggestions on culture, personal observations, or other information. The notes include a star rating as a guide based on these general criteria. With the strictly herbaceous species, the ratings indicate the quality of the plant with the understanding that such plants would be minimally ornamental in winter and usually not have a fall color, and therefore the four-season criterion would not apply.

Also in the Notes section, I mention any cultivars worth highlighting, based on personal experience and advice from other professionals. Because a cultivar is developed in a particular region, it may have requirements or limitations that make it less suitable in your region. Check with other growers in your area to see what works well. Otherwise, chose plants that have traits you like and try them. It is important to note that some cultivars may not look much like the wild species. For example, the cultivar *Symphyotrichum ericoides* 'Snow Flurry' is a very tight-growing, mound-forming aster with tiny leaves, while the wild form is an upright plant with spreading stems.

Finally, I mention other related species and their cultivars that are similar to the main entry and which are worthy of consideration. These descriptions will be brief, just enough to indicate a plant's value and to differentiate it from the main entry; you may need to look up additional information elsewhere.

PLANT RATINGS

★ Useful plant. Has limited ornamental appeal. At least one good trait and one negative trait. You might not plant it, but you might enjoy it if you already have one on your property.

★★ Good plant. Has at least two ornamental traits, one of which may be outstanding; may have a negative trait, but not of great concern. The plant may be hard to find or difficult to grow.

★★★ Very good plant. Has many uses. Very desirable with two or more good traits; no negative traits, though some aspects might not be as ornamental, such as lack of fall color on a flowering shrub. At least moderately easy to grow and frequently available. Worth some trouble to acquire and establish.

★★★★ Outstanding plant. A must-have for the garden. Has three or four seasons of interest and no negative traits. Commonly available and easy to grow. Worth some trouble to acquire and establish.

Ferns and Clubmosses

Ferns are found in virtually every habitat in the Southeast and are always fascinating to encounter. Children walking in the winter woods may notice the dark green Christmas ferns dotting the otherwise brown forest floor and, even without examining the plants closely, can know that they are different from other plants. Ferns are ancient survivors, exploiting niches that flowering plants cannot, and almost define the "look" of many natural landscapes. They may seem a bit mysterious since they grow and reproduce so differently from flowering plants, yet they add an important element of texture to our lives.

In the 16th century people did not understand ferns and thought the plants possessed invisible seeds, which if gathered on Midsummer's Eve would allow the finder to become invisible. In truth, ferns are primitive plants that reproduce solely by microscopic spores, which are produced inside saclike sporangia packed into distinctive clusters called sori. They are located on the underside of the leaves and may resemble brown, tan, or yellowish scale insects that you might want to scrape off, thinking you are ridding the plant of a pest. Sometimes the sori are covered by a thin protective covering (the indusium) that dries up as they mature. Observing the size, shape, and location of the sori and their indusia are critical for accurate identification. Botanists and naturalists use a small 10× magnifying hand-lens (or loupe), which is helpful for looking at fern sori, locating plant hairs, viewing seeds, and removing splinters.

The fern leaf, or frond, has a more-or-less triangular blade that is usually divided into smaller divisions, making it a compound leaf. It may first be divided into rows of segments called pinnae, with possible further divisions of these known as pinnules. These divisions form along the rachis, or axis, of the frond. If the rachis has only one set of completely separate divisions, it's said to be "one time compound." If these divisions, or pinnae, are further divided into a set of separate pinnules, the frond is said to be "two-times compound." And so on.

If the blade of the frond is not divided, it's said to be simple. This very rare condition is known in the Southeast's flora only in Hart's tongue fern and walking fern. If the blade is merely deeply indented from the edge, but the tissue is not completely separate from adjacent tissue on the rachis (as in sensitive fern), it's described as being deeply lobed, but not divided. The highly divided fronds add to the delicate "ferny" appearance of the frond.

The shape of the frond and the nature of the divisions help identify the species; and, of course, it's the beauty of the form and color of the fronds that gives ferns their amazing appeal. The fern frond always develops as an uncoiling fiddlehead. The stalk of the frond is called the stipe, and it may be green or dark, even black, and may be naked, or have hairs or scales to some degree towards the base. These are important diagnostic features and should be noted. The stipe is analogous to the stalk, or petiole, of a tree leaf that attaches it to the twig.

Fronds arise from the growing tips of underground stems, or rhizomes, which produce many brown-black, tough, wiry roots. The rhizome may be elongated or creeping, growing from several inches up to a foot long per year and forming a large colony, or it may be slow growing and form a tight clump. It is important to know whether a fern is a creeper or a clumper, for that will determine how it mixes in with other plants. It is also the first step in identification. Creeping ferns can be as invasive as any fast-growing perennial,

A fernery at Southern Highlands Reserve, Lake Toxaway, North Carolina.

and you may regret planting them in a bed of less-sturdy wildflowers. Yet, creepers can make useful groundcovers in the right location. Clumpers are best used in a fernery (display fern garden) or rockery (shady rock garden for ferns and other woodland perennials), or just planted throughout the shady perennial garden.

Ferns are available in a broad palette of sizes, shapes, colors, and behaviors. While most ferns are deciduous (their leaves die down in winter), some are evergreen and add interest to the winter landscape. Some ferns are not sensitive to frost, while others come up late in the spring to avoid it. Unfurling fern fronds are a most glorious sight and worth special attention in spring, so site your ferns carefully so you can enjoy them at their best.

As a group, ferns are not particularly drought tolerant (though some are) and they are shade loving, though others can tolerate full sun if kept moist. Generally, during a drought they will hold up well until they reach a point of wilting, and then the fronds turn brown and die down—an all-or-none response to drought. Dried fronds never recover well, but new ones will grow quickly if moisture is replenished and the rhizomes remain alive. All else being equal in your garden, plan to water the ferns first during a dry spell because they are such important components of the garden structure for the entire growing season, whereas many spring wildflowers die down in summer.

Most ferns prefer rich, moist soil, that is, really good "woods-dirt" or "topsoil" with ample organic matter for holding some moisture, but by no means do ferns require regular watering to "keep them wet." Other ferns might prefer a drier well-drained soil in the crevices on rock ledges and walls. Knowing the natural habitat of your fern will help you place it in the most appropriate soil in your garden.

Ferns are generally pest-free, though the occasional leaf-rolling or chewing insect may be seen. They are tougher than you think and are usually easy to propagate: just divide the clump or separate a portion of the colony and keep them moist during reestablishment. But—and this is an important

The common evergreen Christmas fern (*Polystichum acrostichoides*) is often called fiddlehead fern in the South, a name that is also used for ostrich fern (*Matteuccia struthiopteris*) in the North. All ferns display fiddleheads as their leaves unfold.

point—don't roughly disturb the fern roots *and* cut all the fronds off at the same time during active growth, as they often will not recover as would a flowering plant. Even slightly broken and bruised fronds will remain active and help a fern recover, so keep as many fronds as you can when transplanting and dividing ferns.

Another way to grow ferns is from spores, but this method requires careful preparation of the proper spot. Fern spores must land in a suitable micro-niche where there is sufficient moisture for them to grow and enough water for sexual fertilization. Under these conditions, the spores will slowly grow into mature plants that can tolerate harsh conditions. You may discover the perfect spot in your garden, probably on the north side of a wall, log, or mossy steps, where some ferns will self-sow their spores and establish as volunteers that you can relocate as they become larger.

In cultivation, you may start ferns in an enclosed plastic food container. Sow dried spores on sterile potting soil and seal the container to keep the spores moist. Tiny plants will form, and perhaps after more than a year, following repeated transplantations, they will become large enough to place in the garden. For adult ferns, soil acidity (pH) and moisture levels are usually more important than fertilizer in fostering their growth.

Ferns are important components of the shady woodland perennial garden and may be used to provide delicate charm and beauty, convey a feeling of lushness and serenity, or produce bold structures and focal points. Ferns and many wildflowers just naturally go together, and ferns provide strong visual elements that retain their presence throughout the growing season. Ferns are also interesting and diverse enough to be grown in garden areas unto themselves. They can be cultivated in any normally shaded garden site, in a raised bed or terraced situation, on a hillside, or in a location where a moist area or creek bank gives rise to a slope. Just as there are characteristic ferns in every wild habitat, there is a place for ferns in every native garden.

Two rows of round sori (with white indusia) dot the underside of a Dixie wood-fern (*Dryopteris* ×*australis*) frond.

Adiantum capillus-veneris

Adiantum pedatum

Asplenium rhizophyllum

Adiantum capillus-veneris ★★★★

Southern maidenhair fern, Venus maidenhair fern

Pteridaceae (maidenhair fern family)

HABITAT & RANGE limestone walls and woods, scattered throughout the Southeast **ZONES** 7–10 **SOIL** moist, well drained, slightly acidic or limey **LIGHT** shade.

DESCRIPTION Deciduous clumper developing into a slowly spreading, tight colony, 12–18 in. tall. Stipes black, becoming branched as the rachis of the frond. Fronds delicate, generally gently arching or pendant, twice divided. Sori form along the blade margins by flaps of tissue that look like fingers of a hand folded back onto the palm.

PROPAGATION spores; division **LANDSCAPE USES** specimens in ferneries, rockeries, woodland gardens **EASE OF CULTIVATION** easy **AVAILABILITY** common.

NOTES A delicate and beautiful fern that forms clumps and groundcover in neutral or acidic soils, or among limestone or acidic rocks and walls; it may naturalize in rock walls. Mixes well with other small plants and can fit under open shrubs.

Adiantum pedatum 

Northern maidenhair fern, five-finger fern

Pteridaceae (maidenhair fern family)

HABITAT & RANGE rich woods throughout the Southeast **ZONES** 2–8 **SOIL** moist, rich, well drained **LIGHT** shade, or morning sun.

DESCRIPTION Deciduous clumper developing into a slowly spreading, tight colony, 1–2 ft. tall. Stipes black. Fronds stiffly erect, forming a unique horseshoe-shaped arrangement of the pinnae, which are held in a horizontal symmetrical circle. Sori form along the blade margins by flaps of tissue that look like the fingers of a hand folded back onto the palm.

PROPAGATION spores; division **LANDSCAPE USES** specimens in spacious ferneries or rockeries, mixed with larger wildflowers and small shrubs in woodland gardens **EASE OF CULTIVATION** easy **AVAILABILITY** common.

NOTES The loveliest, most distinctive fern for American gardens. From the time the pinkish fiddleheads begin to unfurl and expand into the very attractive fronds, you will find this plant a joy to behold. No garden should be without it. Because it spreads, it will eventually overtake nearby small plants; but it's easy to thin out periodically. Withstands heat and drought fairly well, but once it wilts, there is no recovery and new fronds must be produced.

Adiantum pedatum 'Miss Sharples', a variant selected in Britain, is commonly available and offers shorter, lime-green fronds that are distinctly more fluffy.

Asplenium platyneuron ★★★

Ebony spleenwort

Aspleniaceae (spleenwort family)

HABITAT & RANGE upland woodlands and rocky wooded slopes throughout the Southeast **ZONES** 4–9 **SOIL** moist to dryish, well drained **LIGHT** shade. to part sun.
DESCRIPTION Evergreen clumper. Stipes dark reddish brown, not really black like the common name implies. Fronds brittle, once divided, 6–22 in. tall, somewhat dimorphic (of two different forms); sterile frond shorter and slightly arching; fertile fronds taller and straight. Sori straight, along the middle in two rows on each blade.
PROPAGATION spores; division **LANDSCAPE USES** rockeries, rocky sites, path edges, walls **EASE OF CULTIVATION** easy **AVAILABILITY** infrequent.
NOTES A delicate and interesting little fern that can become quite robust. Able to fend for itself in the woodland garden—you will be surprised where it turns up—and in ideal growing conditions, will form dividable small clumps on a mossy log or path edge. Plant a few and then leave them alone (without fertilizer or excessive water) to spread into their own niches.

Asplenium rhizophyllum ★★

Walking fern

Syn. *Camptosorus rhizophyllus*
Aspleniaceae (spleenwort family)

HABITAT & RANGE well-drained mossy limestone cliffs and boulders on wooded slopes, mostly in the mountains and adjacent Piedmont, west through Alabama and Tennessee **ZONES** 4–8 **SOIL** thin, moist, among (or on) limestone rocks **LIGHT** shade.
DESCRIPTION Evergreen clumper, forming a colony by the elongated fronds, which root at the tips like strawberry runners. Stipes green. Fronds also green, somewhat leathery, gently arching, to 10 in. long, undivided, tapering to long "tails" in a unique fashion for vegetative reproduction. Sori linear, scattered along the middle of each blade.
PROPAGATION spores; plantlets **LANDSCAPE USES** specimens on mossy boulders or walls, or in rockeries **EASE OF CULTIVATION** difficult **AVAILABILITY** infrequent.
NOTES A very interesting and diminutive fern. Attempt establishment in mossy mats or thin, moist soil in a limestone or slightly acidic rockery. A marvelous sight to see on a rock wall. I kept several plants alive for thirty years with minimal care among limestone rocks. I have seen robust little colonies in the right cultural conditions.

Asplenium platyneuron

Athyrium asplenioides ★★★★

Southern lady fern

Athyriaceae (lady fern family)

HABITAT & RANGE moist acidic woods, swamps, and thickets throughout the Southeast to New Jersey **ZONES** 4–9 **SOIL** moist to average **LIGHT** shade to part sun.
DESCRIPTION Deciduous clumper, becoming large, 2–3 ft. tall and as wide. Stipes typically reddish, with a few scales and hairs. Fronds brittle, large, erect, triangular, two-times compound (with pinnules usually lobed). Sori slightly curved to hooked, in two rows along the middle of each blade.
PROPAGATION spores; division **LANDSCAPE USES** specimens in ferneries or beds, borders, groupings, naturalized in woodland gardens **EASE OF CULTIVATION** easy **AVAILABILITY** common.
NOTES One of the commonest ferns throughout the South. In the mountains where it's cooler and wetter, it may form massive swaths along open sunny road banks and ditches with hay-scented fern, where both appear stiffly erect and a lighter green. In the shade, lady fern is open and more delicate and lacy. Clumps can be rejuvenated by carefully removing older and broken fronds, or cutting back altogether if the clump becomes disheveled. It makes a bold landscape statement in woodland gardens and borders. Easy to grow in almost any situation; somewhat drought tolerant and will wilt when dry, to perk up if watered soon. A very agreeable and reliable fern.

A close relative, the common northern lady fern (*Athyrium angustum*), is hardly distinguishable, but the Southern lady fern has reddish stipes, as southern ladies characteristically wear red lipstick. That last part may not be patently true, but it will help you remember the fern.

Athyrium asplenioides

Botrychium virginianum ★★

Rattlesnake fern

Syn. *Botrypus virginianus*
Ophioglossaceae (adder's-tongue family)

HABITAT & RANGE very rich deciduous woods throughout the Southeast **ZONES** 4–8 **SOIL** moist, well drained **LIGHT** shade.
DESCRIPTION Deciduous, solitary frond from a brittle fleshy root, 6–18 in. tall. Frond dimorphic; the sterile "leafy" portion is triangular and three-times divided into numerous smaller blades, up to 1 ft. wide; the fertile portion branches off near the blade to form an erect stalk, bearing small round sori that turn yellow with spores in late spring. The overall height of the fertile spike may reach 2 ft.

Botrychium virginianum

PROPAGATION spores **LANDSCAPE USES** specimens and groupings in rich woodland gardens **EASE OF CULTIVATION** moderate **AVAILABILITY** rare.
NOTES Fiddleheads unfurl in spring and fronds die down in autumn, which allows this fern to mix beautifully when interplanted with spring wildflowers. In fact, it may look more at home with wildflowers than with other ferns. It is

usually found with ginseng, and thus is referred to as "sang sign" by gatherers. It is difficult to propagate, as it does not form clumps and spores are very slow, but it transplants well and can be rescued and easily established.

Closely related are the true grape ferns, such as cut-leaf grape fern (*Botrychium dissectum*, syn. *Sceptridium dissectum*), which are also not dividable. Nevertheless, they are well worth seeking on plant rescues, being aware that the roots are deep and brittle and are best moved in fall when next year's growth bud has already formed at the roots. Unlike rattlesnake fern, grape ferns are wintergreen, usually dying down in summer.

Cheilanthes lanosa

Cheilanthes lanosa ★★

Hairy lipfern

Pteridaceae (maidenhair fern family)

HABITAT & RANGE exposed granitic rock outcrops throughout the Upper Southeast **ZONES** 5–8 **SOIL** thin, well drained **LIGHT** part sun to sun.

DESCRIPTION Deciduous clumper, gray-hairy, forming a small colony, 6–12 in. tall. Fronds erect to spreading, twice divided, with lobed pinnae; stipes and fronds are copiously hairy, especially underneath the frond where sori are along the in-rolled margins of the pinnae.

PROPAGATION spores; division **LANDSCAPE USES** exposed rockeries and rock gardens, walls, ledges **EASE OF CULTIVATION** moderate **AVAILABILITY** infrequent.

NOTES One of the easiest rock-dwelling ferns to grow, making attractive slow-growing clumps. The fronds may curl up in dry weather, but revive upon watering. This fern likes plenty of bright light or direct morning sun. Don't overwater it once established.

The related woolly lipfern (*C. tomentosa*) is very attractive and equally worthwhile to try to grow in a dry rockery. It is also hairy, but more green than gray, with fronds 6–12 in. long.

Cystopteris protrusa

Cystopteris protrusa ★★★

Southern fragile fern, southern bladder fern

Cystopteridaceae (brittle fern family)

HABITAT & RANGE rich woods, moss-covered boulders, and alluvial flats, scattered throughout the Upper Southeast, mostly nearer the mountains **ZONES** 5–9 **SOIL** humus rich, well drained **LIGHT** part shade to part sun.

DESCRIPTION Deciduous creeper forming an extensive

colony from a vigorous creeping rhizome whose growing tip extends well beyond the last formed frond, 6–18 in. tall. Fronds are delicate and brittle, bright green, gently erect, narrowly triangular, two-times compound (with pinnules deeply lobed). Sori round, scattered over the undersurface, covered by short-lived inflated (hence bladder fern) indusia cut-off on one side to appear half-round.

PROPAGATION spores; division **LANDSCAPE USES** groundcover in moist woodland or beds **EASE OF CULTIVATION** easy **AVAILABILITY** rare.

NOTES A delicate and beautiful fern that quickly spreads to form a soft carpet without posing a competition problem with other plants. It is easy to remove if it wanders where not wanted. It is atypical of its rock-dwelling relatives in that it naturally grows on moist soil. Not particularly drought tolerant, but regrows quickly from dry spells.

The related bulblet fern (*C. bulbifera*) grows on more northern limestone rocks. In areas cooler than Zone 8, it may become a nuisance by proliferating from asexual bulblets (pea-sized detachable propagules) produced on the elongate, delicate fronds.

Dennstaedtia punctilobula ★★

Hay-scented fern

Dennstaedtiaceae (bracken fern family)

HABITAT & RANGE moist to dry, rocky forests, outcrops, roadbanks, and clearings, mostly in the mountains throughout the Southeast north of Florida **ZONES** 3–7 **SOIL** rocky, well drained, acidic **LIGHT** shade to mostly sun.

DESCRIPTION Deciduous creeper, spreading vigorously and densely, 1–2 ft. tall. Stipes shining red-brown, distinctly short-hairy. Fronds erect to arching, yellow-green, narrowly triangular, two times pinnate, the whole frond copiously covered with minute gland-tipped hairs emitting the odor of fresh-mowed hay and feeling clammy to the touch. Sori are borne in distinctive cuplike indusia along the margins of the pinnules.

PROPAGATION spores; division **LANDSCAPE USES** groundcover **EASE OF CULTIVATION** moderate to difficult **AVAILABILITY** common.

NOTES One of our most beautiful and delicate-looking ferns. Its tongue twister of a scientific name honors an early German botanist and the nature of the sori. I have tried for thirty years to establish it in our University botanical gardens, but with no luck. I will keep trying, as I love it. I may rue any success, however, as this rampantly spreading fern has no place in the small garden or with other less formidable plants. The presence of the abundant hairs helps to distinguish this from southern lady fern (which is a clumper) with which it grows characteristically in dense swaths along roadbanks in the southern Appalachian Mountains. Captivatingly beautiful *en masse*.

Dennstaedtia punctilobula

Deparia acrostichoides

Deparia acrostichoides ★★★★

Silvery glade fern, silvery spleenwort

Syn. *Athyrium thelypteroides*
Athyriaceae (lady fern family)

HABITAT & RANGE rich, moist forests, mostly throughout the southern Appalachian Mountains into Alabama and across Tennessee **ZONES** 4–9 **SOIL** moist, well drained **LIGHT** shade or filtered morning sun.

DESCRIPTION Deciduous clumper, forming a tight colony, 1–3 ft. tall. Stipes green, cloaked with hairs and scales. Fronds erect, light green, narrowly triangular, once compound but with pinnae deeply cut into oblong segments to appear almost two-times compound, covered with fine scales that lighten the green color. Sori short and narrow, along the veins, with striking silvery indusia, as reflected in the common name.

PROPAGATION spores; division **LANDSCAPE USES** specimens in ferneries, groupings in woodland gardens **EASE OF CULTIVATION** easy **AVAILABILITY** common.

NOTES Silvery glade is a robust, erect fern but not massive. It mixes very well with wildflowers and other ferns and is a favorite. It is somewhat drought tolerant and not fussy about soil. Because it's upright, it may fit into tighter spots, but then it lacks the fullness of other ferns. One of the easiest ferns to grow.

Diplazium pycnocarpon ★★★★

Glade fern

Syn. *Athyrium pycnocarpon, Homalosorus pycnocarpos*
Athyriaceae (lady fern family)

HABITAT & RANGE woodland glades, rocky slopes, and alluvial thickets, frequently over limestone bedrock, mostly in the southern Appalachian Mountains west of the Carolinas and into Alabama, and across Tennessee **ZONES** 4–9 **SOIL** humus rich, neutral to slightly acidic **LIGHT** shade or filtered morning sun.

DESCRIPTION Deciduous clumper, forming a tight colony, 2–3 ft. tall. Stipe green with a few scattered scales and hairs. Frond erect, narrowly triangular, dark green, once divided, with unlobed tapering slender pinnae. Sori are long and narrow, along the main veins, and mostly on the upper half of the frond where the fertile pinnae may be a bit narrower.

PROPAGATION spores; division **LANDSCAPE USES** specimens in ferneries, clumps or groupings in woodland gardens **EASE OF CULTIVATION** easy **AVAILABILITY** frequent.

NOTES Glade fern is a most charming fern with its once-divided leaves and distinctive long tapering pinnae without teeth or divisions (a rare trait in our ferns). It is one of my favorites, graceful but not delicate. Large and attractive, it slowly forms larger clumps. It likes moisture and will show signs of wear as summer heat and drought take their toll, but it holds up fairly well. Not fussy about soil pH even though it likes less acidic soils in the wild and does fine in average humus-rich garden soil. Here is a fern whose common name is more stable and well known than the changing scientific names.

Diplazium pycnocarpon

Dryopteris intermedia

Dryopteris ×*australis* ★★★★

Dixie wood-fern

Dryopteridaceae (wood-fern family)

HABITAT & RANGE swamp forests and floodplains, scattered through the mid-South where both parents are found **ZONES** 5–9 **SOIL** moist to average, humus rich **LIGHT** shade to morning sun.

DESCRIPTION Semievergreen clumper, forming a lush, very robust colony, 3–5 ft. tall. Stipes greenish, sturdy, densely covered with stout brown scales. Fronds erect, dark green, narrowly triangular, twice compound. Sori large, covering the upper portions of the fronds, round with distinctive kidney-shaped indusia.

PROPAGATION spores; division (of colonies) **LANDSCAPE USES** specimens in ferneries, moist depressions, background screening, masses in woodland gardens **EASE OF CULTIVATION** easy **AVAILABILITY** common.

NOTES Dixie wood-fern is one of the great ferns for Southeastern gardens, where it must be given room to spread in exercising its hybrid vigor. It loves moisture, but is quite tolerant of drought and sun. While neatly standing at organized attention all summer, by winter the huge semievergreen fronds will come to lie conspicuously on the ground, as a battalion of soldiers might simply fall over one-by-one at duties' end, forming a subtle seasonal groundcover. This "species" is a natural hybrid between two other southern ferns, log fern (*D. celsa*) and Southern wood-fern (*D. ludoviciana*), and thus we use the "times sign" in the scientific name and read it as "*Dryopteris* hybrid *australis*." As with many hybrids, it's basically sterile, but it produces some fertile spores. It has become more available as a popular fern since it was first introduced to horticulture through the University of North Carolina Botanical Gardens about 1985 after being discovered in a creek swamp near Darlington, South Carolina. This hybrid may be confused with either of its parents, both of which make good, large, dark green garden ferns, and are hereby endorsed enthusiastically. Log fern is deciduous with a little wider frond; Southern wood-fern is evergreen, with a narrower and very shiny green frond. Neither is as robust as their remarkable hybrid offspring.

Dryopteris intermedia 

Fancy fern, evergreen wood-fern

Dryopteridaceae (wood-fern family)

HABITAT & RANGE rich woods in eastern North America excluding Florida **ZONES** 3–8 **SOIL** moist, humus rich, acidic **LIGHT** shade to part shade.

DESCRIPTION Evergreen clumper from a single crown, 1–2 ft. tall. Stipes greenish, clothed in thick brown scales. Fronds erect-arching, arranged symmetrically to form an open funnel shape, light green, thin-textured, three-times compound, the ultimate pinnules further toothed or lobed to produce the characteristically finely dissected blade (hence fancy fern). The frond is distinctive in having tiny gland-tipped hairs all over it, especially the underside; these hairs are visible with a ten-times magnifying hand lens.

PROPAGATION spores, division of multicrowned older clumps **LANDSCAPE USES** specimens in ferneries, groupings in woodland gardens **EASE OF CULTIVATION** easy **AVAILABILITY** rare.

Dryopteris ×australis

Dryopteris marginalis

NOTES Fancy fern is perhaps the most beautiful, delicate, and symmetrical fern for the garden. It and other dissected-leaved *Dryopteris* ferns can be confused with lady ferns, but *Dryopteris* species have round sori with kidney-shaped indusia and very scaly stipes, literally shaggy toward the base, with a tight crown and one set of leaves. Lady ferns have short, straight sori, very few, if any, scales or hairs on the stipe, and will produce many fronds in an asymmetrical multicrowned clump. All *Dryopteris* species can readily be identified by looking for these three distinctive traits found together. Fancy fern likes to be kept moist during droughts, but well drained. Shows good heat tolerance.

Dryopteris marginalis ★★★★

Marginal wood-fern

Dryopteridaceae (wood-fern family)

HABITAT & RANGE rich woods, often on rocks, mostly in the Upper South from the Carolinas westward **ZONES** 2–8 **SOIL** humus rich, acidic, well drained **LIGHT** shade or part shade.
DESCRIPTION Evergreen clumper from a single crown, 1–2 ft. tall. Stipes green and clothed with brown scales. Fronds erect to arching, arranged symmetrically to form an open funnel-shaped array, dark green, leathery, narrowly triangular, twice compound, the segments quite rounded. Sori round with kidney-shaped indusia located diagnostically along the edges of the frond segments.
PROPAGATION spores **LANDSCAPE USES** specimens in ferneries, groupings in rocky woodland gardens **EASE OF CULTIVATION** moderate **AVAILABILITY** common.
NOTES This very attractive fern is formal in its growth habit. It is a bit fussy about moisture as it does not want to be dry, but it must be well drained, a situation difficult to perfect in the warmer parts of the Southeast. This is a most distinctive fern with its single set of symmetrical dark green leathery leaves, not highly dissected, and with marginal sori.

Somewhat similar, though distinctly different, is the large and stately Goldie's fern (*D. goldiana*) with a broader more robust frond exhibiting a light-greenish sheen. It is more northerly, coming southward only in the mountains where it's cooler and wetter and therefore not as suitable south of Zone 7, even though it's widely available.

Huperzia lucidula ★★

Shining clubmoss

Syn. *Lycopodium lucidulum*
Lycopodiaceae (clubmoss family)

HABITAT & RANGE acidic deciduous and coniferous forests, mostly in the southern Appalachian Mountains **ZONES** 2–8 **SOIL** humus rich, well drained, acidic **LIGHT** shade.
DESCRIPTION Evergreen creeper, forming a loose colony, 6 in. tall, rhizomes creep loosely through rotting leaf mold, turning up at their tips to form upright shoots that fork dichotomously. Leaves 1/4 in. long, thick and narrow, covering the shoots and standing out like the bristles of a bottlebrush. Sori are borne in yellow spore cases at the bases of the upper leaves. Distinctive 1/4-in. vegetative asexual propagules called gemmae are produced late in the summer, then fall off and grow into new plants.
PROPAGATION division; specialized asexual propagules **LANDSCAPE USES** groundcover in a fernery, rockery, or rich woodland garden **EASE OF CULTIVATION** difficult **AVAILABILITY** rare.
NOTES This shining clubmoss would make a fabulous addition to any garden as a groundcover, but, alas, it seems short-lived in cultivation. Try it in highly acidic (pH 4–4.5) moist-but-well-drained peat and leaf-mold mix. It does best in Zones 6 and 7.

Clubmosses, which are not true mosses but fern allies, are fascinating in the wild. Most have small leaves, often scalelike, and, except for shining clubmoss, form erect cones for spore production. As a group, they are difficult to propagate, transplant, and keep alive in cultivation. Apparently, they need very acid conditions. We have grown several species for decades, but rarely do they thrive. They are

Huperzia lucidula

Lycopodium clavatum, upright branching cones

most abundant in harsh conditions in the wild where competition is less, such as exposed rocks, low-nutrient acidic soils, and places where fires burn regularly. Note that all clubmosses were once placed in the classic genus *Lycopodium*, but you will see they have been split up into various genera according to certain traits of leaves and cones.

Running clubmoss (*Lycopodium clavatum*) is a fascinating groundcover on rocks and thin, acidic soils in the mountains. The rhizomes grow flat on the ground, the branching cones rise 6–8 in., with spores ripening yellow in late summer.

Ground-pine (*Lycopodium obscurum*, syn. *Dendrolycopodium obscurum*) is very cute like a small dark-green pine tree with horizontal branches and a single cone at the tip.

Running-cedar (*Lycopodium digitatum*, syn. *Diphasiastrum digitatum*) forms dense mats with branches like fan palms and terminal erect cones on branched yellowish stalks. It is very widespread in and outside the mountains, perhaps our most common clubmoss.

Bog clubmoss (*Lycopodiella appressa*, syn. *Lycopodium appressum*) creates a loose groundcover of flat, creeping strands in moist sandy soils in coastal pitcherplant bogs and pine flatwoods. Easiest to grow in a bog garden.

Matteuccia struthiopteris ★★★

Ostrich fern

Syn. *Matteuccia pensylvanica*
Onocleaceae (sensitive fern family)

HABITAT & RANGE wet forests and swamp edges along the northern edge of Virginia **ZONES** 2–8, with caution **SOIL** moist, well drained to seasonally wet **LIGHT** part shade to sun.
DESCRIPTION Deciduous, rampant creeper, forming an extensive robust colony, 3–4 ft. tall, with three to seven subterranean runners sending up far-ranging new growth. Fronds are strongly dimorphic, the sterile ones larger than the fertile ones, narrowly triangular, tapering at the tip and the base, twice divided, the pinnae deeply lobed; a separate cluster of highly divided fertile fronds forms in the center of each mature growth in late summer, to 2 ft. tall, remaining dark brown and starkly persistent all winter, like tough feathers, shedding spores in early spring.
PROPAGATION separation of runners **LANDSCAPE USES** masses **EASE OF CULTIVATION** easy **AVAILABILITY** frequent.
NOTES Ostrich fern is a remarkable gigantic fern of wet woodlands in the Northeast, barely reaching Virginia. Its formal growths look like giant shuttlecocks, with runners forming dense colonies. The young fiddleheads are a choice edible delicacy. The maturing fronds are indescribably beautiful; and while these are deciduous, the separate, brown, spore-bearing fronds become hardened and persist through the winter, making them an outstanding landscape feature. Keep this fern under control by digging the unwanted offsets (runners) in early spring. It is a magnificent showstopper of a fern, and if you have room, is well worth the maintenance to keep it in check. Plants will be

Matteuccia struthiopteris

shorter and spread less under drier conditions; they can be killed by heat and drought. Typical ostrich fern is not heat tolerant and does not persist south of Zone 7. Two heat-tolerant strains are available and must be utilized in the Southeast: 'The King' and 'Fanfaire'.

Onoclea sensibilis ★★

Sensitive fern

Onocleaceae (sensitive fern family)

HABITAT & RANGE marshes, ditches, and swamps throughout the Southeast **ZONES** 2–10 **SOIL** not particular, moist to wet, acidic **LIGHT** shade or part sun.

DESCRIPTION Deciduous creeper with widely spaced fronds on long-running rhizomes, forming a dense colony, 1–2 ft. tall. Stipes green, slightly scaly near the base. Fronds markedly dimorphic, the sterile ones green, not compound but so deeply lobed as to appear once compound with smooth-margined pinnae that are opposite each other; separate fertile fronds produced in summer have leaf tissues formed into round beadlike structures harboring sori, blackish brown and firm in texture, persisting through the winter or longer.

PROPAGATION spores; division **LANDSCAPE USES** groundcover in moist or wet sites **EASE OF CULTIVATION** easy **AVAILABILITY** common

NOTES Sensitive fern grows quickly to make a coarse-textured mass, and, while stately, should not be used with other plants. It could have a place in a very moist or wet situation around a pond or other isolated situation. It is very sensitive to drought and cold, and new fronds are often killed by late frosts—hence the common name. Sensitive fern can be distinguished from the very similar netted chain fern (*Woodwardia areolata*) by its smooth-margined lobes that are opposite each other on the rachis and persistent almost-woody fertile fronds; netted chain fern has minutely toothed margins, alternate lobes, and fertile fronds that wither and die down in winter.

Osmunda cinnamomea 

Cinnamon fern

Syn. *Osmundastrum cinnamomeum*
Osmundaceae (royal fern family)

HABITAT & RANGE swamps, wooded stream banks, pitcher-plant bogs, and wet situations generally, throughout the Southeast **ZONES** 2–10 **SOIL** moist, acidic, well drained to seasonally wet **LIGHT** part shade to sun.

Onoclea sensibilis

Osmunda cinnamomea

DESCRIPTION Deciduous clumper, with very large and stiffly erect fronds arising via attractive hairy fiddleheads from a crown of tough wiry roots, often forming an open colony of growths, 3–5 ft. tall. Stipes green, soft tan cottony-hairy. Fronds dimorphic; sterile frond light green, narrowly triangular, once divided and then deeply lobed into somewhat sharply pointed segments, the rachis clothed with soft tan hairs that can be rubbed off, also with a tuft of tan hairs at the base of each pinna; fertile fronds not quite as long, twice pinnate, arising in spring, bearing a thick plume of globular sporangia without leafy tissue, turning cinnamon-brown, and withering by summer.

Osmunda cinnamomea

Osmunda claytoniana

PROPAGATION spores; division when young **LANDSCAPE USES** specimens in borders, ferneries, pond borders, wetlands, or woodland gardens **EASE OF CULTIVATION** easy **AVAILABILITY** common.

NOTES This well-known gorgeous big fern makes prominent and stately specimens or masses, and is suitable for almost any location. The unfurling fiddleheads are as attractive as they come, and the variable textures add much interest. In autumn, the fronds turn a striking bronze-yellow. The tough root mass is nearly impossible to divide without a root saw. It can crowd out smaller plants, so watch your spacing. Exposure to persistently dry soil will cause this fern to grow smaller and smaller. Plants in full sun must be kept moist for best results; you can't overwater them.

Osmunda claytoniana ★★

Interrupted fern

Osmundaceae (royal fern family)

HABITAT & RANGE rich wooded mountain slopes and seeps, mostly in the southern Appalachian Mountains **ZONES** 2–7 **SOIL** moist, well drained **LIGHT** shade to part shade.

DESCRIPTION Deciduous clumper, with very large and stiffly erect to spreading fronds arising via attractive fiddleheads from a crown of wiry roots, 2–4 ft. tall. Stipes green, with a few soft hairs. Fronds dimorphic, the sterile ones light green, narrowly triangular, once divided and then the pinnae deeply lobed into somewhat round-tipped segments that are wider than those of cinnamon fern, the rachis with only a small amount of soft tan hairs, and without a tuft of hairs at the base of each pinna; fertile fronds characterized by dimorphic pinnae, where only two to four pairs of pinnae in the middle of the frond become fertile, bearing loosely arranged globular sporangia without leafy tissue, turning dark brown and withering by summer. Only well-grown plants in brighter light make fertile pinnae.

PROPAGATION spores; division when young **LANDSCAPE USES** specimens in woodland gardens **EASE OF CULTIVATION** moderate **AVAILABILITY** rare.

NOTES This fascinating big fern produces a large specimen, but overall does not possess as much subtle character as cinnamon fern, although it's still majestic and almost as attractive in autumn. Adaptable to humus-rich woodland soil, interrupted fern does not like heat or drought, but will persist for years in less-than-ideal conditions. Keep it in open shade in the Southeast, with supplemental watering in summer.

Osmunda regalis var. *spectabilis* ★★★★

Royal fern

Osmundaceae (royal fern family)

HABITAT & RANGE swamps, marshes, stream valleys, and moist woods throughout the Southeast into tropical America **ZONES** 2–10 **SOIL** moist, well drained to seasonally wet **LIGHT** shade to sun.

DESCRIPTION Deciduous clumper, with very large and stiffly erect to spreading fronds arising via attractive fiddleheads from a single crown, 2–5 ft. tall, often growing together in open colonies and forming dense masses of tough black roots around each crown, only slightly similar to the other *Osmunda* species. Stipes green to reddish with whitish waxy coating, lacking hairs and fuzz when mature. Fronds dimorphic, broadly triangular; sterile fronds twice compound, pinnae elongate, widely spaced and somewhat pointed; fertile fronds with only the terminal pinnae replaced by clusters of globular sporangia turning brown and then withering by summer.

PROPAGATION spores; division when young **LANDSCAPE USES** specimens in borders, beds, ferneries, pond edges, open woodlands, wet sites **EASE OF CULTIVATION** easy **AVAILABILITY** common.

NOTES Royal fern rounds out the triumvirate of big *Osmunda* ferns, and seems to adapt to any situation, but is especially good in moist-to-wet areas and in bright shade to full sun (with supplemental watering). It can become overwhelming in scale when grown well, forming larger multigrowth colonies. If kept on the dry and shady side, it will remain smaller while still providing the coarser texture for which it's renowned. Has good fall color as do other *Osmunda* species. The more sun and moisture it receives, the larger it becomes; it has been reported to 10 ft. tall.

Phegopteris hexagonoptera ★★★

Broad beech fern

Syn. *Thelypteris hexagonoptera*
Thelypteridaceae (marsh fern family)

HABITAT & RANGE rich wooded slopes and swamp margins, mostly throughout the Southeast **ZONES** 5–9 **SOIL** moist, acidic, well drained **LIGHT** shade to part sun

DESCRIPTION Deciduous creeper, forming a large, dense colony, 6–18 in. tall, from long-creeping shallow rhizomes. Stipes green to dark with a few scales or hairs. Fronds arching, the blades becoming almost horizontal, thin-textured, rich green, almost triangular, twice deeply lobed to appear twice deeply divided, the pinnae displaying hexagonal wings at their bases that seem to connect them along the rachis. Sori are small and round near the blade margins, lacking indusia.

Osmunda regalis var. *spectabilis*

Phegopteris hexagonoptera

PROPAGATION spores; division **LANDSCAPE USES** masses in woodland gardens **EASE OF CULTIVATION** easy **AVAILABILITY** infrequent.

NOTES Broad beech fern is beautiful to behold because of the architectural structure of the triangular leaves. It needs room to spread, however, and the more you water it, the larger it grows and the faster it colonizes. Don't plant it with other wildflowers. This fern is easily transplanted or thinned out, and is heat but not drought tolerant. The fronds are brittle.

Polypodium virginianum ★★

Rock-cap fern

Polypodiaceae (polypody family)

HABITAT & RANGE capping dry to moist rocks, boulders, and ledges, and in rock crevices, mostly in the Piedmont and mountains of the Upper Southeast **ZONES** 2–8 **SOIL** humus rich, moist but well drained **LIGHT** shade.

DESCRIPTION Evergreen creeper, forming dense colonies on boulders and ledges from scaly rhizomes. Stipes without scales. Fronds thick-textured, bright green, 4–8 in. long, deeply lobed but not quite compound. Sori are large, round, and yellow, without indusia.

PROPAGATION spores; division **LANDSCAPE USES** specimens on rocks, in rockeries, walls, raised beds, and rock gardens **EASE OF CULTIVATION** difficult **AVAILABILITY** infrequent.

NOTES Rock-cap fern is charming in the wild, covering large boulders like a head of curly hair. Desirable to grow, but difficult to get the moisture and drainage just right. If you could just bring home one of those big boulders with ferns already on it! Best as a crevice plant for the shady rockery. The mainstay of the boulder fields and shady rock crevices along the Blue Ridge Parkway, it's rarely as vigorous in cultivation. I love to turn the fronds over and look for the conspicuous yellow sori late in the growing season.

A curious and common relative throughout the Southeast is the resurrection fern (*Pleopeltis polypodioides*, syn. *Polypodium polypodioides*), found frequently towards the coast on large rocks and boulders, and especially characteristic of forming dense mats on the trunks and branches of large live oak trees toward the coast. Its common name comes from its striking behavior of curling up when dry. The leaves and stipes are covered with gray scales that efficiently absorb water when it rains to bring the dried plants quickly back to life. Very difficult to establish in the garden, but may be introduced by bringing in pieces of fallen tree limbs upon which it's growing. Definitely don't keep wet, but let dry a wee bit between frequent waterings.

Polystichum acrostichoides ★★★★

Christmas fern

Dryopteridaceae (wood-fern family)

HABITAT & RANGE shaded slopes, especially north-facing, throughout the Southeast **ZONES** 3–9 **SOIL** moist, well drained to dryish **LIGHT** shade to part sun.

DESCRIPTION Evergreen clumper, slowing forming a robust dense mass from multiple crowns, 1–2 ft. tall. Stipes covered in light brown scales. Fronds very narrowly triangular, erect-arching, once divided, then usually just sharply toothed, with a bristle at the tip of each tooth. Most pinnae have a characteristic single larger lobe at the base. Sori abundant, covering whole pinnae, but only on the terminal third of the fertile fronds where the pinnae are much reduced, thus creating dimorphic fronds.

PROPAGATION spores; division **LANDSCAPE USES** specimens in ferneries, beds, borders, masses, walls, around rocks and tree trunks, naturalized in woodland gardens **EASE OF CULTIVATION** easy **AVAILABILITY** frequent.

NOTES Christmas fern is one of the most recognized ferns, known by many from childhood walks in the woods, and often called "fiddle-head fern." One of the favorite stories to help children recognize Christmas fern is to hold a single pinna vertically to resemble a Christmas stocking with heel and toe (the basal lobe). This very beautiful and durable fern grows in almost any condition and is easily divided and transplanted. It can provide backgrounds and companions for many woodland wildflowers and shrubs, and may even be used as path edging since it forms such dense, discrete clumps. It is one of the commonest ferns in eastern North America and should be in every garden. It could be the most prominent evergreen plant in your wildflower garden in winter. Christmas fern is tolerant of sun and shade, moist and dry, heat and cold. It is common, even in the drier parts of the Piedmont region, to find whole hillsides of Christmas ferns on the north-facing slopes of ravines and steep creekbanks. The plants seem to find the smallest niche and take a foothold.

Thelypteris kunthii ★★★

Southern shield fern

Syn. *Thelypteris normalis*
Thelypteridaceae (marsh fern family)

HABITAT & RANGE swamps, low woods, creek banks, and limestone seeps, throughout the Deep Southeast, not in North Carolina or Tennessee **ZONES** 8–10 **SOIL** moist, well drained to damp, or dryish **LIGHT** part shade to sun.

DESCRIPTION Deciduous (or semi-evergreen southward) creeper, forming a very dense fast-growing colony, 1–3 ft. tall. Stipes light green, fuzzy. Fronds large, gracefully arching-erect, light green to whitish-grayish green, hairy on all parts, once divided and then the pinnae deeply lobed and strikingly long tapering to a pointed tip. Sori small, near the margins, with delicate indusia.

PROPAGATION spores; division **LANDSCAPE USES** masses for borders, backgrounds, or patches with larger shrubs **EASE OF CULTIVATION** easy **AVAILABILITY** common.

NOTES Southern shield fern has taken the gardening world by storm. It is wonderfully full, robust, and graceful, forming whitish green dense masses in light shade. or sun (if kept moist). It is very aggressive and fast growing, adaptable, drought tolerant, even volunteering readily from spores in normal garden situations. Don't overwater and it will be less aggressive. Don't plant with other plants, as this

Polypodium virginianum

Polystichum acrostichoides

Pleopeltis polypodioides

fern will choke them out. This striking species was less well known in cultivation (it's rare in the wild) before the early 1990s and is truly a denizen of the Deep South. Now, it's a widely adaptable garden mainstay, and is very good at persisting, spreading, and sporing around. It is most worthwhile in the right location—all by itself with room to grow.

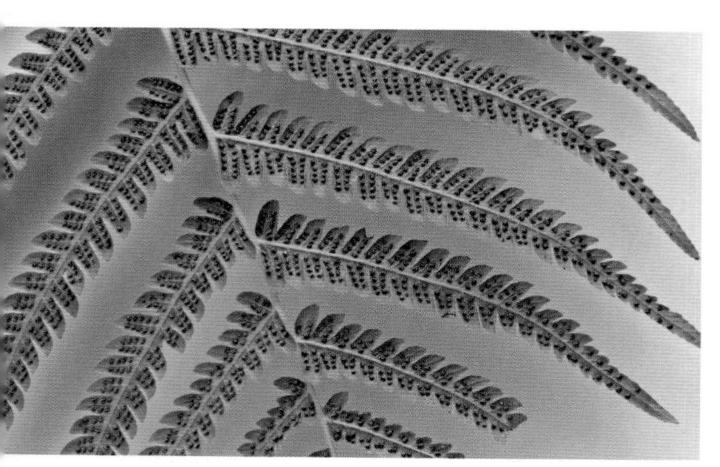

Thelypteris kunthii, frond underside

Thelypteris kunthii, indusium detail

Thelypteris kunthii

Thelypteris noveboracensis ★★★

New York fern

Thelypteridaceae (marsh fern family)

HABITAT & RANGE moist woods, thickets, and road banks, mostly throughout the Upper South **ZONES** 4–8 **SOIL** moist, well drained to dryish **LIGHT** shade. to part sun.
DESCRIPTION Deciduous creeper, forming a delicate but extensive colony, 1–2 ft. tall. Stipes green and slightly hairy (or not). Fronds erect, somewhat hairy, light green, brittle, narrowly triangular, once divided, the pinnae deeply lobed and gradually becoming shorter towards both the tip and the bottom. Sori small, near the margins, lacking indusia.
PROPAGATION spores; division **LANDSCAPE USES** masses in woodland gardens, edging within bounds **EASE OF CULTIVATION** easy **AVAILABILITY** common.
NOTES New York fern can make a beautiful and graceful fast-growing groundcover. Because it's not tough, just relentless, it may blend with other sturdy woodland plants, but I like to keep it away from most wildflowers. In nature, it can form the most extensive colonies that literally fill up the woods with an endless sea of delicate green, especially in situations where other plants will not grow. I have seen no other fern (or plant for that matter) capable of carpeting the woods so uniformly. Its delicate frond with wide-spaced pinnae diminishing toward the bottom is distinctive.

The wetland counterpart is the very similar marsh fern (*T. palustris*) found all over eastern North America, where it spreads and forms dense colonies in swampy woods to standing water in sunny marshes. Its delicate fronds can be attractive when it makes dense colonies in moist to very wet soil. Its pinnae don't taper down the rachis.

Thelypteris noveboracensis

Woodwardia areolata ★★★

Netted chain fern

Syn. *Lorinseria areolata*
Blechnaceae (deer fern family)

HABITAT & RANGE bogs and swamps, throughout the Southeast **ZONES** 3–9 **SOIL** moist, well drained to seasonally wet **LIGHT** shade to part sun
DESCRIPTION Evergreen creeper, forming an extensive colony of scattered fronds, 1–2 ft. tall. Stipes dark and shiny. Fronds erect, strongly dimorphic; sterile frond dark green, broadly triangular, deeply once-lobed, the divisions alternating on each side of the rachis, their margins wavy and finely toothed; fertile fronds form in autumn, usually longer than the sterile, deeply divided into very narrow, widely diverging projections like disheveled hair, dying down in winter.
PROPAGATION spores; division **LANDSCAPE USES** coarse groundcovers in average to wet shady gardens **EASE OF CULTIVATION** easy **AVAILABILITY** common.
NOTES This robust, spreading fern is very handsome in its fine details. The blade is pinkish when young, becoming bronzy-green and glossy as it matures. The plant is rather attractive in winter when little else is green. While you would not plant such an aggressive fern with other delicate plants, it might blend well with other moisture-loving giants like royal and cinnamon ferns. Although this fern is found in sunny, wet conditions in the wild, it's perfectly happy—and actually less full-sized and less aggressive—when grown in average garden soil in light shade. Netted chain fern is perhaps more handsome and interesting than its look-alike, sensitive fern (*Onoclea sensibilis*), which differs in having deciduous leaves, opposite pinnae without tiny marginal teeth, and a "woody" fertile frond with little globular units that persists through the winter. Named for a distinctive trait, netted chain fern has veins along the midribs of the frond and its lobes that appear to follow the pattern of a chain stitch in sewing, forming narrow enclosed areas.

A related species that does not look like netted chain is Virginia chain fern (*W. virginica*, syn. *Anchistea virginica*). This coarse, rapid colony-producer with beautiful bold featherlike fronds (that have great fall color) to 3 ft. tall is much too aggressive for the typical ornamental garden. While adapted for wetlands, growing it in drier situations might make it manageable for certain garden situations.

Thelypteris palustris

Woodwardia virginica

Woodwardia areolata

Grasses and Grasslike Plants

The Southeast has distinctive native grass species in every habitat from the shifting sand dunes of the coastal Carolinas to the stormy tops of the highest mountains, from wettest to driest, and in every form from bold and striking to diminutive and obscure. Even with this wealth of flora, the region has not viewed its own grasses as worthy of cultivation until rather recently. Indeed, most grasses are best appreciated for their unparalleled ecological roles in the Midwestern prairies, the highly productive coastal salt marshes of the Inland Waterway, the so-called sea of grass in the Florida Everglades, and in waving sweeps of broom-sedge so familiar in stabilizing old fields across the Southeast. Since the late 1990s, however, many native grasses have taken on a new life in the ornamental garden by virtue of their sparse need for water and their outstanding textural manifestations whose contrasting forms and colors work well with other perennials and last well into the winter months.

Among the dozens of native grasses in the Southeast that are worthy of cultivation are most of the famous prairie species that tiptoe into the edges of the region in suitable sites. In addition, countless introduced species have naturalized and seemingly become part of the regional flora, though often they thrive only in lawns and disturbed places. Many of them are seen as weeds, and certainly some of our worst garden pests are introduced grasses such as Bermuda grass, crabgrass, and the vigorous and competitive Johnson grass.

Grasses are members of the Poaceae, whose ancient name is Gramineae. Closely related to grasses are sedges in the family Cyperaceae and rushes in the family Juncaceae. The term *graminoid* is useful in referring to these plants as a group. All are monocots and generally have wiry stems, narrow leaves with parallel veins, and nonshowy flowers whose parts are in sets of threes. Most of them are perennials.

The leaves of true grasses wrap around hollow, jointed stems to form a sheath, which is split down one side and is easy to remove. The grass blade diverges from the sheath near the top of the sheath at a distinct region called the collar. On the inside of the collar is a flange of tissue called the ligule. It can have various distinctive features, such as hairs or lobes, which can be used to recognize different grass species.

Another unique feature of grasses is their ability to grow from special embryonic tissue located above each of their joints. This tissue allows the stems to continue to elongate at those points, even after they are seemingly hardened and mature. Because of this unique trait, grasses can be mowed, as in a lawn, or grazed regularly as in a prairie, and the stems continue to grow above the nodes or joints.

Grasses also regularly branch from the base of the clump, or along creeping rhizomes, adding to their enhanced capacity to renew and spread after mowing. Their ability to tolerate heat and drought also contributes to their improved competitiveness in sunny, dry conditions. Thus, their fast-growing stems and leaves are good at competing in dense, sunny grasslands and harsh environmental conditions. This helps explain why there are relatively few woodland grasses.

In contrast to true grasses, sedges have sheaths that are not open along one side and there is no ligule. They usually have triangular stems in cross-section with sheathing leaves diverging in a three-way pattern. Rushes usually have sheaths without leaf blades and their stems are solid. By using your sense of touch, you can recognize these differences in many grasslike plants, especially if you remember this famous mnemonic ditty:

Sedges have edges,
And rushes are round,
Grasses are hollow,
Right up from the ground.

Grasses fill an old field.

Andropogon elliottii in autumn.

The flowers (and seed heads) of graminoids are small and rarely showy, at least not as tiny individual units. In cut bouquets and as long-lasting dried vegetation to leave standing in the garden, however, their myriad reproductive clusters explode with beauty and can occupy space as a burst of fireworks. In their sex life, these structures function just like the familiar large and showy flowers. Each tiny flower has the requisite three to six dangling male stamens for making pollen, and two to three feathery female stigmas to catch the pollen, accomplish fertilization, and produce fruits with generally a single seed inside a grainlike fruit. (Rushes are the exception with several seeds per mature seedpod.) The tiny flowers are produced in short or long clusters called spikelets, and these may be arranged into ever-complex aggregations such as spikes, plumes, or airy panicles. When taken in mass, they can be quite stunning.

All graminoid flowers are wind pollinated and thus don't need showy petals to attract insects or birds; however, they must produce pollen in overabundance to achieve adequate dispersal via the capricious action of the wind. This leads to hay fever in humans during summer when some of this wind-blown grass pollen lands in your nose. Successful cross-pollination results in an abundance of seeds that may be wind distributed or dropped by animals as they scurry and fly about gathering the seeds for food. This makes these plants above average in usefulness for naturalizing in wild or managed areas and in informal garden situations such as meadows and open woodlands. Many of these plants thrive on disturbance and may disappear if woody vegetation is allowed to take over. Many native and naturalized graminoids can be found in disturbed areas that we wryly call "waste areas."

Grasses are generally sun loving. Even a half day of sun may not be enough to bring out the best in most grasses. While adapted to a wide variety of soil types, grasses generally will better tolerate well-drained and dry soils. All Southeastern grass species are warm growing, that is, they grow during the warm-to-hot seasons and go dormant in cold weather. A few warm-growers may have some winter appeal, as the evergreen rosettes of some species of *Dichanthelium*, but they flower and set seed in summer. Cool-season grasses are typically nonnative lawn grasses such as bluegrass and fescue.

Most grasses can be grown easily from seeds, though they will take longer to mature than plugs or plants. Some species may be available only as seeds. Grasses are also propagated by division in spring as the weather warms and growth starts. Don't plant or divide warm-season grasses during very cold weather.

Although grasses often self-sow in the garden and thrive with abandon in situations where there is less competition, we would not call them invasive. Their reproduction may be curtailed by deadheading (removing ripening fruits) before they mature, but you may miss some of the show from seed heads late in the season.

The greatest landscape features of graminoids are in their abilities to appeal to our senses as we enjoy the various aspects of plants in the garden—graceful movements in the wind, the rustle of dry grasses in the breeze, striking colors in the fall, vertical structure in the winter snow, blue-green sedge leaves on a shady forest floor. Thus, they act as textural enhancements in the perennial garden, providing stiffly upright or willowy growth, usually enhanced by backlighting, in contrast to their broad-leaved garden companions that are grown for their bold leaves or showy flowers.

THE AMERICAN LAWN

There has been much talk about the unnaturalness and energy consumption of typical American lawns. You sow, mow, weed, feed, mow, water, spray . . . and then mow some more. In times of drought, the lawn turns brown, although most grasses can survive, even going dormant for a period. When it's too wet, diseases crop up, and the grasses die.

Despite these drawbacks, lawns are not going away. Too many people like them, and they do have ornamental appeal. Almost all lawns are produced by nonnative turf-forming grasses, but if you want to try a native lawn, St. Augustine grass may be viewed as a native species that could be considered a turf grass. It makes a coarse-textured lawn that does not require heavy mowing. It apparently originally grew in brackish marshes. Additionally, I could suggest common carpet grass (*Axonopus fissifolius*) that is native from Virginia on southwards as a potential candidate, with reports of it growing in less than ideal situations of low nutrition and acidic pH. Both grasses spread by creeping stems.

Andropogon gerardii ★★★★

Big bluestem, turkey-foot grass

Poaceae (grass family)

HABITAT & RANGE prairielike roadside sites and open woodlands in the Southeast **ZONES** 3–9 **SOIL** wide range, moist to dry **LIGHT** sun.

DESCRIPTION Robust, strongly upright, clump forming, 5–8 ft. tall. Leaves to 18 in. long and to 1/4 in. wide, green to blue-green in summer, turning reddish brown in autumn, persisting into winter. Inflorescence a cluster of 4-in.-long stiff spikes, usually occurring in threes at the tips of the stems, diverging like the toes of turkey feet, flowering late August to early September.

PROPAGATION seeds (slow); division in spring **LANDSCAPE USES** screenings, perennial borders, meadows, prairie restorations **EASE OF CULTIVATION** easy **AVAILABILITY** common.

NOTES This elegant grass is unmistakable in its tight, upright clumps topped with three-clustered spikes. It offers an element of formality to the garden and stalwart majesty to the prairie-meadow. The tallest of the *Andropogon* species, it can be quite dramatic, especially in autumn and winter.

Andropogon gerardii

Andropogon glomeratus ★★★

Brushy bluestem

Poaceae (grass family)

HABITAT & RANGE fields, roadsides, open woods, savannahs, and bogs, mostly throughout the Southeast **ZONES** 6–10 **SOIL** moist, well drained to seasonally wet **LIGHT** sun.

DESCRIPTION Upright, strictly clump forming, 4–5 ft. tall, with round stems and ribbonlike leaves that are green in summer, turning orange-brown in autumn, persisting well into winter. Inflorescences much branched and enlarged towards the top, becoming showy as they fluff up with the ripening seeds, in autumn.

PROPAGATION seeds; division in spring **LANDSCAPE USES** meadows, natural areas, embankments **EASE OF CULTIVATION** easy **AVAILABILITY** common.

Andropogon glomeratus

Andropogon ternarius

Aristida stricta

NOTES Brushy bluestem is commonly seen throughout eastern North America in the lower edges of old fields and roadside ditches and is an indicator of soil that is a bit more moist than the surrounding soil. The robust clumps of plumes may form eye-catching stands along a roadside or embankment seepage. It is a worthwhile addition to the ornamental grass garden or adjacent areas if it is not allowed to self-seed.

Elliott's broom-sedge (*A. elliottii*, syn. *A. gyrans*), with also enlarged inflorescences and broader leaves in the inflorescence, grows to 3 ft. tall and tolerates drier conditions.

These are related to the far more commonly seen broom-sedge (*A. virginicus*), an abundant grass of old fields, meadows, and roadside waste areas. While beautiful with its expansive masses of waving stems, turning brown in autumn and winter, it's too aggressive and "ordinary" for the garden. It is also famous for leaching allelopathic chemicals into the soil, inhibiting the growth of other plants; therefore, its cut stems should not be used as mulch.

Andropogon ternarius ★★★

Split-beard bluestem

Poaceae (grass family)

HABITAT & RANGE dry to moist roadsides and meadows, mostly throughout the Southeast **ZONES** 6–10 **SOIL** sandy, well drained **LIGHT** sun.

DESCRIPTION Strictly clump forming, to 3 ft. tall, less stiffly erect stems often diverging. Leaves ribbonlike, often purplish, producing white fluffy discrete seed heads along the stems in autumn.

PROPAGATION seeds; division in spring **LANDSCAPE USES** specimens in grass collections, meadows, adjacent areas, embankments **EASE OF CULTIVATION** easy **AVAILABILITY** frequent.

NOTES Split-beard bluestem is a smaller, less robust and less plumelike plant than brushy bluestem, with very attractive white fluffy seed heads. It should be more widely planted. Because it's less stiffly upright, it might mix better with other kinds of perennials in the border or bed. It holds its color well into winter.

Aristida stricta ★★

Carolina wiregrass, pineland three-awn

Poaceae (grass family)

HABITAT & RANGE open pine forests in coastal North and South Carolina **ZONES** 7–10 **SOIL** sandy, well drained **LIGHT** sun to light shade.

DESCRIPTION Strictly clump forming to 3 ft. tall, with narrow stems and leaves diverging symmetrically to form a large starburst 3–5 ft. across. Flowers and fruits somewhat inconspicuous in autumn.

PROPAGATION seeds; division in spring **LANDSCAPE USES** groundcover and scattered masses in dry, sandy situations **EASE OF CULTIVATION** difficult **AVAILABILITY** rare.

NOTES Wiregrass species are very conspicuous in the Southeast in sandy, open or piney woods. They form strong clumps and are tolerant of fire. They may be grown only in very well drained sunny situations.

Southern wiregrass (*Aristida beyrichiana*) is a very similar species that replaces Carolina wiregrass from central South Carolina to Florida and west to Mississippi. It is just as ubiquitous in ecologically important long-leaf pine stands.

Arundinaria gigantea ★★

Giant cane, river cane

Poaceae (grass family)

HABITAT & RANGE low woods, bogs, savannahs, and dry woods throughout the Southeast **ZONES** 6–9 **SOIL** wide range, moist to dry to seasonally wet **LIGHT** sun to shade

DESCRIPTION Long-creeping rhizomes with tough, greenish woody stems to 30 ft. tall, often shorter, forming extensive colonies. Leaves evergreen, to 1 ft. long and 2 in. wide, formed into coarse featherlike arrangements. Flowering branches nonshowy, about 1 ft. long, appearing sporadically in summer.

PROPAGATION division of rhizome **LANDSCAPE USES** dense habitat cover in natural areas, or for floodplains and land restorations **EASE OF CULTIVATION** easy **AVAILABILITY** frequent.

NOTES You may not purposely plant giant cane unless you want a good slow-growing, soil-holding species for erosion control, but it's found widely in the Southeast and is a critical element of our flora, often forming extensive canebrakes of impenetrable growth. It is very important to many species of wildlife. It is fire dependent, often flowering after burning. Some believe that the colony dies after flowering and must be reestablished by seeds. This native cane is not to be confused with the widely naturalized and highly invasive golden or fishpole bamboo (*Phyllostachys aurea*) from China and Japan.

Arundinaria gigantea

Carex flaccosperma ★★★

Blue wood sedge

Syn. *Carex glaucodea*
Cyperaceae (sedge family)

HABITAT & RANGE low woods and meadows, mostly in the Piedmont and coastal plain of the Southeast **ZONES** 6–9 **SOIL** wide range, moist to dry **LIGHT** part shade to shade

DESCRIPTION Evergreen, loosely clump forming, to 1 ft. tall, spreading slowly by seed. Leaves up to $\frac{1}{2}$ in. wide, blue-green. Flowers not showy.

PROPAGATION seeds; division in spring **LANDSCAPE USES** groundcover, groupings, massing, naturalizing **EASE OF**

Carex flaccosperma

Carex glaucescens

CULTIVATION easy **AVAILABILITY** common.
NOTES Useful and attractive in formal and informal settings, especially in poorer woodland soils where other plants may not thrive. Self-sows but not invasive.

A similar species of the Upper Southeast, spreading sedge (*Carex laxiculmis*) has an attractive clump-forming cultivar 'Hobb' (trade name Blue Bunny).

Carex glaucescens ★★★

Wax sedge, blue sedge

Cyperaceae (sedge family)

HABITAT & RANGE pinelands, swamps, and savannahs throughout the Southeast **ZONES** 3–9 **SOIL** moist, well drained to poorly drained **LIGHT** part sun to sun.
DESCRIPTION Evergreen, large, clump forming, to 3 ft. tall. Leaves arching, lax, to ½ in. wide. Flowers not showy. Fruits are small nutlets clustering in pendulous spikes, appearing whitish blue-green in summer.
PROPAGATION seeds; division in spring **LANDSCAPE USES** specimens, groupings, coarse groundcover, wetland naturalizing **EASE OF CULTIVATION** easy **AVAILABILITY** infrequent.
NOTES Showier than most sedges, this interesting plant is suitable for a moist sun garden, or a moist woodland setting in part sun. The striking fruits are eye-catching.

The similar fringed sedge (*C. crinita*) is a large evergreen clumper, 2–4 ft. wide that makes a dramatic statement in an open moist site or in moist to dry woodland gardens; it may be just the right informal plant for some special setting. Can become massive and will self-sow a bit.

Carex lupulina

Carex lupulina ★★★

Hop sedge

Cyperaceae (sedge family)

HABITAT & RANGE moist to wet woods throughout the Southeast **ZONES** 3–9 **SOIL** wide range, moist to wet **LIGHT** part sun to part shade
DESCRIPTION Loosely clump forming, to 3 ft. tall. Leaves to ½ in. wide. Flower stems bear elongated spikes of inflated hopslike fruit clusters near the top of the stalk in summer.
PROPAGATION seeds; division in spring **LANDSCAPE USES** naturalizing in wet to moist woods and pond margins **EASE OF CULTIVATION** easy **AVAILABILITY** frequent.
NOTES This is a nifty sedge for the informal moist-to-wet garden or wetland edge. The enlarged female spikes are dramatic in summer.

Gray's sedge (*Carex grayi*) is similar but with more rounded spikes.

Carex pensylvanica ★★★

Pennsylvania sedge

Cyperaceae (sedge family)

HABITAT & RANGE dry to moist woodlands, mostly throughout the Piedmont and mountains of the Upper Southeast **ZONES** 4–8 **SOIL** wide range, average to dryish **LIGHT** part sun to shade
DESCRIPTION Delicate, loosely clump forming, from elongate rhizomes, with stems to 1 ft. tall. Leaves delicate, green, may turn yellowish in autumn. Spring flowers not showy.

Carex pensylvanica

PROPAGATION seeds; division in spring **LANDSCAPE USES** groundcover, groupings, naturalizing **EASE OF CULTIVATION** easy **AVAILABILITY** frequent.

NOTES Forms a delicate groundcover in dryish woods. The spreading clumps always look a bit disheveled, but pleasing.

A similar species is the Appalachian sedge (*C. appalachica*).

Carex plantaginea

Carex plantaginea ★★★★

Seersucker sedge, plantain-leaf sedge

Cyperaceae (sedge family)

HABITAT & RANGE rich woods, often in less-acidic soil, mostly in the mountains, and in Valley and Ridge, and Cumberland Plateaus **ZONES** 5–8 **SOIL** moist, well drained **LIGHT** part shade to shade

DESCRIPTION Evergreen, loosely clump forming, to 1 ft. tall and wider. Leaves grass green, ribbonlike to 1 in. wide, somewhat crinkled and becoming gracefully lax. Flowers small, worth noticing on graceful stems, blooming in spring.

PROPAGATION seeds; division in spring **LANDSCAPE USES** groundcover, groupings, naturalizing **EASE OF CULTIVATION** easy **AVAILABILITY** common.

NOTES This most wonderful sedge was one of the first species to adorn my shady wildflower garden. It mixes well with all kinds of herbaceous plants and ferns, and looks soothing on gentle slopes and ledges in a woodland setting. It is long-lived, and if sufficiently moist will naturalize readily.

A similar species is broadleaf or silver sedge (*C. platyphylla*), whose leaves are more upright, not crinkled, and often more bluish, and should prove to be equally attractive and desirable.

Carex platyphylla

Chasmanthium latifolium 

River-oats, wild-oats

Syn. *Uniola latifolia*

Poaceae (grass family)

HABITAT & RANGE bottomland woods, streambanks, and glades, throughout the Southeast **ZONES** 6–9 **SOIL** wide range, moist to dry **LIGHT** sun to shade

DESCRIPTION Strongly upright, clump forming, to 4 ft. tall. Leaves green, to ¾ in. wide. Flowering begins in midsummer, producing conspicuous ¾ in. flattened spikelets on dangling stalks, the whole plant more lax in shade, lasting through the winter.

PROPAGATION seeds; division in spring **LANDSCAPE USES**

Chasmanthium latifolium

perennial borders, groupings, masses, containers, naturalizing **EASE OF CULTIVATION** easy **AVAILABILITY** common.
NOTES This widely known and attractive grass is especially suitable for informal woodlands and shaded settings. Seedlings readily self-sow, but are easily removed.

Cymophyllus fraserianus ★★★★

Fraser's sedge, flowering sedge

Syn. *Carex fraseriana*
Cyperaceae (sedge family)

HABITAT & RANGE rocky ledges and seeps in shady, acidicv woods, mostly restricted to the southern Appalachian Mountains **ZONES** 6–8 **SOIL** moist, well drained **LIGHT** shade to part shade
DESCRIPTION Evergreen, clump forming, with numerous growing points, to 1 ft. tall. Leaves are basal, up to 1 in. wide, dark green with very fine teeth. Flowers are showy, in tight clusters on long green stems, blooming in spring. Fruits are small nutlets, somewhat showy, maturing in summer.
PROPAGATION seeds, sown fresh; division of clump **LANDSCAPE USES** specimens in woodland rock gardens or slopes **EASE OF CULTIVATION** moderate to easy **AVAILABILITY** infrequent.
NOTES Fraser's sedge is an amazingly beautiful plant. Its broad leaves are dark green, and the striking pure white flowers are much showier than those of any other species of *Carex*, or most any sedge or graminoid for that matter. This sedge is tough and long-lived in average garden soil. I have had a small clump for thirty-five years that flowers every year. It requires two different plants to cross-pollinate and make seeds. Well worth growing.

Cymophyllus fraserianus

Cyperus strigosus ★★

Umbrella flatsedge

Cyperaceae (sedge family)

HABITAT & RANGE marshes and wet ditches throughout the Southeast **ZONES** 4–10 **SOIL** moist to wet **LIGHT** sun.
DESCRIPTION Coarse, clump forming, with several stems to 30 in. tall. Leaves greenish yellow, about ½ in. wide. Flowers and fruits in clustered spikelets in late summer on the tops of stems, with wide leaves beneath them, giving an umbrella effect.
PROPAGATION seeds; division in spring **LANDSCAPE USES** specimens, naturalizing in pond margins and wetlands **EASE OF CULTIVATION** moderate to easy **AVAILABILITY** infrequent.

Cyperus strigosus

NOTES This is one common representative of a group with many species across the region. They are architecturally distinctive, but often freely self-sowing. Great for waterfowl food.

Elymus hystrix ★★

Bottlebrush grass

Syn. *Hystrix patula*
Poaceae (grass family)

HABITAT & RANGE moist woodlands and floodplains, mostly throughout the Piedmont and mountains of the Upper Southeast through Alabama **ZONES** 3–8 **SOIL** moist, well drained **LIGHT** shade.
DESCRIPTION Erect, clump forming, to 3 ft. tall. Flowering in summer, inflorescences upright, producing bottlebrush-like spikes with prominent needlelike awns.
PROPAGATION seeds **LANDSCAPE USES** informal groupings and naturalizing in woodlands **EASE OF CULTIVATION** easy **AVAILABILITY** infrequent.
NOTES The unusual needlelike heads are distinctive and attractive. This grass makes a great addition to the informal woodland garden because it makes you want to reach out and feel it. It is not dangerously sharp.

A similar species is Virginia wild-rye (*E. virginicus*). Robust and striking, it perhaps is more available.

Elymus hystrix

Muhlenbergia capillaris ★★★★

Pink muhly

Poaceae (grass family)

HABITAT & RANGE woodlands, savannahs, and rock outcrops throughout the Southeast **ZONES** 6–10 **SOIL** wide range, moist to dry **LIGHT** sun.
DESCRIPTION Informal clump forming, to 3 ft. tall. Flowering September to November, with masses of tiny pink flowers and fruits, lasting into winter.
PROPAGATION seeds; division in spring **LANDSCAPE USES** perennial borders, groundcover, groupings **EASE OF CULTIVATION** easy **AVAILABILITY** common.
NOTES Pink muhly is one of the most distinctive grasses for fall color and is especially useful in masses in difficult places such as traffic medians and dry, rocky soils. The effect is a dense cloud of pink spray that waves with the slightest breeze.

The very similar dune hairgrass or sweetgrass of coastal Carolinas (*M. sericea*) comes from the outer coastal fringes of the Carolinas. The related savanna hairgrass (*M. expansa*) is from the damp pinelands of the Southeast. Each would produce a beautiful autumn billowy pink "puffusion."

Another pink-to-purple billowy grass is purple lovegrass (*Eragrostis spectabilis*). It is a clumper to 3 ft. tall and

Muhlenbergia capillaris

Panicum virgatum

Rhynchospora latifolia

Saccharum giganteum

wide with late summer airy masses. It is excellent when used to stabilize disturbed soil on roadbanks and ditches, and in garden drifts; it does not last well into winter.

Panicum virgatum ★★★

Switch grass

Poaceae (grass family)

HABITAT & RANGE marshes, pinelands, and open ground throughout the Southeast **ZONES** 4–10 **SOIL** wide range, moist to dry **LIGHT** sun.
DESCRIPTION Variously shaped, clump forming, 4–8 ft. tall, may slowly spread. Leaves green to bluish. Flowering in mid to late summer, producing attractive plumes of tiny flowers.
PROPAGATION seeds; division in spring **LANDSCAPE USES** specimens, perennial borders, groupings **EASE OF CULTIVATION** easy **AVAILABILITY** common.
NOTES In the wild, switch grass is large and informal, distinctive in producing tall airy flowering stalks. In cultivation, more than a dozen named cultivars emphasize various features from stiffly upright growth and steel-blue foliage ('Heavy Metal'), more lax and broadly billowing greenish growth ('Hänse Herms'), to informal red foliage ('Shenandoah').

A number of related species in the Southeast are now recognized as the witch grasses. These have overwintering rosettes of green leaves that could have some merit as attractive groundcovers in garden situations. One example is deer-tongue witch grass (*Dichanthelium clandestinum*).

Rhynchospora latifolia 

White-top sedge, white-bracted sedge

Syn. *Dichromena latifolia*
Cyperaceae (sedge family)

HABITAT & RANGE moist ditches, sand flats and savannahs, in the outer coastal plain from North Carolina to Texas, disjunct into Coffee County, Tennessee **ZONES** 7–10 **SOIL** moist to wet **LIGHT** sun.
DESCRIPTION Colony forming from creeping rhizomes, stems to 1 ft. or more tall. Leaves basal, to ½ in. wide. Flower clusters whitish, topping the stems, in summer, with five or more distinctive long white bracts immediately beneath them.
PROPAGATION division anytime **LANDSCAPE USES** groundcover, naturalizing in moist meadows or ditches **EASE OF CULTIVATION** easy **AVAILABILITY** frequent.
NOTES A strikingly beautiful species, but aggressively forms colonies and thus not compatible with other plants in moist soil or in a bog garden.

A slightly smaller almost identical species is narrow-bracted sedge (*R. colorata*).

Saccharum giganteum 

Sugarcane plume grass

Syn. *Erianthus giganteus*
Poaceae (grass family)

HABITAT & RANGE savannahs, ditches, and woodland borders throughout the Southeast **ZONES** 7–10 **SOIL** moist to wet to dryish **LIGHT** sun.
DESCRIPTION Robust, upright, clump forming, to 12 ft. tall. Leaves green, to 3 ft. tall, turning reddish in autumn. Flowering in late summer into autumn, with showy plumes.
PROPAGATION seeds; division of clump **LANDSCAPE USES**

specimens, groupings, naturalizing in rough areas **EASE OF CULTIVATION** easy **AVAILABILITY** frequent.
NOTES One of our largest native grasses, sugarcane plume grass is distinctive and attractive in the wild, a bit aggressive in self-sowing in cultivation. It makes a bold statement, perhaps a good native substitute for pampas grass. It can grow in wet or dry sites. Be aware that it produces a very tough specimen.

Schizachyrium scoparium

Schizachyrium scoparium ★★★

Little bluestem

Syn. *Andropogon scoparius*
Poaceae (grass family)

HABITAT & RANGE dry woods and prairielike roadside sites throughout the Southeast **ZONES** 3–10 **SOIL** wide range, average to dry **LIGHT** sun.
DESCRIPTION Strictly clump forming, to 3–4 ft. tall. Leaves green to bluish, turning orangish to reddish in fall and into winter. Flowering in late summer, producing delicate silvery tufts of plume along the drying stems.
PROPAGATION seeds; division in spring **LANDSCAPE USES** perennial borders, groupings, meadows, prairie restorations **EASE OF CULTIVATION** easy **AVAILABILITY** common.
NOTES Along with big bluestem, little bluestem is one of the hallmarks of the original American tallgrass prairies. Several cultivars have better bluish stems (for example, 'The Blues'). May have a wider formal landscape appeal than once thought and works well as a big clumper in a perennial border or bed.

Scirpus cyperinus

Scirpus cyperinus ★★

Wool-grass

Cyperaceae (sedge family)

HABITAT & RANGE marshes and wet ditches throughout the Southeast **ZONES** 4–9 **SOIL** dry to wet **LIGHT** sun.
DESCRIPTION Coarse clump forming, to 6 ft. tall. Leaf blades green, to about ¾ in. wide. Flowering in late summer, with striking loose-to-dense floppy heads of fluffy fruits.
PROPAGATION seeds; division in spring **LANDSCAPE USES** naturalized in wetlands or moist meadows and ditches **EASE OF CULTIVATION** easy **AVAILABILITY** frequent.
NOTES Wool-grass is a beautiful and distinctive roadside indicator of wet areas in late summer. While bold and attractive, it's coarse and self-sowing, making it best used in naturalizing.

Sorghastrum nutans

Sorghastrum nutans ★★★★

Indian grass

Poaceae (grass family)

HABITAT & RANGE dry and alluvial woods, prairielike roadsides, and fields throughout the Southeast **ZONES** 4–10 **SOIL** moist to dry, well drained **LIGHT** sun.
DESCRIPTION Strictly clump forming, to 7 ft. tall. Leaves green to bluish, turning yellow to orange in autumn and persisting into winter. Flowering in late summer, with showy clusters of small yellowish flowers.
PROPAGATION seeds; division in spring **LANDSCAPE USES** perennial borders, meadows, prairie restorations **EASE OF CULTIVATION** easy **AVAILABILITY** common.
NOTES This is my favorite big grass, elegantly robust in form but delicate in flower. It is perhaps the showiest grass for its conspicuous yellow stamens in late summer into early autumn. With the bluestems, this is a characteristic species of the American tallgrass prairie. I love to walk among the roadside clumps of Indian grass and think on how majestic a mature prairie must have been in late summer, and am glad we have this representative so common in the Southeast. A great choice in grasses. It has blue-stemmed cultivars 'Sioux Blue' and 'Indian Steel'.

Stenotaphrum secundatum ★★★

St. Augustine grass

Poaceae (grass family)

HABITAT & RANGE brackish marshes in the outer coastal areas of the Southeast **ZONES** 8–10 **SOIL** sandy, moist to dry **LIGHT** sun.
DESCRIPTION Somewhat evergreen, spreading, rhizomatous groundcover to 6 in. tall. Leaves green, 4 in. long and ⅜ in. wide. Flowers inconspicuous on short, dense spikes in summer.
PROPAGATION division of rhizomes in summer **LANDSCAPE USES** lawn and informal groundcover **EASE OF CULTIVATION** easy **AVAILABILITY** common.
NOTES Here is your choice for a native grass lawn if you live in the mid- to Deep South. It is especially good in sandy soil but is quite tolerant. Better yet, this grass rarely needs mowing. It has a much coarser texture than any other southern lawn grasses and is drought tolerant.

The similar-looking carpet grass (*Axonopus fissifolius*) might be worth trying in acidic, low-nutrient situations.

Tridens flavus ★★

Purple-top grass

Poaceae (grass family)

HABITAT & RANGE fields, roadsides, and open woods throughout the Southeast **ZONES** 4–10 **SOIL** wide range, wet to dry **LIGHT** sun.
DESCRIPTION Upright, clump forming, to 4 ft. tall. Flowering in late summer with loose, open purple plumes, turning tan and lasting all winter
PROPAGATION seeds; division in spring **LANDSCAPE USES** naturalizing in meadows, ditches, disturbed areas **EASE OF CULTIVATION** easy **AVAILABILITY** frequent.
NOTES This is one of the most characteristic roadside grasses of late summer. Because it self-sows readily, it's best used outside the maintained garden.

Stenotaphrum secundatum

Tridens flavus

Aquatic Plants

Native aquatic plants are fun to grow because they attract interesting wildlife such as dragonflies and frogs, are prone to flower over a long period, are easy to propagate, and only require that you keep them sunny and very wet. They are important because they help make a backyard water feature more stimulating and natural, provide shelter for animals, and help keep the water clean and clear. These plants thrive in standing water, but can survive if the water dries up to mud for short periods.

Typically, aquatic plants have abundant air-filled tissues in their roots, stems, and leaves that help them survive in the mucky substrate of the underwater environment. True aquatic plants (the subject of this chapter) are not adaptable to the average perennial border, unlike some wetland plants that are (see box). Some additional species may be more suitable for the merely moist or wet soil (but not constant standing water) of a bog garden (see separate chapter).

Aquatics can be divided into four categories depending on their position in the water.

1. **Floating-leaf plants**, like water-lilies and water shield, root in the mucky soil below while their leaves rise on elongating petioles to float on the surface of the water.
2. **Free-floating plants**, like duckweeds and water-meal, ride the surface of the water without attaching to the bottom, though they may strike root in moist muck if the water level goes way down.
3. **Emergent plants,** like cattails and pickerel-weed, root in mucky substrate in shallow water usually 1–12 in. deep, and then grow mostly well-above water level. These are often called "marginals," as they grow at the edges of bodies of water.
4. **Submerged plants**, like cabomba and parrot-feather, are entirely below water, though always the flowers and often some vegetative parts may break the surface.

The primary plants for landscaping belong to the first three groups. Plants in the fourth category, being underwater, are rarely showy in landscape settings and therefore are not described in this book.

The magnificent shoals spider-lily (*Hymenocallis coronaria*) survives in the Catawba River of north-central South Carolina by producing rapid-growing roots during low water times and attaching in the rock crevices.

All aquatic plants described here are herbaceous, rhizomatous, and perennial. Unless otherwise noted in the descriptions, all are deciduous. Generally, water plants are heavy feeders and benefit from a little slow-release (tabs or stakes) fertilizer during the growing season if more flowers or darker green leaves are desired. This is especially true of nursery selections with larger flowers, such as water-lilies and lotus. If a water feature has naturally accumulated organic matter and a variety of live-in wildlife, less fertilizer is necessary or desired as it just makes the plants grow more abundantly.

The most common aquatic habitats are ponds, marshes, slow-flowing streams, and swamps. All are mostly sunny sites except the swamp, which is, by definition, a type of forest. All water plants play important roles in native aquatic ecosystems and have specific niches based on water depth around the margin of a lake or slow-moving stream. These plants provide spawning beds for aquatic animals, create places for animals to hide and hunt, and prevent erosion.

Beautiful and stately, the flowers of pickerel-weed (*Pontederia cordata*), an emergent aquatic, are as handsome as those of any nonaquatic species.

Alas, they may give trouble to people who like to fish in small ponds, and they become aggressive in taking over part of the habitat.

Aquatic plants usually occupy distinct zones in water, with emergent plants in the shallow water's edge and floating plants in deeper water. Often species will segregate themselves in the shallower water based on precise depth, with, for example, blue-flag iris and lizard-tail preferring 6 in. or less of water, arrow-head and pickerel-weed at 1 ft. deep, and cattails at 18 in. Don't worry about providing exact water levels, as most aquatics are quite adaptable to various water depths within their range in cultivated settings.

Virtually all aquatic plants are competitive and fast growing, forming large clumps or spreading colonies by underground rhizomes. Keep this in mind as you plant them. Give them plenty of room to spread and don't be afraid to thin them out ruthlessly as necessary—they will normally grow back. In addition to their propensity to spread, most aquatics are self-fertile and readily produce copious fruit and seeds. Birds and mammals benefit from this abundance, and in their activities carry seeds on muddy feet to new habitats. You may end up with more new seedlings than you wish.

Be diligent in recognizing and removing any before they become firmly established.

The flowers of many aquatic plants are sun loving and warm loving. In general, aquatic plants start growing later in the spring than terrestrial plants, as they need the water to warm up. Thus, it's best to wait for warm weather to divide plants or start seeds.

In the landscape, aquatics are typically grown in containers and pools. A variety of backyard vessels can be used, such as barrels, tubs, troughs, wading pools, or other large watertight containers. Ponds can be fabricated, perhaps as part of a rubber-sheet-lined excavation edged with stones and featuring a waterfall or a constructed system of flowing water. In fabricated ponds, grow emergent (marginal) and floating-leaved plants in broad, squat pots with clay-based garden soil covered with 1 in. of gravel. This keeps the plants heavy and sunk to the bottom, and from "washing out" of their pots if water movement or animal activity is severe. Never use peat-based potting soils since peat and perlite float so well. Aquatic plants love water movement and aeration; hot, stagnant water in a cramped or pot-bound situation will not allow for best performance.

Natural ponds, lakes, marshes, or other wetlands, of course, would be places to naturalize aquatic plants in various ways. Be sure to get some hardy local native "mosquito fish" called gambusia (catch them with permission from a local pond) to eradicate mosquito larvae if you have no ornamental fish.

Finally, enjoy your water feature or wetland. It can be more dynamic than a perennial border by providing beautiful flowers, vibrant foliage, water movement, gurgling waterfalls, shimmering reflections, and abundant animal activity and sounds. I am reminded every evening that my neighbor has a bullfrog that came to live in his 3-ft.-square patio pool.

WETLAND PLANTS THAT ALSO GROW IN SOIL

The following common species of wetland plants can grow in regular garden soil and are described elsewhere in this book:

WILDFLOWERS

Asclepias incarnata (swamp milkweed)
Asclepias purpurascens (purple milkweed)
Eutrochium fistulosum (Joe-pye weed)
Helianthus angustifolius (swamp sunflower)
Hibiscus moscheutos (marsh mallow)
Lobelia cardinalis (cardinal flower)

VINES

Ampelaster carolinianus (climbing aster)
Decumaria barbara (climbing hydrangea)
Smilax smallii (Jackson-vine)

SHRUBS

Aronia spp. (chokeberry)
Cephalanthus occidentalis (buttonbush)
Itea virginica (Virginia-willow)
Ilex verticillata (winterberry holly)

TREE

Taxodium ascendens (pond-cypress)

Azolla caroliniana ★

Eastern mosquito fern, water fern

Azollaceae (waterfern family)

HABITAT & RANGE bodies of water throughout the Southeast **ZONES** 5–10 **SOIL** floats on water or forms mats on wet soil **LIGHT** part sun to sun.

DESCRIPTION Floating, evergreen fern with unfernlike little fronds with scalelike leaves, growing rapidly to form a mat on any water surface. Floating mats turn a beautiful reddish color in winter. Nitrogen-fixing blue-green bacteria (*Anabaena* spp.) live mutualistically in the frond flaps and are a source of fertilizer in Asian rice paddies.

PROPAGATION division **LANDSCAPE USES** novelty specimens in water gardens **EASE OF CULTIVATION** easy **AVAILABILITY** common, usually comes with other aquatic plants.

NOTES May be effective in any small contained water feature as a soft uniform cover that turns reddish in winter sun. But be warned: this species is aggressive and invasive and multiplies so rapidly as to choke the water surface and obscure any view of underwater features (keeps algae down though). Don't put it in large ponds as it will be difficult to eradicate, and any small pieces will quickly regrow. On the other hand, it cleanses the water; koi will eat it; and a little bit looks good with other plants. It is curious that if you push it underwater, it will bob back up, unwetted. Great for hiding frogs.

Bacopa caroliniana ★★

Blue water-hyssop

Plantaginaceae (plantain family), formerly in Scrophulariaceae (figwort family)

HABITAT & RANGE wet shores, tidal mud flats, and marshes, mostly in the coastal plain from South Carolina to Mississippi **ZONES** 7–10 **SOIL** clay-loam, sandy to mucky, wet **LIGHT** sun.

DESCRIPTION Creeping or floating perennial, deciduous, much-branched, rooting along the stem, up to 1 ft. tall, aromatic when bruised. Leaves somewhat succulent, opposite, simple, with three to nine veins per leaf, untoothed, ovate, 1 in. long. Flowers showy, blue, one at each leaf, with five petals about ¾ in. wide, blooming all summer. Fruit a tiny pod maturing sequentially.

PROPAGATION cuttings anytime **LANDSCAPE USES** floating aquatic or rooted in shallow water in large tubs and ponds, naturalized in natural pond edges and seeps **EASE OF CULTIVATION** easy **AVAILABILITY** common.

NOTES This bacopa is rarely noticed in the wild, but has a beautiful flower and grows so rapidly it has become used as a hanging basket plant in the florist trade. It adapts to growing in water or wet soil, and might be worth trying if you can keep its rapid spread under control. The lemony fragrant leaves are unique and interesting. It is often grown submerged in aquariums, nonflowering.

Its similar relative, white water-hyssop (*B. monnieri*), is even more widely used as a hanging basket and pot plant. The one-veined oblong leaves are not fragrant and the flowers are white.

Brasenia schreberi ★★

Water shield

Cabombaceae (watershield family)

HABITAT & RANGE shallow ponds throughout the Southeast **ZONES** 4–9 **SOIL** clay-loam, sandy to mucky, wet **LIGHT** sun.

DESCRIPTION Submerged perennial rooting in the mud with an expanse of floating elliptic leaves 1–4 in. long. Flower solitary, 1 in., reddish purple, on a stalk above the leaves, blooming all summer. All young underwater stems, leaves, and buds are heavily coated with unique slimy mucilage.

PROPAGATION division in late spring **LANDSCAPE USES** floating aquatic in large tubs and ponds, naturalized in ponds and sluggish streams **EASE OF CULTIVATION** easy **AVAILABILITY** frequent.

NOTES The football-shaped leaves of this miniature water-lily-like plant can cover the surface of still water in any situation, especially where a real water-lily would be too large. While the flowers are not striking, they are interesting; and I like to finger the slime-encased buds, especially in an accessible waist-high container.

Azolla caroliniana

Bacopa caroliniana

Brasenia schreberi, flower

Canna flaccida

Canna flaccida ★★★

Golden canna, bandana-of-the-Everglades

Cannaceae (canna family)

HABITAT & RANGE marshes, lake margins, and flooded pine flatwoods in the outer coastal plain from South Carolina to South Florida, westward to southwestern Louisiana **ZONES** 7–10 **SOIL** clay-loam, sandy to mucky, moist to wet **LIGHT** sun.
DESCRIPTION Emergent rhizomatous perennial, forming small stands, 2–4 ft. tall. Leaves yellow-green, bold, 12–20 in. long and 5 in. wide. Flowers pure yellow, 4–6 in. long, borne one to four on erect spikes, blooming in summer. Fruits are attractive, plump, warty seedpods about 2 in. long.
PROPAGATION seeds; division in late spring **LANDSCAPE USES** specimens for containers or naturalized in damp soil or shallow water **EASE OF CULTIVATION** easy **AVAILABILITY** frequent.
NOTES We are fortunate to have such an attractive, free-flowering, native canna that also makes a handsome foliage plant.

Brasenia schreberi, buds

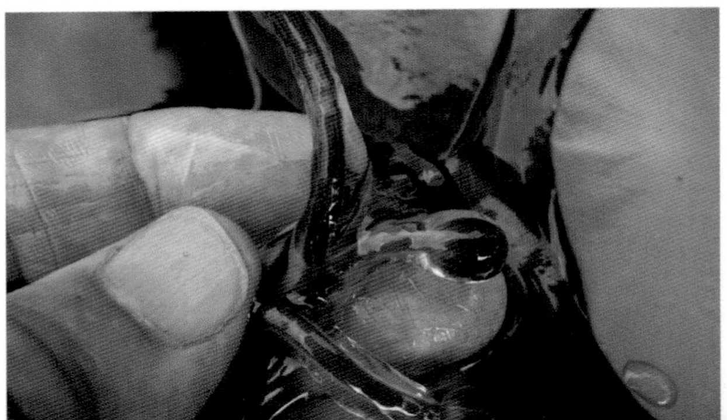

Dulichium arundinaceum ★★★

Three-way sedge

Cyperaceae (sedge family)

HABITAT & RANGE streambanks, marshes, bogs, and ditches throughout the Southeast **ZONES** 4–9 **SOIL** clay-loam, sandy to mucky, moist to wet **LIGHT** sun.
DESCRIPTION Emergent rhizomatous perennial spreading in shallow water, 1–2 ft. tall. Leaves light to dark green, narrow, 3 in. long and 1/4 in. wide, conspicuously three-ranked, spreading to erect. Flowers and fruits inconspicuous, late summer.

Dulichium arundinaceum

PROPAGATION division in late spring **LANDSCAPE USES** large containers and pond margins **EASE OF CULTIVATION** easy **AVAILABILITY** common.
NOTES Unique to North America, this species has a wonderful texture and tight growth habit that allows it to mix well with a variety of floating-leaved aquatic plants such as water-lilies and with shorter emergents around pond edges.

Equisetum hyemale ★★

Scouring-rush, horse-tail

Equisetaceae (horsetail family)

HABITAT & RANGE railroad embankments, roadsides, and streambanks, mostly throughout the Southeast **ZONES** 2–10 **SOIL** any type, wet to dryish **LIGHT** shade to sun.
DESCRIPTION Creeping evergreen, forming extensive colonies, with hollow, jointed, upright clustering stems that are ribbed, about ½ in. in diameter, with embedded sand crystals in the outer skin, 2–4 ft. tall. Leaves are tiny black toothed sheaths at each node. This being a primitive fern relative, it has no sori *per se*, but spores are borne in ½-in.-long yellowish cones at the tips of the larger stems.
PROPAGATION division **LANDSCAPE USES** container plants for patio and shallow pool **EASE OF CULTIVATION** easy **AVAILABILITY** common.
NOTES Wow, scouring-rush is a most striking plant when you see it growing in dense colonies and hear the stems rustle like dry parchment as you brush by. It might remind you of pan-pipes or hollow reeds with the wind whistling through its broken-off stems. Everyone wants to grow it, but watch out; it spreads rapidly and there is virtually no known herbicide for effective control. We have used the listed herbicide Casoron (active ingredient dichlobenil) with some success. We planted a gallon-sized clump outdoors in Charlotte in 1970 and it spread to over 800 square feet in mucky wet to dry rocky soil in sun and shade. Pieces of rhizomes and stems can reroot if you break them off in pulling. Be careful what you wish for, and perhaps enjoy these fascinating plants in containers.

Field horsetail (*E. arvense*) produces fascinating multibranched softer, brittle stems up to 18 in. tall. Don't plant that one in the ground either.

Equisetum hyemale

Habenaria repens ★★

Water-spider orchid

Orchidaceae (orchid family)

HABITAT & RANGE marshes, wet meadows, bogs, ponds, stream margins, and wet ditches, in the coastal plain from South Carolina to South Florida and westward to East Texas **ZONES** 7–10 **SOIL** clay-loam, sandy to mucky, moist to wet **LIGHT** mostly sun.
DESCRIPTION Emergent (or terrestrial) single-stemmed leafy perennial with somewhat weak tuberous roots, to 1 ft. tall (or more) in typical situations. Leaves 2–4 in. long, somewhat succulent. Flowers about ½ in. long, greenish to

Habenaria repens

whitish, many crowded on erect stalks, sepals and petals narrow and arching, spiderlike, blooming all summer. Fruits are elongate capsules.

PROPAGATION division of leafy proliferations from base **LANDSCAPE USES** novelty specimens for up-close viewing **EASE OF CULTIVATION** easy **AVAILABILITY** rare.

NOTES Here is your chance to grow a wild orchid in a relatively easy, self-maintaining aquatic setting, being careful not to let the rather delicate, brittle plant become overgrown by neighbors that are more vigorous.

Hymenocallis crassifolia ★★★★

Coastal spider-lily

Amaryllidaceae (amaryllis family)

HABITAT & RANGE marshes, wet meadows, bogs, ponds, stream margins, and wet ditches, in the coastal plain from South Carolina to South Florida and westward to East Texas **ZONES** 6–9 **SOIL** clay-loam, sandy to mucky, moist to wet **LIGHT** sun.

DESCRIPTION Emergent clump-forming perennial from a bulb. Leaves basal, narrow, erect, up to 2 ft. long, thick, shiny green. Flowers fragrant, white, two or three on a hollow stalk to 20 in. tall, with narrow sepals and petals, and six long stamens uniquely connected with a thin white veil-like membrane (corona), blooming in late spring. Inflated fruits have up to three large green grapelike seeds in summer.

PROPAGATION seeds as soon as ripen; division after flowering **LANDSCAPE USES** specimens or groupings in shallow water or damp soil **EASE OF CULTIVATION** easy **AVAILABILITY** rare.

NOTES The spider-lily's flowers are uniquely beautiful but short-lived. Each flower lasts one night, but there may be multiple stalks flowering over a few weeks.

The more robust shoals spider-lily, or Cahaba-lily (*Hymenocallis coronaria*), occurs in several special Fall Line (the distinct break between the Piedmont and the coastal plain) rocky river habitats in several large rivers in the Piedmont of South Carolina, Georgia, and Alabama. A mass of these in full bloom is a breathtaking sight in late spring. It blooms at night under the full moon with an alluring scent, attracting hawk moths and giddy onlookers. You wonder how these magnificent plants can survive in the middle of a major river, but the big clumps produce rapid-growing roots during low water times and attach in the rock crevices. They are worth trying in a man-made flowing stream.

Iris virginica

Southern blue flag

Iridaceae (iris family)

HABITAT & RANGE wetlands, ditches, and margins of lakes and streams from eastern Virginia, southward to South Florida and westward to East Texas **ZONES** 5–9 **SOIL** clay-loam, sandy to mucky, dryish to wet **LIGHT** part sun to sun.

DESCRIPTION Emergent perennial forming dense clumps, to

Hymenocallis crassifolia

Iris virginica

Iris fulva

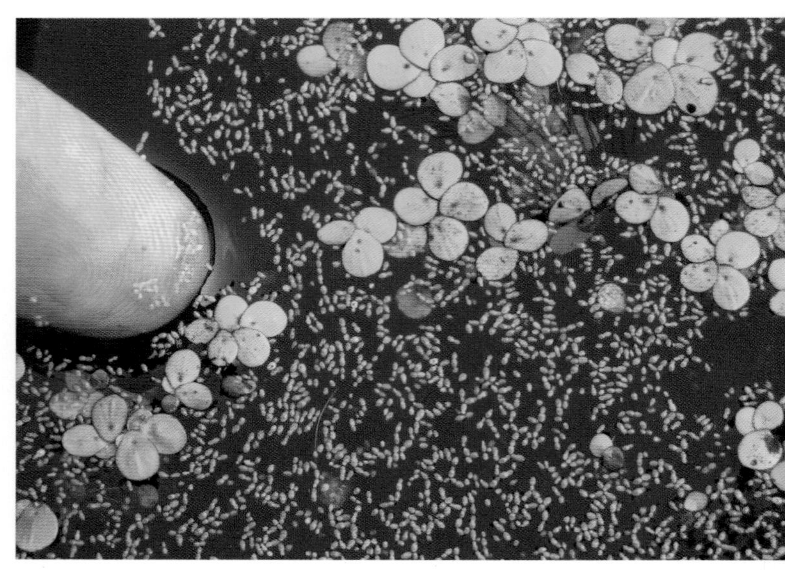

Lemna perpusilla with smaller water-meal (*Wolffia* sp.)

about 3 ft. tall. Leaves basal, narrow, gray-green to bright green. Flowers lavender to violet, with yellow spot on sepals (nonbearded), about 3 in. long, two or three on each tall stalk, blooming May to June. Fruits are fat green pods about 2 in. long.

PROPAGATION seeds sown when ripe (not stored dry); division in fall **LANDSCAPE USES** specimens, groupings, or masses in moist to wet soil or in rain gardens, naturalized in wetlands **EASE OF CULTIVATION** easy **AVAILABILITY** common.

NOTES Blue flag makes impressive clumps and slowly spreads by rhizomes to form masses. It can take some shade and drought, and may be a good choice in rain gardens if not too dry. There are several named horticultural selections, including a choice white-flowered form.

Louisiana iris, or copper iris (*Iris fulva*), is another showy southern iris found mostly along the Mississippi River corridor south from southern Illinois to the Gulf Coast of Alabama and extreme western Florida. It is less clump forming than blue flag, but otherwise growth requirements are similar. It has coppery-orange to reddish flowers on tall stems. Many selections are available.

Lemna perpusilla ★

Tiny duckweed

Araceae (arum family), formerly in Lemnaceae (duckweed family)

HABITAT & RANGE ponds scattered across the eastern United States **ZONES** 4–10 **SOIL** floats on water **LIGHT** part sun to sun.

DESCRIPTION Free-floating plant that reproduces by branching and rapidly covers a pond with green growth. Leaves rounded, floating, about 1⁄32 in. long, producing two or more roots per leaf. Flowers are miniscule and are rarely produced or seen.

PROPAGATION multiplies asexually on its own **LANDSCAPE USES** pleasant green pond cover **EASE OF CULTIVATION** easy **AVAILABILITY** common, often comes with other aquatic nursery plants; rarely sold *per se*.

NOTES Duckweeds are the world's smallest flowering plants. Perhaps you view them as the undesirable misnomer "pond scum" and don't want them on your water feature. While I would not purposely acquire any duckweed, I do find them pleasant to enjoy on a wild pond, and they do find their way into many a pond, brought by migrating waterfowl, acquiring nursery-grown aquatic specimens, or dumping of aquarium plants.

Another native, greater duckweed (*Spirodela polyrrhiza*), has leaves to 1⁄4 in. long (see *Taxodium distichum* photo on p. 21).

Ludwigia alternifolia ★★

Seedbox

Onagraceae (evening-primrose family)

HABITAT & RANGE marshes, ditches, savannas, and low woods, mostly throughout the Southeast **ZONES** 6–10 **SOIL** clay-loam, sandy to mucky, moist to wet **LIGHT** sun.

DESCRIPTION Perennial, single branched, 2–3 ft. tall. Leaves alternate, narrowly elongate, to 6 in. long. Flowers showy, yellow, with four petals, about 1 in. wide. Fruit is a dry, cubical seedpod with a hole in the middle, maturing in summer, persisting into winter.

PROPAGATION seeds; cuttings; division of clump **LANDSCAPE USES** delicate specimens in wetland association **EASE OF CULTIVATION** easy **AVAILABILITY** infrequent.

NOTES Ludwigias are classic wetland plants. They are not always showy, but the seedpods are downright charming.

Ludwigia alternifolia

Kids love the little square pods with a hole to let out the tiny seeds. The genus is divided into two distinct groups: some are the low, creeping and floating ones that can take over a pond. Then there are the upright, 2- to 3-ft. tall herbaceous or somewhat woody species called seedboxes. While only marginally ornamental, they are delightful to observe, and attractive to pollinators.

Nelumbo lutea ★★★

American lotus, yellow lotus

Nelumbonaceae (lotus-lily family)

HABITAT & RANGE ponds and natural lakes, mostly throughout the nonmountainous Southeast **ZONES** 4–9 **SOIL** clay-loam, sandy to mucky, moist to wet **LIGHT** sun.

DESCRIPTION Massive floating or emergent herbaceous perennial from very long creeping rhizomes. Leaves huge, round, 2 ft. or more across, with a petiole up to 8 ft. long (or more in deeper water) attached in the middle. Flower large, solitary on a stalk exceeding the leaves, up to 10 in. across, pale yellow, all parts numerous. Fruit's an enlarged pyramidal receptacle housing individual hard "seeds" in little chambers.

PROPAGATION division of rhizomes almost anytime during growing season **LANDSCAPE USES** large containers or ponds, but will invade any body of water **EASE OF CULTIVATION** easy **AVAILABILITY** frequent.

NOTES These impressive plants are exciting to behold, as they can become impressively large. Because they can also spread to unbelievable lengths and depths, they perhaps are best kept in very large containers or small man-made ponds where they do very well as long as the root system does not freeze hard. Don't plant one unless you don't plan to try and get rid of it; it will take over any body of water up to 4 ft. or more deep. The sacred or oriental lotus from Asia with beautiful pink flowers is more often seen in cultivated situations. Several selections of lotus are available.

Nymphaea odorata ★★★★

White water-lily

Nymphaeaceae (water-lily family)

HABITAT & RANGE ponds, lakes, slow streams, and ditches throughout the Southeast **ZONES** 4–10 **SOIL** clay-loam, sandy to mucky, moist to wet **LIGHT** sun.

DESCRIPTION Floating-leaf herbaceous perennial from thick

Nelumbo lutea

Nymphaea odorata

slowly creeping rhizomes, forming extensive stands. Leaves arising in close succession, blades orbicular, up to 1 ft., with a slit cut on one side (lotus and water shield leaves are completely circular). Flowers floating, 3–8 in. wide, white or rarely pink, all parts numerous, fragrant, opening and closing each day for (typically) two days, blooming all summer. Fruits are slime-filled seedpods that mature underwater.
PROPAGATION seeds; division of rhizomes in late spring
LANDSCAPE USES containers and ponds **EASE OF CULTIVATION** easy **AVAILABILITY** common.
NOTES Water-lilies are well-known and beautiful additions to larger home ponds. The more sun and fertilizer, the more flowers. Tropical water-lily flowers arise above the water, while temperate (native) ones float directly. The leaves can cover small ponds; a single mature plant can easily cover an 8-ft. diameter in one season. Alternately, I have seen water-lilies become affective leafy groundcover in wet mud as the water level shrinks. White water-lily is the only common native water-lily in the Southeast (outside of South Florida). The pink form is rarer. Several cultivars are available.

Nymphoides cordata

Orontium aquaticum

In many wild ponds the spatterdock or yellow pond-lily (*Nuphar advena*) can dominate with its large orbicular to oval leaves crowding the water surface. You may think they are white water-lilies until you see the not-so-showy 1 in., solitary, globose yellow flowers emergent on long stalks. Not nearly as ornamental as water-lilies.

Nymphoides cordata ★★★

Little floating-heart, banana plant, water snowflake

Menyanthaceae (buckbean family), formerly in Gentianaceae (gentian family)

HABITAT & RANGE ponds, streams, and swamp forests in the coastal plain of the Southeast **ZONES** 7–10 **SOIL** clay-loam, sandy to mucky, moist to wet **LIGHT** sun.
DESCRIPTION Perennial from stocky rhizomes. Leaves floating, orbicular with a rounded slit to the petiole, green above, purple beneath, with a smooth texture; flowers white, 3⁄4 in. wide, in clusters on stems fused to the leaf petiole such that they appear to arise closely from the floating leaf. After flowering, a cluster of 1-in. dark green tubers forms on the flowering stem.
PROPAGATION division of rhizome, or tubers **LANDSCAPE USES** specimens in containers or ponds for up-close inspection **EASE OF CULTIVATION** easy **AVAILABILITY** common.
NOTES This is a curiosity as the floating leaves appear to bear a cluster of flowers. Furthermore, the unique bunching of the tubers makes all the floating-hearts popular as floating aquarium plants.

Big floating-heart (*Nymphoides aquatica*) has larger flowers at 1¼ in. wide. Neither species is as invasive as other aquatic plants.

Orontium aquaticum 

Golden-club

Araceae (arum family)

HABITAT & RANGE bogs, marshes, swamps, streams, and wet ditches throughout the Southeast, rarer in Tennessee **ZONES** 6–9 **SOIL** clay-loam, sandy to mucky, moist to wet **LIGHT** part sun to sun.
DESCRIPTION Emergent perennial, tightly clump forming. Leaves blue-green usually with reddish petioles, 6–16 in. long. Flowers tiny, bright yellow, lacking conspicuous sepals and petals, embedded in a fleshy spike to 6 in. long, blooming in early spring. Fruits are one-seeded, berrylike, ripening soon after flowering.
PROPAGATION seeds sown when ripe; division after flowering **LANDSCAPE USES** marginal specimens in containers, ponds, and slow-moving streams **EASE OF CULTIVATION** moderate **AVAILABILITY** frequent.
NOTES A very distinctive and attractive aquatic plant with blue-green leaves and yellow spikes, golden-club likes slowly moving water best and does not like to be pot bound. This member of the arum family does not have a leafy bractlike spathe surrounding the flower stalk like Jack-in-the-pulpit. In the wild, it can grow in conditions ranging from 3 ft. of water to barely wet muck, but is typical in shallow water.

Peltandra virginica 

Green arrow-arum, tuckahoe

Araceae (arum family)

HABITAT & RANGE bogs, swamps, marshes, ponds, rivers, and ditches throughout the Southeast **ZONES** 5–10 **SOIL** clay-loam, sandy to mucky, moist to wet **LIGHT** shade to sun.
DESCRIPTION Emergent perennial, strongly clump forming. Leaves light green to 4 ft. tall, blades arrowhead shaped, to 2 ft. long. Flowers tiny, yellowish, in a spike tightly

Peltandra virginica

surrounded by a single green spathe, on a stalk 2 ft. tall, blooming all season. Fruits are slimy green berries in a cluster tightly enclosed by the spathe, ripening soon after flowering.

PROPAGATION seeds sown when ripe; division after flowering **LANDSCAPE USES** specimens or groupings in shallow water containers or edges of ponds, naturalized in moist to wet soil **EASE OF CULTIVATION** easy **AVAILABILITY** common.

NOTES Green arrow-arum forms a stately upright plant that is never flashy, its architectural form worthwhile. It does not spread by rhizomes and so does not colonize aggressively, but it does self-sow.

A striking related species is spoonflower or white arrow-arum (*P. sagittifolia*). It is smaller, with mottled dark green and purple leaves, a white, more open spathe, and bright red berries. It would be a choice water garden or wetland plant if you can grow it by providing very acidic soil (pH 5).

There can be confusion when identifying aquatic plants with arrowhead-shaped leaves. Arrow-arum (*Peltandra*) leaves have a strong midvein with side parallel veins running to the margins, and more pointed basal lobes. Pickerel-weed (*Pontederia*) leaves lack a midvein and have parallel veins and rounded basal lobes. Arrowhead (*Sagittaria*) leaves have parallel veins with a strong midvein that forks into the pointed basal lobes.

Pontederia cordata

Pontederia cordata ★★★★

Pickerel-weed

Pontederiaceae (pickerel-weed family)

HABITAT & RANGE wet ditches and marshy margins of ponds and lakes throughout the Southeast **ZONES** 4–10 **SOIL** clay-loam, sandy to mucky, moist to wet **LIGHT** sun.

DESCRIPTION Emergent perennial, forming extensive colonies. Leaves light green, elongate arrowhead-shaped with rounded basal lobes, 2–3 ft. tall. Flowers numerous, 3⁄4 in., mauve or violet-blue with a two-lobed yellow spot, borne in tight clusters on a spike, opening sequentially and lasting one day each, flowering stalk to 4 ft. tall. Fruits are ribbed nutlets, ripening soon after flowering.

Sagittaria lancifolia

PROPAGATION seeds sown when ripe; division anytime **LANDSCAPE USES** clumps in containers and ponds, naturalized in sunny wetlands **EASE OF CULTIVATION** easy **AVAILABILITY** common.

NOTES This beautiful and stately species blooms all summer with attractive flowers that insects adore. The rounded leaves are delightful with the stiff flower spikes. There are dwarf and white-flowered forms. Pickerel-weed is an aggressive colonizer, taking over all space, and is a zealous self-sower. The floating seeds will clog pond filters. Grow it anyway.

Sagittaria latifolia ★★★

Common arrowhead, duck-potato

Alismataceae (water-plantain family)

HABITAT & RANGE wet ditches and marshy margins of ponds and lakes throughout the Southeast **ZONES** 3–10 **SOIL** clay-loam, sandy to mucky, moist to wet **LIGHT** sun.

DESCRIPTION Emergent perennial forming tube-like corms, colonizing by spreading runners. Leaves dark green strongly arrowhead-shaped, 1–2 ft. tall. Flowers white, 1½ in., in progressive whorls, males higher and females lower, on a stalk 1–2 ft. long or longer. Fruits are clusters of tiny dry seeds.

PROPAGATION seeds sown when ripe; digging plantlets **LANDSCAPE USES** specimens in containers or ponds, naturalized in wetland settings **EASE OF CULTIVATION** easy **AVAILABILITY** common.

NOTES This old familiar aquatic species is known for its (reportedly) edible tubers. I tried to eat one as a boy scout but found it too rubbery. Arrow-head makes a distinctive structural plant with its bold triangular leaves, but will take over whatever aquatic space is available. Keep plants potted or thin them ruthlessly. The flowers are long blooming and attractive.

There are many species with different shapes and sizes of leaves. Lance-leaved arrow-head (*S. lancifolia*) is similar but the leaves are long and narrow, not arrow-shaped.

Saururus cernuus ★★

Lizard-tail

Saururaceae (lizard-tail family)

HABITAT & RANGE marshes, swamps, and wetlands, in water up to 2 ft. deep, throughout the Southeast **ZONES** 6–10 **SOIL** clay-loam, sandy to mucky, moist to wet **LIGHT** shade to sun.

DESCRIPTION Emergent perennial, leafy stems to 4 ft. or more tall from long creeping rhizomes, forming extensive colonies. Leaves alternate, heart-shaped, 1–6 in. long. Flowers white, tiny, on a nodding terminal spike, blooming all summer. Fruits are dry, brown seeds.

PROPAGATION fresh seeds germinate at 75°F; division of rhizomes **LANDSCAPE USES** potted in container or pond, naturalized in wetland gardens **EASE OF CULTIVATION** easy **AVAILABILITY** common.

NOTES Lizard-tail has a uniquely attractive form with its tall leafy stems and nodding flowers. A colony of plants is impressive, but is too aggressive for the small garden. Growth can be somewhat suppressed when plants are sited in shade and drier soil.

Thalia dealbata ★★★

Powdery thalia, alligator-flag

Marantaceae (arrowroot family)

HABITAT & RANGE swamps, marshes, ditches, and pond margins in the coastal plain of South Carolina and Georgia, then Alabama to East Texas **ZONES** 6–9 **SOIL** clay-loam, sandy to mucky, moist to wet **LIGHT** sun.

DESCRIPTION Robust emergent perennial, stout rhizomes creep to produce large colonies. Leaves 3–6 ft. tall with long petioles bearing spear-shaped blades 1–2 ft. long, coated with a white powder. Flowers many, an intriguing blue

Saururus cernuus

and purple color, on a crowded panicle 3–6 ft. tall, blooming over a long period. Fruits are round with dark brown or black seeds.
PROPAGATION seeds after cold stratification; division anytime
LANDSCAPE USES colonies in large containers or ponds, naturalized **EASE OF CULTIVATION** moderate **AVAILABILITY** common.
NOTES Powdery thalia may be our largest and boldest aquatic for ornamental plantings. In a large pond or lake in an open setting, it makes a delightful scene along the shore, attracting many insects. It is even more striking as a large island clump in shallow water. Avoid crowding with other plants. I love this plant, but watch out—dragonflies will sit on the flowers.

Typha latifolia ★

Cattail

Typhaceae (cattail family)

HABITAT & RANGE swamps, marshes, ditches, and pond margins throughout the Southeast **ZONES** 3–10 **SOIL** clay-loam, sandy to mucky, moist to wet **LIGHT** shade to sun.

DESCRIPTION Emergent perennial from long-creeping rhizomes to make vast colonies. Leaves crowded, green to bluish green, to 6 ft. tall, about 1 in. wide. Flowers yellowish, minute, countless, in a very tight 1-in.-thick spike. Fruits are brown clublike structures that fluff in autumn and winter to release wind-borne seeds.
PROPAGATION seeds sown when ripe; division of rhizome
LANDSCAPE USES containerized clumps, naturalized around a large pond **EASE OF CULTIVATION** easy **AVAILABILITY** common.
NOTES Cattails are very familiar and are indicative of wetlands. Patches of them always seem to harbor a red-wing blackbird or two. The seeds can find the smallest wetland site and grow rapidly to form extensive impenetrable colonies that are not compatible with other aquatic plants. Cat-tails are beautiful in their own right—and in their own space.

Thalia dealbata

Typha latifolia

Utricularia inflata ★★

Floating bladderwort

Lentibulariaceae (bladderwort family)

HABITAT & RANGE wet ditches, marshes, and ponds in the coastal plain from New Jersey to South Florida, westward to East Texas **ZONES** 7–10 **SOIL** floats in water **LIGHT** sun.
DESCRIPTION Free-floating carnivorous short-lived perennial. Stems elongate, multibranched with tiny forked leaves bearing small dark bladders that trap minute aquatic insects. Inflorescence an unbranched stalk supported centrally on a radiating pontoon of four to nine inflated leaves, blooming in spring. Flowers up to fifteen, yellow, two-lipped, about 1 in. long.
PROPAGATION grows from vegetative pieces **LANDSCAPE USES** novelty specimens in containers or ponds, naturalized in ponds or wet ditches **EASE OF CULTIVATION** moderate **AVAILABILITY** rare.
NOTES These curious carnivorous plants are like no others. Their tiny bladders can be seen best with 5–10× magnification. Several related species are easy to grow in acidic water, from truly aquatic to terrestrial in wet sand. The very similar *U. radiata* is smaller. *Utricularia gibba* has much smaller flowers on nonbuoyant 2- to 3-in. spikes and can fill a small pool with a mass of interwoven bladder-bearing strands in no time. *Utricularia purpurea* has attractive lavender-red flowers and blooms sporadically all summer. *Utricularia subulata* is a very tiny species of wet sand with upright wiry stems and little yellow flowers, common in many aquatic plant containers. The aquatic forms are best grown in separate containers from other plants.

Utricularia inflata

Bog Plants

Venus flytrap is the most wondrous plant in the world. Its carnivorous behavior and limited range in the coastal Carolinas pique our curiosity and light up our imagination with countless questions. Everybody wants to grow one, but it will never become a mainstream landscape plant any more than the enchanting Oconee bells will come to replace hostas in the shade garden. However, such a dynamic and charismatic plant as Venus flytrap should be grown as it provides a great entry into a specialized but attainable gardening niche—bog gardening.

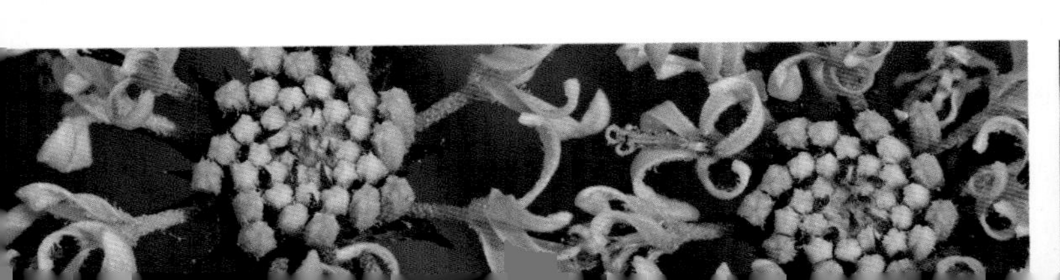

Recognition of bog plants as a category represents a movement towards increased interest in growing them and conserving their natural wetland habitats. While carnivorous plants are prominent in native bogs, many other interesting species coexist there as well.

Bog plants differ from aquatic plants in that they don't grow in standing water, but the soil must remain saturated at all times. Bog plants don't like to wilt from water stress. In addition, bog plants naturally come from low-nutrient peaty-sandy habitats, and as such, they neither require nor can tolerate much fertilizer. In contrast, aquatic plants are heavy feeders and languish in sterile soil. Neither group can survive in stagnant soil conditions; aquatic plants are especially common in slowly flowing water of streams and river edges. Bog plants especially benefit from good aeration in their otherwise moist soils.

The sunny wetland environments where bog plants grow are generally uncomfortable to modern humans as they are often muggy, buggy, mucky, and prickly. At the same time, they are captivating when you consider the beauty and diversity of the plants that grow there. The surprising reasons for this diversity are low nutrients and periodic fires, which go hand in hand to prevent any one species from dominating, in contrast to the nutrient-rich marsh or swamp forest habitats.

Southeastern bog wetlands are relatively flat, poorly drained areas where peat mosses and sedge peats accumulate. These wetlands are thus different from the famous

Boasting the largest hood in the genus, the often-grown yellow pitcherplant (*Sarracenia flava*) comes in several attractive color forms with red-flushed tubes and red veining.

An aboveground bog garden featuring various pitcherplants.

northern cranberry bogs where thick floating mats of sphagnum mosses accumulate in small lakes left by the glaciers. Up north, such spots are called bogs because the mosses are spongy and buoyant. As the mosses die and partly decompose, the accumulated substrate, termed *peat*, is mined to create horticultural soil mixes.

In the Southeast, the term *bog* is carried over, and the various habitats are called pitcherplant bogs, grass-sedge bogs, savannahs, wet meadows, shrubby pocosins, or pine flatwoods, depending on their degree of wetness and the dominant species. These kinds of habitats were once more abundant in the southeastern coastal plain and functioned ecologically to help absorb much of the abundant rainfall in the region. They have been extensively drained for agriculture and development, and in many cases have been kept from burning for decades such that the diversity they once knew has diminished.

EASY BOG PLANTS FOR A BEGINNER'S DISH GARDEN

Dionaea muscipula (Venus flytrap)
Drosera filiformis (thread-leaf sundew)
Drosera intermedia (intermediate sundew)
Drosera tracyi (dew-threads)
Pogonia ophioglossoides (rose pogonia)
Sabatia dodecandra (marsh-pink)
Sarracenia hybrids (pitcherplant hybrids)
Sarracenia purpurea (purple pitcherplant)
Sarracenia rubra (red pitcherplant)

Bog Habitats

All bog habitats are wet but with no standing water, and may grade into wetter cypress swamps or marshes on the one hand, or drier pine woods or sand ridges on the other. In the wild, there is a gradient from very wet to very dry soil. Carnivorous plants and other bog plants prefer sites where it's never one extreme or the other for long periods.

A pond edge is quite wet and may be in shallow standing water. Next is the coastal plain savanna or meadow, a treeless tract with only grasses and other herbaceous plants. Wiregrass and toothache grass can be common in some coastal habitats. A pine savanna is similar to a pine flatwoods, an area with more sandy soil and scattered longleaf pines, but still mostly very sunny and wet.

Pocosins are strange wet places. The term *pocosin* means "swamp-on-a-hill" and refers to an area dominated by mostly evergreen shrubs. Tall pocosins are quite impenetrable and shady with shrubby vegetation 15–25 ft. tall. Short or shrubby pocosins are more open, and are transitions between the denser tall pocosins and the less dense savanna or flatwood vegetation containing longleaf and loblolly pine. The pocosin is the source of a surprising number of ornamental woody plants including coastal sweet-pepperbush, titi, witch-alder, loblolly bay, inkberry holly, shining fetterbush, Atlantic azalea, and honey-cups.

Creating a Backyard Bog Garden

While there are many types of intergrading bog habitats in nature, the best way to recreate a suitable site at home is to make a bog garden containing a mixture of peat and sand that holds plenty of moisture. A small bog garden that is easily accessible to children and adults will provide hours of pleasure in watching the insects visit and be caught by the traps. An array of bog plants in addition to carnivorous plants can be colorful throughout the growing season.

Creating a bog garden can take several routes, depending on the desired size. One option is to use a container such as a plastic wash tub, a water trough, a child's wading pool, a wooden half-barrel, even a leaky old boat—anything that will retain some water. For something smaller, you can make a manageable bog dish garden in a 12- to 14-in. plastic bowl. Alternatively, if you wish to go larger but don't want to dig out a hole in your backyard, you can construct a raised bed of 2-by-8-in. lumber, larger railroad ties, or with concrete wall-blocks arranged in a rectangle or oval. Line the space with a high-quality rubber or PVC pool liner that has been punched with a few holes for drainage. Remember, you want it to just retain moisture, not hold standing water like a pond. If you have a backyard waterfall and pool feature for aquatic plants, you might just add a large container with bog plants on a ledge near the edge of the pool. Above all else, the bog "soil" must remain moist and not be stagnant or anaerobic.

A good planting medium for a bog garden is a 50:50 mix of Canadian (brown) peat and white quartz sand. At UNC Charlotte Botanical Gardens, we use pool filter sand, or sand-blaster's sand, as it's coarse and clean. We usually have to hand-mix the dry out-of-the-bag peat (as in kneading dough for bread) using very warm water to cause it to become wetted. Once wet, it will hold moisture better than other soils, but it has to be kept constantly moist. The final mixture should be wet enough such that you can squeeze a little water out of the medium with your hand, but not so wet that water leaks out constantly when you fill the garden container with the medium. You might call it very moist, or constantly moist, or saturated, or wet as I have done below.

Soil for a bog garden should not be anaerobic (stagnant and without oxygen), hence the usefulness of sand or other coarse materials (such as perlite, milled sphagnum) in the peaty mix. Since the medium should never dry out, it may require daily watering in a dish garden or freestanding container. If you are going to be away for several days in hot weather, set a small dish garden in a container with several inches of water. It is better for the bog to be a little too wet than too dry for a short time.

You may need to replenish the soil in your bog garden every three to five years. This means deconstructing the bog garden and putting in fresh soil to replace that which will have deteriorated over time.

AGGRESSIVE-GROWING PLANTS

These are better avoided or used only in large gardens.

Hypericum species (St.-John's-worts)
Osmunda cinnamomea (cinnamon fern)
Rhexia mariana (meadow beauty)
Rhexia virginica (meadow beauty)
Rhynchospora latifolia (white-top sedge)
Utricularia subulata and other semi-terrestrial spp. (bladderwort)
Viola pallens and *V. lanceolata* (white violets)
Woodwardia virginica (Virginia chain fern)

The final soil medium will be acidic and very low in nutrients. You may need to fertilize some noncarnivorous plants to get them to grow properly, but do this sparingly. I use ½ tablespoon of acid fertilizer per gallon of water, and apply twice a month during the growing season. Don't fertilize Venus flytraps—feed them freshly caught small insects (and no hamburger meat).

Plants for a Home Bog Garden

Many beautiful and easy-to-grow species are found in Southeastern bogs. Those described in this chapter have been grown by me and others, and have proved to be manageable in relatively small containers. That is, these species are compatible plants for the home bog garden because they have similar requirements, exhibit slow (or manageable) growth and attain relatively short stature (less than 2 ft.), and look good all summer.

I have been testing wild bog plants for years, looking for candidates suitable for the small bog garden. I have done quite a bit of work with pitcherplants and even created several garden hybrids, and thus they are more thoroughly described here. After all, they are among the premiere bog plants and likely will form the foundation for your collection.

Some very attractive species mentioned here and elsewhere are very aggressive growers and thus I suggest you avoid using them in a small bog garden. In general, any bog species that spread by underground rhizomes or roots are better in large gardens than in small ones. Some bog plants, such as most yellow-eyed grasses, some species of meadow-beauties, and certain violets, also reproduce vigorously by seeds such that the seedlings take over and crowd out species that are more desirable.

As you read the plant descriptions, keep in mind that when I say a plant may be grown as a landscape specimen in a natural bog, that implies you live in an area where you have a proper savanna or pine-dominated wetland on your property and you want to enhance it with particular species. In most cases, however, you will be dealing with a manufactured containerized bog garden of some kind. Don't get bogged down in the particulars. Give bog gardening a try and put a bog dish garden on your sunny deck.

Aletris farinosa ★★

Colic-root

Nartheciaceae (bog-asphodel family), formerly in Liliaceae (lily family)

HABITAT & RANGE roadsides, ditch banks, meadows, seeps, and shrubby pocosins from Virginia to Florida and Texas **ZONES** 5–9 **SOIL** sandy-peaty, wet **LIGHT** sun.
DESCRIPTION Clump-forming herbaceous perennial to 3 ft. tall, with a basal rosette of leaves 2–3 in. long. Flowers white, pebbly-textured, about ½ in. long, numerous, borne on leafless racemes more than 1 ft. tall, blooming from late spring to midsummer.
PROPAGATION seeds, cold stratified for one month; division **LANDSCAPE USES** specimens or groupings in bog gardens **EASE OF CULTIVATION** easy **AVAILABILITY** frequent.
NOTES This long-flowering species doesn't take up much room, unless seedlings appear. Old flower stalks will last as dried ornaments, and leaves may remain somewhat evergreen.

Two beautiful yellowing-flowering species are found in the coastal plain of the Southeast: golden colic-root (*Aletris aurea*) flowering May–July, and yellow colic-root (*Aletris lutea*) flowering March–May.

Aletris farinosa

Asclepias lanceolata

Few flower milkweed

Apocynaceae (dogbane family), formerly in Asclepiadaceae (milkweed family)

HABITAT & RANGE savannahs, swamps, and brackish marshes in the southeastern coastal plain from Virginia to Florida and Texas **ZONES** 7–9 **SOIL** sandy-peaty, wet **LIGHT** sun.
DESCRIPTION Slender, unbranched, flexuous-stemmed perennial to 5 ft. tall. Leaves opposite, very narrow, to ½ in. wide. Inflorescence an umbel with rarely more than eight or nine flowers, each ¾ in. long, red-orange to sometimes a little yellow, rarely all yellow, blooming all summer; each flower has a conspicuous corona consisting of five nectar horns in the center. Fruits are slender pods with many seeds; each seed has a tuft of fluffy hairs for wind dispersal.
PROPAGATION seeds, cold stratified for one month **LANDSCAPE USES** specimens in bog gardens **EASE OF CULTIVATION** easy **AVAILABILITY** frequent.
NOTES This species is tall and striking, and is a grand addition to larger dish gardens. Another beautiful species rare in cultivation is red milkweed (*A. rubra*). It has broader leaves and flowers in pastel colors of cream-pink-red and occurs scattered in bogs and savannas.

Asclepias lanceolata

Calopogon tuberosus ★★★★

Common grass-pink

Syn. *Calopogon pulchellus*
Orchidaceae (orchid family)

HABITAT & RANGE bogs, meadows, savannahs, and seeps from Virginia to Florida and Texas **ZONES** 4–10 **SOIL** sandy-peaty, wet **LIGHT** sun.
DESCRIPTION Slender herbaceous perennial, from a 1-in. long ovate corm. Leaves one or two, to about 1 ft. long. Flowering stems solitary, 1–2 ft. tall, with several long-lasting flowers opening in sequence, flower color pale to deep rich pink (rarely white), blooming in early summer. Fruits are 1-in. fat green pods with thousands of dustlike seeds maturing in late summer.
PROPAGATION separation of corms **LANDSCAPE USES** specimens or groupings in containers **EASE OF CULTIVATION** moderate **AVAILABILITY** infrequent.
NOTES One of our showiest bog orchids, common grass-pink slowly multiplies to form a colony. Bees love to visit the flower, where the showy upper petal flops down to dash the surprised-but-unharmed visitors against the pollen-holding column.

The pale grass-pink (*C. pallidus*) is very similar with rose pink to white flowers.

Calopogon tuberosus

Dionaea muscipula ★★★★

Venus flytrap, meadow clam

Droseraceae (sundew family)

HABITAT & RANGE savannas, wet sandy roadside and shrubby pocosin margins in extreme southeastern North Carolina and adjacent South Carolina **ZONES** 6–9 **SOIL** sandy-peaty, wet **LIGHT** sun.
DESCRIPTION Carnivorous evergreen herbaceous perennial, from a bulblike rhizome, with six to eight tightly grouped leaves in a rosette. Leaves 1–3 in. long with a two-lobed snap-trap to 1 in. long on the end, often red inside, and with long firm bristles around the edges; there are three triggers hairs on the inside of each lobe that must be touched twice in rapid succession for the trap to snap shut on a small animal victim, which is then digested over a period of several days. Flowers white, 1 in. in diameter, borne in an umbel-like cluster of six to twelve on a solitary stem to 10 in. tall, blooming in late spring. If cross-pollinated, the plant produces tiny black seeds that mature in early summer.
PROPAGATION seeds sown when ripe; careful clump divisions **LANDSCAPE USES** specimens or groups in containers for close observation **EASE OF CULTIVATION** easy **AVAILABILITY** common.
NOTES This famous plant is a classic subject for a bog container. It should be grown in a sterile medium, never fertilized, but it grows faster and more robustly when fed small insects. It does not do well indoors, and needs abundant light and moisture. It must never dry out nor be kept sopping wet. Under good conditions, a plant will live for years and develop into a large clump that can be divided. Grows easily from seeds. There are several named cultivars.

Dionaea muscipula

Drosera intermedia ★★★

Intermediate sundew

Droseraceae (sundew family)

HABITAT & RANGE bogs, savannahs, and wet ditches in the coastal plain from Virginia to Florida and eastern Texas **ZONES** 4–9 **SOIL** sandy-peaty, wet **LIGHT** sun.
DESCRIPTION Carnivorous, herbaceous perennial (often annual), from very delicate roots, with a short stem 1–2 in. tall, or taller when growing in standing water. Leaves alternate, numerous, 1–2 in. long, very slender, with a spatula-like apex bearing gland-tipped hairs that catch and curl over small prey. Flowering stalks 3–4 in. tall, arising from the base of the leafy stem and bearing several small white flowers in sequence as they curiously uncoil, blooming all summer. Fruits are small pods ripening shortly after the flowers bloom.
PROPAGATION seeds as soon as ripe **LANDSCAPE USES** single or, more commonly, groupings in smaller bog gardens for close examination **EASE OF CULTIVATION** moderate **AVAILABILITY** frequent.
NOTES This interesting carnivorous plant can have a distinct presence in the small bog garden, forming a striking mass of glistening leaves. It will overwinter as dormant buds, or return annually from self-sown seeds.

Of the other similar rosette-forming Southeastern sundews, pink sundew (*D. capillaris*) would be worth growing, often acting as an annual and forming a groundcover of small red-leaved rosettes rarely more than 1 in. wide. Round-leaf sundew (*D. rotundifolia*) is a northern species that is very rare in the Southeast and is not recommended because it does not survive well in the heat.

Drosera intermedia

Drosera tracyi ★★★★

Dew-threads

Droseraceae (sundew family)

HABITAT & RANGE savannahs and wet sandy roadsides in the Gulf Coastal Plain from southwestern Georgia to eastern Louisiana **ZONES** 7–9 **SOIL** sandy-peaty, wet **LIGHT** sun.
DESCRIPTION Carnivorous, herbaceous perennial, clump forming, from an enlarged cormlike base. Leaves light green, slender, 6–14 in. long, uncoiling as they elongate, covered with greenish gland-tipped hairs that catch tiny insects. Flowering stalks 6–12 in. tall in early summer, bearing a succession of bright pink ½-in. wide flowers. Fruits are small pods, maturing in midsummer.
PROPAGATION seeds; division of clump in spring or after flowering **LANDSCAPE USES** specimens or groupings in containers for close observation **EASE OF CULTIVATION** easy **AVAILABILITY** frequent.
NOTES This species has the most beautiful of sundew flowers and the most abundant profusion of sticky leaves. It is vigorous all summer and is stunning to see the dewy leaves backlit by the sun.

The similar thread-leaf sundew (*D. filiformis*) occurs from North Carolina to New Jersey. It is smaller (4–8 in. tall) and somewhat less vigorous, with red leaves bearing red-purple gland-tipped hairs and fewer bright pink flowers.

Drosera tracyi

Eriocaulon decangulare ★★★

Pipewort, bog-hatpins

Eriocaulaceae (pipewort family)

HABITAT & RANGE bogs, savannahs, and wet pinelands from New Jersey to Florida and Texas **ZONES** 6–10 **SOIL** sandy-peaty, wet **LIGHT** sun.
DESCRIPTION Clump-forming herbaceous perennial with fibrous roots and a basal rosette of leaves 3–6 in. long. Flowers tiny, numerous, borne in very tight, hard, whitish knoblike heads ½ in. across, on numerous thin, stiff, leafless flowering stalks 1–2 ft. tall, blooming all summer.
PROPAGATION seeds; division into smaller clumps (not individual plants) after flowering **LANDSCAPE USES** specimens allowed to form larger masses in containers or natural bogs **EASE OF CULTIVATION** moderate **AVAILABILITY** rare.
NOTES This is an intriguing plant that adds an unusual note to the bog garden. It does not like disturbance, so be careful when dividing it and keep at least three growths together.

A much smaller and easy relative with grayish heads on 3- to 4-in. delicate stalks is bog-buttons (*Lachnocaulon anceps*).

Eriocaulon decangulare

Helonias bullata ★★

Swamp-pink

Heloniadaceae (swamp-pink family), formerly in Liliaceae (lily family)

HABITAT & RANGE bogs and sandy seepage swamps in New Jersey and Virginia, very rare in the mountains of the Carolinas and Georgia **ZONES** 5–7 **SOIL** peaty-sandy, wet **LIGHT** part sun to shade.
DESCRIPTION Clump-forming herbaceous perennial, from a stout rootstock. Leaves numerous, 4–8 in. long, evergreen, arranged in a rosette. Flowering stalk single (per rosette), hollow, leafless, up to 1 ft. tall in flower, much taller in maturity, bearing numerous ½ in. wide pink flowers in very early spring. Fruits are small seedpods maturing in early summer.
PROPAGATION seeds, cold stratified for three months; separation of larger clumps after flowering **LANDSCAPE USES** specimens in container gardens **EASE OF CULTIVATION** moderate **AVAILABILITY** frequent.
NOTES This is one of the most charming bog plants and is extremely rare (and endangered) in the southern part of its range. Propagates readily from seeds, though grows slowly at first. See it at its best at Garden in the Woods, Framingham, Massachusetts. It hates heat, but will survive in Zone 8 and bloom some years. Although it's not a typical southern

Helonias bullata

pitcherplant bog denizen, but rather grows in shaded mountain bog sites, I am treating it with bog plants because it needs wetness and is so worth growing.

Iris tridentata ★★★

Savanna iris

Iridaceae (iris family)

HABITAT & RANGE wet savannahs, pine flatwoods, and shrubby pocosins in the coastal plain from the Carolinas to the western part of the Florida Panhandle and Alabama **ZONES** 7–10 **SOIL** sandy-peaty, wet **LIGHT** part sun to sun.
DESCRIPTION Herbaceous perennial, from a slender rhizome, forming a small grouping. Leaves to 12 in. long and ½ in. wide. Flowering stems to 20 in. tall, producing one flower at a time from inside tight bracts; flowers 2–3 in. long, sepals up to 1½ in. wide, violet marked with yellow, the three petals rudimentary, blooming for a short period in early summer. Fruit's a dry pod 1½ in. long.
PROPAGATION seeds; division **LANDSCAPE USES** specimens in containers or natural bogs **EASE OF CULTIVATION** moderate **AVAILABILITY** frequent.
NOTES This is the primary iris you can grow with pitcherplants and other slow-growing bog plants because most other irises are so vigorous. It is showy and easy to recognize because of its smaller size and lack of conspicuous petals, which are uniquely much shorter than the showy sepals.

Liatris spicata var. *resinosa* ★★★

Savanna blazing-star, savannah gayfeather

Asteraceae (aster family)

HABITAT & RANGE bogs, ditches, and pine savannahs in the coastal plain from New Jersey to Florida, westward to Louisiana **ZONES** 6–9 **SOIL** peaty-sandy, wet **LIGHT** sun.
DESCRIPTION Clump-forming herbaceous perennial from a globose tuberlike rootstock. Leaves very narrow, to 12 in. long and ½ in. wide. Flowering heads (few-flowered clusters with tight set of surrounding bracts) numerous, arranged on a spikelike stem up to 5 ft. tall, the individual flowers lavender, and the bracts purple, blooming from top to bottom in mid to late summer. Seeds mature with tufts of bristles for dispersal.
PROPAGATION seeds collected only when fully mature, cold stratified for one month; division of larger clumps in spring **LANDSCAPE USES** specimens or groupings in larger containers or natural bogs **EASE OF CULTIVATION** easy **AVAILABILITY** frequent.
NOTES A strikingly tall plant, it may get too large for the smaller container garden. However, it provides a nice compact addition for late-summer color if kept on the lean side nutritionally. It readily self-sows, making grasslike seedlings with tiny tubers.

While this variety is native to pitcherplant bogs, its more robust upland relative (*L. spicata*) is more typical of ordinary soils and makes a good garden perennial.

Iris tridentata

Liatris spicata var. *resinosa*

Lilium catesbaei

Macbridea caroliniana

Marshallia graminifolia

Lilium catesbaei ★★

Pine lily, Catesby's lily

Liliaceae (lily family)

HABITAT & RANGE bogs, pine savannahs, and sandhills seeps in the coastal plain from southeastern North Carolina to Florida and westward to eastern Louisiana **ZONES** 7–9 **SOIL** sandy-peaty, wet **LIGHT** part sun to sun.

DESCRIPTION Herbaceous perennial with a single stem from a true bulb. Leaves alternate, grasslike, 3 in. long and ½ in. wide. Flower solitary, pointing upwards, the three sepals and three petals similar, narrowing abruptly at their bases, bright orange (rarely yellow) with spots on a yellow basal blotch, blooming mid to late summer, short lasting. Fruit is a 2-in. pod, maturing in autumn.

PROPAGATION seeds, cold stratified for three months, but very slow growing; division of bulb scales **LANDSCAPE USES** specimens in containers and natural bogs **EASE OF CULTIVATION** difficult **AVAILABILITY** rare.

NOTES The pine lily has unmatched beauty, but brings great challenges to cultivation. It is difficult to propagate and grow, requiring just the right moisture (fairly wet). The main reason for including it here is to encourage enjoyment in the wild, conservation of its habitats, and experimentation with propagations.

Macbridea caroliniana ★★★

Carolina birds-in-a-nest

Lamiaceae (mint family)

HABITAT & RANGE swamp forests in seepage areas and also in edges of boggy savannas; uncommon in the coastal plain from North Carolina through Georgia **ZONES** 7–8 **SOIL** peaty-sandy, wet **LIGHT** part shade to sun.

DESCRIPTION Herbaceous perennial with underground stolons forming a colony, leafy stems square in cross section, erect to 2 ft. tall. Leaves opposite, simple, slightly toothed, 4–5 in. long. Flowers showy, pink to lavender, with darker stripes, two-lipped, 1–2 in. long, tightly clustered at the tip of the stem, blooming in late summer into autumn. Fruits are small nutlets, maturing in late autumn.

PROPAGATION seeds, cold stratified for three months; division in spring **LANDSCAPE USES** specimens and groundcover in natural boggy areas **EASE OF CULTIVATION** easy **AVAILABILITY** frequent.

NOTES This beautiful and unusual species is becoming more widely grown because it propagates well. It forms a colony of delightful flowering stems in moist, bright situations; it does not like drought or competition. It is not for the small containerized bog garden, but more for the open woodland seep or larger bog garden. Worth growing for the conspicuous "up-top" flowers. Oh, the common name refers to the opened flower and the unopened round buds, which look like birds and eggs in a green leafy nest.

The related white birds-in-a-nest (*M. alba*) is from the Florida Panhandle and is striking.

Marshallia graminifolia ★★★★

Grassleaf Barbara's-buttons

Asteraceae (aster family)

HABITAT & RANGE shrubby pocosins, savannas, and wet pinelands in the coastal plain from North Carolina to northeastern Georgia **ZONES** 7–9 **SOIL** peaty-sandy, wet **LIGHT** sun.
DESCRIPTION Clump-forming herbaceous perennial becoming robust, stems to 30 in. tall. Leaves grasslike, evergreen, 8 in. long. Flower heads numerous atop slender stems, about 1 in. wide with numerous fragrant pink flowers, blooming late summer. Fruits are nutlets with fluff, maturing in autumn.
PROPAGATION seeds, cold stratified for three months; division in spring **LANDSCAPE USES** specimens and grouping in larger containers and natural bogs **EASE OF CULTIVATION** easy **AVAILABILITY** infrequent.
NOTES Beautiful and rarely cultivated, this species is fragrant and compatible with the larger bog garden. It definitely should be grown more and is an easy subject for nursery production. The intricate detail of the individual flowers is intriguing.

Some consider this species the "northern" form of the broader distributed narrowleaf Barbara's-buttons (*M. tenuifolia*), which continues on further south from northeastern Georgia to southern Florida.

Pinguicula lutea ★★

Yellow butterwort

Lentibulariaceae (bladderwort family)

HABITAT & RANGE savannahs, pine flatwoods, and sandhills seeps in the coastal plain from southeastern North Carolina to South Florida and westward to eastern Texas **ZONES** 7–10 **SOIL** sandy-peaty, wet **LIGHT** part sun to sun.
DESCRIPTION Carnivorous herbaceous perennial, often behaving as an annual, forming a single basal rosette of leaves, from delicate roots. Leaves yellowish green, greasy-textured, broadly triangular, growing flat against the ground. Flower solitary, yellow, 1 in. diameter, on an erect

Pinguicula lutea

stalk 4–6 in. tall, plant may produce several flowers during the spring blooming period. Fruits are small pods with tiny seeds, maturing in early summer.
PROPAGATION seeds, sown fresh; gently divide clumps **LANDSCAPE USES** specimens or groupings in small container gardens **EASE OF CULTIVATION** moderate **AVAILABILITY** infrequent.
NOTES This unusual plant has cheerful yellow flowers and basal leaves that catch small invertebrates in a surface slime and curl up over them around the edges. Difficult to maintain in dish gardens.

In similar habitat is the blue-flowered butterwort (*P. caerulea*), from North Carolina to Florida. Two species from the western part of the Florida Panhandle to Louisiana are Chapman's butterwort (*P. planifolia*), with large bronzy-green rosettes 3–4 in. wide, and clearwater butterwort (*P. primulifolia*), with rosettes 2–3 in. wide, the latter growing along streams and seeps, and proliferating nicely from leaf-tip plantlets.

Platanthera ciliaris ★★★

Yellow fringed-orchid

Syn. *Habenaria ciliaris*
Orchidaceae (orchid family)

HABITAT & RANGE savannahs, meadows, bogs, and moist road banks in the coastal plain from Virginia to Florida and westward to eastern Texas **ZONES** 5–9 **SOIL** sandy-peaty, moist **LIGHT** part sun to sun.

DESCRIPTION Stout herbaceous perennial, from a thickened fleshy branched root, single leafy stem 1–2 ft. tall. Inflorescence an unbranched stem of many long-lasting flowers, lower petal distinctly fringed, orange (sometimes yellow), about 1½ in. long including a distinctive slender nectar spur protruding backwards from each flower, blooming in late summer. Fruits are slender pods with thousands of dustlike seeds maturing in autumn.

PROPAGATION seeds in sterile culture; separated roots rarely regenerate new plants **LANDSCAPE USES** specimens or groupings in containers or natural bogs **EASE OF CULTIVATION** moderate **AVAILABILITY** infrequent.

NOTES Finding wild orchids is a great pastime for naturalists and photographers, and this is one of our showiest. It is even more thrilling to grow one. This is by far the commonest wild orchid across the region in various wetlands, especially moist roadbanks and ditches, from the mountains to the coast, and it's becoming more available as a nursery-propagated plant from sterile seed culture. Strangely, plants don't multiply from their fleshy roots; instead, a single new bud forms on a root each summer to regenerate the plant the next spring. The proper moisture level and non-stagnant soil are critical for growing this orchid.

Platanthera ciliaris

Pogonia ophioglossoides ★★★★

Rose pogonia

Orchidaceae (orchid family)

HABITAT & RANGE bogs and savannahs in the coastal plain from Virginia to South Florida and westward to Texas **ZONES** 4–9 **SOIL** sandy-peaty, wet **LIGHT** sun.

DESCRIPTION Herbaceous perennial, from a creeping rhizome, colony forming, stems slender 6–10 in. tall, with a single leaf and a single (sometimes two) 1-in. long-lasting flower with a pink and yellow lower petal, blooming in early summer. Fruit's a slender pod with thousands of dustlike seeds, maturing in late summer.

PROPAGATION division of new growths from the rhizome **LANDSCAPE USES** colonies spreading in containers or natural bogs **EASE OF CULTIVATION** moderate **AVAILABILITY** frequent.

NOTES This charming species creeps around in the bog garden, often producing dozens (eventually hundreds) of flowers opening at once—quite a sight. It is one of the easiest orchids to grow and multiply in a bog garden.

Pogonia ophioglossoides

Polygala lutea ★★★

Orange milkwort

Polygalaceae (milkwort family)

HABITAT & RANGE wet savannahs, ditches, and bogs in the coastal plain from Virginia to South Florida and westward to eastern Louisiana **ZONES** 6–9 **SOIL** sandy-peaty, wet **LIGHT** sun.

DESCRIPTION Herbaceous biennial, from delicate roots, forming a small basal rosette of evergreen, fleshy ovate leaves. Flowering stems sparingly branched, 6–10 in. tall, weak, brittle, each branch topped with a single globose cluster of very small bright orange flowers, blooming all season. Seedpods are quite small, two-seeded, and fall off sequentially as the flowers fade.

PROPAGATION seeds only, cold stratified for one month **LANDSCAPE USES** informal groupings of self-sowing specimens in containers or natural bogs **EASE OF CULTIVATION** moderate **AVAILABILITY** rare.

NOTES This brightly colored plant fits in with pitcherplants and other species and is a welcome addition to the bog garden because it blooms all season. We have learned that the plant must be left to flower and self-sow, as it's a biennial and will die after dropping all its spent flowers containing the tiny seedpods. Young seedlings can be left *in situ* or carefully pricked out and moved to improve spacing. They will grow and develop over winter to bloom the following summer. The species is easy to propagate once this is known and should become more widely available. Orange milkwort is an indicator of Venus flytrap habitat in the wild, but is widespread far beyond that range and is a conspicuous roadside wetland wildflower.

Polygala lutea

Rhexia alifanus

Rhexia alifanus 

Smooth meadow-beauty

Melastomataceae (melastome family)

HABITAT & RANGE pine flatwoods, savannahs, and pocosin borders, mostly in the coastal plain of the Southeast, south of Virginia **ZONES** 7–9 **SOIL** sandy-peaty, wet **LIGHT** sun.

DESCRIPTION Herbaceous perennial, from an enlarged spongy root, unbranched stems to 4 ft. tall. Leaves opposite, to 3 in. long. Terminal inflorescence of several flowers opening one or two at a time, petals four, 1 in. long, bright pink or purple, eight conspicuous curved yellow stamens, blooming all summer. Fruits are ¾ in. seedpods, persistent, juglike.

PROPAGATION seeds, fresh or dried, with or without stratification, sown on wet soil surface **LANDSCAPE USES** specimens in larger containers or natural bogs **EASE OF CULTIVATION** easy **AVAILABILITY** infrequent.

NOTES The striking large flowers on tall wandlike stems make this a distinctive species and worthy of growing.

Of the dozen species in the Southeast, other good non-spreading ones are the yellow meadow-beauty (*R. lutea*) and the ciliate meadow-beauty (*R. petiolata*); both are much shorter plants with straight anthers. It is vital to avoid the common aggressively spreading species *R. mariana* and *R. virginica*.

Sabatia dodecandra ★★★★

Marsh-pink

Gentianaceae (gentian family)

HABITAT & RANGE savannahs and marshes in the coastal plain from Virginia to northeastern Georgia **ZONES** 6–9 **SOIL** sandy-peaty, wet **LIGHT** sun.

DESCRIPTION Herbaceous perennial, clump forming, from a short-creeping rhizome, stems erect, normally 6–20 in. tall. Leaves semievergreen, opposite, to 3 in. long. Inflorescence terminal branching clusters, flowers 1½–2 in. wide, typically with ten petals, bright pink with yellow center, blooming in midsummer. Seedpod green, surrounded by ten green persistent sepals.

PROPAGATION seeds cold stratified for one month; division after flowering **LANDSCAPE USES** robust clumps in containers or natural bogs **EASE OF CULTIVATION** moderate **AVAILABILITY** frequent.

NOTES This unbelievably beautiful species should be grown more. It is easy to propagate by vegetative division.

The similar Plymouth gentian (*S. kennedyana*) is rare in coastal Carolinas and northward, but is very easy to grow and forms robust evergreen basal rosettes of 3- to 4-in. leaves and taller stems to 3 ft. with numerous large flowers. An excellent bog subject that perhaps becomes a little too big for the small dish garden.

Sabatia dodecandra

Sarracenia flava ★★★★

Yellow pitcherplant

Sarraceniaceae (pitcherplant family)

HABITAT & RANGE savannahs, pine flatwoods, seepage bogs, and shrubby pocosins in the coastal plain from southeastern Virginia to the Florida Panhandle and westward to eastern Alabama **ZONES** 6–9 **SOIL** sandy-peaty, wet **LIGHT** sun.

DESCRIPTION Carnivorous herbaceous perennial, clump forming, from a short-creeping rhizome. Leaves erect, robust, to 3 ft. tall, tubular, bright greenish yellow with a large opening at the top covered by a flaplike hood, the rounded back margins of the hood almost touching each other; one to three evergreen swordlike nontubular leaves are produced in later summer. Flower showy, nodding, smelling of cat urine, 2–3 in. wide, solitary on a 2-ft. leafless stalk arising from the crown, with five thick yellow sepals and five thin yellow petals, numerous stamens, and an umbrella-like style, blooming in early spring before the leaves come up. Fruits are globose seedpods ¾ in. wide, with numerous seeds, maturing in late summer.

PROPAGATION seeds, cold stratified for one month; division of large clumps after flowering **LANDSCAPE USES** specimens and

Sarracenia flava

groupings in containers or large bog gardens **EASE OF CULTIVATION** moderate **AVAILABILITY** frequent.
NOTES This striking species is the most widely distributed pitcherplant in the Southeast and is widely grown. It is robust, has the largest hood, up to 5 in. wide, and comes in several attractive color forms with red-flushed tubes and red veining. It makes its largest pitchers in spring and early summer.

Pitcherplants lure insects with bright colors, spots, veins, and sweet nectar around the mouth of the pitcher. When the visitors fall in, they can't escape due to downward-pointing hairs. The plants digest the insects and absorb the nutrients. This dietary supplement helps pitcherplants survive in nutrient-poor bog soils. Of the eleven *Sarracenia* species in the Southeast, three are endangered. The species hybridize readily, and there are several named hybrids available along with most of the species, such as 'Dixie Lace', 'Doodle Bug', 'Flies Demise', 'Judith Hindle', 'Mardi Gras', 'Red Bug', and 'Scarlet Belle'. Early October is the best time to transplant and divide pitcherplants, but you can do it in summer, too. The plants are quite cold hardy as long as they are kept from drying out in winter. Cut old pitchers off in March before new growth begins.

Sarracenia leucophylla

Sarracenia leucophylla ★★★★

Whitetop pitcherplant

Sarraceniaceae (pitcherplant family)

HABITAT & RANGE savannahs and pine flatwoods in the Gulf Coastal Plain from western Florida to eastern Mississippi **ZONES** 6–9 **SOIL** sandy-peaty, wet **LIGHT** sun.
DESCRIPTION Carnivorous herbaceous perennial, clump forming from a short-creeping rhizome. Leaves erect, robust, to 3 ft. tall, tubular, greenish, with a large opening at the top covered by a flaplike hood with long hairs underneath, with conspicuous white translucent tissue surrounded by red veins in the upper tube and hood; several evergreen swordlike nontubular leaves are produced in later summer. Flower showy, nodding, fragrant, 2–3 in. wide, solitary on a 2-ft. stem arising from the crown, with five thick red sepals and five thin red petals, numerous stamens, and an umbrella-like style, blooming in early spring as the leaves come up. Fruits are globose ¾-in. seedpods, with numerous seeds, maturing in late summer.
PROPAGATION seeds, cold stratified for one month; division of clump after flowering **LANDSCAPE USES** specimens or groupings in containers or natural bogs **EASE OF CULTIVATION** moderate **AVAILABILITY** frequent.
NOTES This is perhaps the showiest species with its flowerlike white-topped pitchers. The best pitchers are made in August–September.

Another charming species is the hooded pitcherplant (*S. minor*), growing from North Carolina to Peninsular Florida. It has prominent silver-white spots on the back of the hood and yellow flowers on 1-ft. stems in late spring. Pitchers are produced all summer, 1–2 ft. tall.

Sarracenia purpurea

Purple pitcherplant

Sarraceniaceae (pitcherplant family)

HABITAT & RANGE bogs, savannahs, and seeps in the Coastal plain from southeastern Virginia to the Carolinas and in the mountains of southwestern Carolinas and northern Georgia **ZONES** 3–9 **SOIL** sandy-peaty, wet **LIGHT** part sun to sun.
DESCRIPTION Carnivorous herbaceous perennial with evergreen leaves, rosettelike clump forming, from a short-creeping rhizome. Leaves resting on the ground, thick-textured, to 8 in. tall, tubular, green red-purple, with open

Sarracenia purpurea

Sarracenia rosea

mouth at the top not covered by the very hairy hood, holding water inside. Flower showy, nodding, fragrant, 2–3 in. wide, solitary on a single 1-ft. stem arising from the leaf crown, with five thick red sepals and five thin red petals, numerous stamens, and an umbrella-like style, blooming in spring as the new leaves come up. Fruits are globose ¾-in. seedpods with numerous seeds, maturing in late summer.
PROPAGATION seeds, cold stratified for one month; division of clump after flowering **LANDSCAPE USES** specimens or groupings in containers or natural bogs **EASE OF CULTIVATION** easy **AVAILABILITY** common.
NOTES The purple pitcherplant is the most commonly seen and the most widespread. It is the only species whose pitcher has a gaping mouth and therefore can receive rainwater. It is a myth that the taller species contain water. Even though it's evergreen, this species is quite cold hardy to well below 0°F.

Along the Gulf Coast near Mobile is the similar rose pitcherplant (*S. rosea*) with more bulging pitchers and charming pink flowers on short 8- to 10-in. stems.

Sarracenia rubra ★★★

Red pitcherplant, sweet pitcherplant,

Sarraceniaceae (pitcherplant family)

HABITAT & RANGE seepage bogs, savannahs, and shrubby pocosins in the coastal plain from North Carolina to the western part of the Florida Panhandle **ZONES** 6–9 **SOIL** sandy-peaty, wet **LIGHT** sun.
DESCRIPTION Carnivorous herbaceous perennial, clump forming, from a short creeping rhizome. Leaves erect, slender, to 1 ft. tall, tubular, red-purple, with opening at the top covered by a narrow flaplike hood (no hairs, no wide hood, and no nontubular leaves). Flower showy, nodding, sweetly fragrant, 1 in. wide, solitary on a 1-ft. stem, two or three arising from the same crown, with five thick green sepals and five thin maroon-red petals, numerous stamens, and an umbrella-like style, blooming in spring after the leaves come up. Fruits are globose ½-in. seedpods with numerous seeds, maturing in late summer.
PROPAGATION seeds, cold stratified for one month; division of clump after flowering **LANDSCAPE USES** specimens or groupings in containers or natural bogs **EASE OF CULTIVATION** moderate **AVAILABILITY** frequent.
NOTES This rather slender species is famous for several fragrant blood-red flowers that are taller than the pitchers.

A somewhat similar species with which it can be confused (when not in bloom) is the beautiful yellow-flowered pale pitcherplant (*S. alata*) of the Gulf Coast from Alabama to eastern Texas; the latter has earlier blooming, larger flowers on stalks that are shorter than the mature pitchers and hoods that are slightly wider than long.

Stenanthium densum ★★

Crow-poison

Syn. *Zigadenus densus*

Melanthiaceae (bunchflower family), formerly in Liliaceae (lily family)

HABITAT & RANGE seepage bogs, pine savannahs, and flatwoods, mostly in the coastal plain of the Southeast **ZONES** 7–9 **SOIL** sandy-peaty, wet **LIGHT** sun.
DESCRIPTION Herbaceous perennial, densely clump forming, with one flowering stalk per growth unit. Leaves basal, narrow, to 16 in. long, gently arching. Flowers showy, white, fragrant,½ in. wide, densely clustered at the end of a tall

Sarracenia rubra

Stenanthium densum

Triantha racemosa

3-ft. unbranched stalk, blooming in late spring. Fruits are small pods, maturing in summer.

PROPAGATION seeds, if you have two plants to cross-pollinate; division after flowering **LANDSCAPE USES** specimens allowed to form large clumps in larger containers or natural bogs **EASE OF CULTIVATION** easy **AVAILABILITY** rare.

NOTES This beautiful species is rarely grown, but with age it produces the magnificent effect of a fountain of grasslike leaves softly arching with tall stately flowers in a striking clump.

Triantha racemosa ★★

Southern bog asphodel

Syn. *Tofieldia racemosa*

Tofieldiaceae (false-asphodel family), formerly in Liliaceae (lily family)

HABITAT & RANGE savannahs, seepage bogs, and shrubby pocosins, mostly in the coastal plain from New Jersey to northwestern Florida and westward to eastern Texas **ZONES** 7–9 **SOIL** sandy-peaty, wet **LIGHT** sun.

DESCRIPTION Herbaceous perennial, slightly clumping, from a short rhizome. Leaves narrow, 3–12 in. long, folded to fit into each other at the base (equitant). Inflorescence glandular and rough in texture, spikelike, 8–24 in. tall, with units of three to seven flowers in short-branched clusters along the axis; flowers about ½ in. wide, with narrow white segments, blooming in summer. Fruits are small pods, maturing in autumn.

PROPAGATION seeds, cold stratified for one month; division **LANDSCAPE USES** groupings in up-close bog gardens **EASE OF CULTIVATION** moderate **AVAILABILITY** rare.

NOTES Although rarely grown, this charming species has intricate detail in the flowers, so plant it where you can examine the flowers when they bloom. The blue-green leaves are stiffly erect, and they may be confused with the commonly associated coastal dwarf iris (*Iris verna*), but the latter has finely toothed leaf edges. Southern bog asphodel has staying power as I have grown if for years in a raised bog bed. It has a very similar mountain form found in a few mountain bogs that might prove better adapted for colder sites.

Trilisia paniculata

Zigadenus glaberrimus

Trilisia paniculata ★★★

Trilisia, hairy chaffhead

Syn. *Carphephorus paniculatus*
Asteraceae (aster family)

HABITAT & RANGE savannas and flatwoods in the coastal plain from North Carolina to Alabama **ZONES** 7–9 **SOIL** sandy-peaty, wet **LIGHT** sun.

DESCRIPTION Herbaceous perennial, with semievergreen basal leaves, clump forming, flowering stems arising 1–4 ft. tall. Leaves narrow, rough-hairy, toothed, 3–6 in. long. Flowers showy, tiny, pink-purple, in few-flowered heads tightly arranged on a floriferous plumelike spike, blooming in late summer into autumn. Fruits are nutlets with fluff, maturing in midautumn.

PROPAGATION seeds, cold stratified for one month; division of clump **LANDSCAPE USES** specimens in bog gardens **EASE OF CULTIVATION** easy **AVAILABILITY** rare.

NOTES This rarely grown but charming species has intricate detail in the flowers. It is one of my favorites as I love the rich flower color and anticipate its blooming each summer. It is interesting to watch the flowering stalk develop and slowly elongate over several weeks. The showy blossoms attract many insects, and add wonderful color to the late season bog garden.

Zigadenus glaberrimus ★★

Large death-camas

Melanthiaceae (bunchflower family), formerly in Liliaceae (lily family)

HABITAT & RANGE bogs, savannas, and pocosins in the coastal plain throughout the Southeast **ZONES** 7–9 **SOIL** sandy-peaty, wet **LIGHT** sun.

DESCRIPTION Herbaceous perennial, with basal leaves, clump forming, slowly spreading, flowering stems arising 1–3 ft. tall. Leaves narrow, with a keeled edge along the back, to 1 ft. long. Flowers showy, white, 1 in. wide, with six tepals, each with a conspicuous pair of yellow spots at the base, arranged spaciously in a large branching inflorescence, long blooming in summer. Fruits are pods, maturing in autumn.

PROPAGATION seeds, cold stratified for one month; division of clump **LANDSCAPE USES** specimens in bog gardens **EASE OF CULTIVATION** moderate **AVAILABILITY** rare.

NOTES This is another rarely grown yet beautiful bog species. There is great beauty in the detail of the thick-textured flowers, which attract many insects to the yellow glands on the tepals. The species is probably too large for the small bog garden, as it tends to want to spread a little.

Wildflowers

When our daughter was young, her mother and I encouraged her to pick wildflowers in roadside meadows. These included Queen Anne's lace, daisies, and bachelor's-buttons, all non-native naturalized "wildflowers." For our daughter, it was simply a thrill. We saw it as a way to help her develop a personal relationship with plants that should help her to "not be afraid" of nature as she grew up.

All children should have a chance to pick a flower and get to know nature. If they can feel like a particular flower is there "just for them," it might encourage them to be more observant and ask good questions.

Wildflowers are the easiest native plants to relate to, perhaps because of their smaller stature (being more human-sized and easier to deal with than trees and shrubs), bright colors, and appealing fragrances, and because they attract fascinating butterflies and hummingbirds. Who does not know about the migratory flight of the monarch butterfly and that its distinctive black-and-white caterpillar eats orange milkweed plants to sequester bitter-tasting chemicals that protect it from hungry birds?

Herbaceous plants may not provide shade, lumber, firewood, or physical landmarks like trees and shrubs, and they rarely have fall color. Yet for their apparent lack of practicality, they are endlessly attractive to us. They appeal to our senses of sight and smell, to our wonderment about nature and the mysterious interactions of plants and animals, maybe even to a subconscious revelation of the real story behind the coming-of-age traditions of the birds and the bees.

What is it about flowers that makes us want to pick them (simple beauty), to give them as gestures of love (for an anniversary), of bereavement (at funerals), of hope (at weddings), of accomplishment (flowers on stage)? I don't think our analytical minds want to examine the architectural structure of a rose bud upon first encounter, nor are we initially fascinated by the mathematics of the spiral patterns of the florets in a sunflower head, nor do we care that the flower must be pollinated to produce fruits and seeds. No, there must be some fundamental trait in the human psyche that draws us to flowers.

Perhaps it's their ephemeral nature, beauty that is here today and gone tomorrow, so that we must hold precious every moment of joy. Could it be that we get a personal thrill to see a newly opened flower in the woods, and think, "That flower is there just for me, I am the only one here to see it." Perhaps also we are concerned about the web of life, and see plants as food and shelter for the myriad insects and birds that share our environment. Whatever it is that attracts us to flowers, more and more people have become interested in growing native wildflowers in their gardens, from formal borders and informal meadows to pleasant woodland walks and naturalized backyard oases.

This chapter is about herbaceous wildflowers. All of them were selected for their ornamental contribution to the garden. A few of them are shy denizens of deep dark forests, many come from the diverse herbaceous layer of spring ephemerals in open deciduous forests, a number are found in colorful prairielike meadows and pastures, and yet some are roadside species of disturbed places and tough environments.

Successful gardening with these plants depends on your placing them in their preferred environments, whether shade or sun, wetter or drier. Herbaceous plants adjust to local conditions even more so than woody species, and so

Under a thick deciduous tree canopy, the cheerful flowers of green-and-gold (*Chrysogonum virginianum*) welcome visitors to the entrance of Van Landingham Glen at UNC Charlotte Botanical Gardens.

your planting might not look like you had envisioned. It could be lusher if the site is rich and moist, or thinner and spotty if the site is drier. The best thing you can do for wildflowers is to give them supplemental water during times of drought, because after all, you have taken them out of their native environment. The sunlight can be the same in the new environment, but water is a variable critical to survival. Don't think you can plant a native and walk away, that just because it's native it will adjust perfectly to conditions in your garden. After the first summer of stressful adjustment, you will be able to tell which species have "taken" to their new home and will likely thrive with less attention.

Be most aware of clumpers versus creepers, and keep the spreaders away from other plants so they don't overgrow them. Some plants will grow more aggressively in cultivation than in the wild, and do it faster than you realize when they find themselves with little competition. It is best not to overwater or overfertilize natives, as this could lead to unnatural growth and aggressiveness.

More than with trees and shrubs, wildflowers offer an opportunity to colonize and establish populations of volunteer seedlings in your garden. This can be both good and bad. Diversity is good, but overcrowding is bad.

To get viable seeds of wildflowers, usually you need two different individuals of the same species that are genetically dissimilar, meaning they were grown from seed. No two seedlings are genetically equal, just like with animal offspring. Clones of cultivars (or even clones of a single wild plant) are usually self-sterile and may not set seed. Sometimes you can't know whether a plant is sterile or not, but you can ask when you acquire plants how they were propagated. If the plants are matched to their environment (woodland or meadow, for example) and given adequate moisture, some will readily cross-pollinate, self-sow seedlings, and create new plants in sites they like best. You can accept them as volunteers, move them, or discard the extras.

Most seeds of natives need a cold winter period to break dormancy. That is, they need to be moist and exposed to one to three months of cold temperature (40°F or below) before they will germinate. The farther south you go, the less cold is required; the farther north you go, the more cold is required. The requirement for a cold period keeps seeds from germinating too soon in winter and then encountering a hard freeze that would kill the seedlings. This explains why seeds collected in Michigan may not germinate well in Georgia, even though the species itself is part of the flora of both regions. Plants are adapted to regional climate along a gradient of cold winters from north to south. Get your nursery-grown seeds, and plants, no more than one state away from where you live, if possible.

Some natives are better propagated by taking cuttings from stems or roots. Generally, herbaceous stem cuttings 3–4 in. long taken before flowering, or of nonflowering shoots, will root, particularly if you use a rooting hormone. When taken during or just after flowering, stem cuttings probably will not root. Root cuttings can be taken usually when you lift and divide a perennial. Place the pieces in a shady, moist environment while they root, perhaps showing results the next spring. Consult a good book on propagation of natives to learn of all the possibilities. Each species has its optimal method, and timing could be critical.

Cultivars are generally more common in herbaceous plants than in woodies, probably due to the greater ease of growing herbaceous plants from seed to flowering in two or three years, allowing for faster breeding and selecting for better traits. It may also be because herbaceous plants make more seeds (for the plant size) and tend to be easier to grow in a variety of conditions, such that unusual seedlings might appear spontaneously in a garden setting more often. Herbaceous plants are also easier to propagate by division of the clump, and by cuttings, than shrubs; and they generally work much better in tissue culture to multiply selected clones by the thousands. Look, for example, at all the color forms of phlox and alumroot (*Heuchera*) available today via tissue culture.

Unless otherwise stated in the descriptions, most of the species described in this chapter are perennials. Perennials take longer to grow and mature from seed before flowering, and will bloom for a specific period during the growing season. They may require dividing and thinning out periodically. Annuals and biennials have a finite life (one or two years, respectively). They tend to bloom more profusely and for a longer period, and then they set seeds and die. These species truly need more than one individual to cross-pollinate for viable seed production; otherwise, there is no continuity from year to year. I recommend you acquire three specimens of each type to be sure to have some that are compatible, and plant them close-by. Fernleaf phacelia is an excellent example of a biennial that is tricky to establish, but once you get the first batch of seeds produced, it will form a breeding population and self-sow, establishing a ground cover of plants to bloom and re-seed themselves every year.

Timing of bloom can be very specific, and vary from region to region depending on how early spring comes. The earliest flowering spring ephemerals come up in the woods before the deciduous tree canopy leafs out. These include the familiar bloodroot, Mayapple, trillium, bleeding-heart, and wild geranium. They have several weeks of abundant sunlight to bloom, set seeds, and then die down for the summer. Before they go, they pre-form next year's flowers and leaves in underground structures called perennating buds, and then they require a cold winter to break dormancy before they will wake up and bloom again. Their pre-formed flowers allow for rapid growth in spring during the relatively brief period of sun exposure.

A mainstay of the sunny border with its boldly colored flower, purple coneflower (*Echinacea purpurea*) is also attractive to goldfinches, who relish its flattened, thistlelike seeds.

Plants that bloom in summer need several months of growth in full sun to develop their mature plant body and make enough energy to produce flowers. Each species blooms when it does, in part, due to synchronization to changes in day length, so that all members of the population can bloom at the same time and cross-pollinate. Fall bloomers, such as goldenrods, asters, and gentians, are thus triggered by the shortening of days in late summer.

Plants that like full sun in the Southeast should be monitored to see how they fare in hot afternoon sun. For northern species, or those that typically live in open woods or forest edges, the full brunt of afternoon summer sun could be devastating. In general, no plant likes to be in hot afternoon sun without some amelioration, such as growing in a mixed meadow where the effect of the sun is broken by growing among sun-tolerant grasses, or being on the shady side of a big tree or house.

In most cases, the specifics of soils may not matter. Of course, woodland species tend to like very humus-rich soils, while sun-loving perennials like mineral soil enriched with some organic matter for water retention. The specific pH (acidity or alkalinity) is less important, except in a few cases. The point is, don't fret initially over soils, but try to use the best you have. Avoid pure red clay and plain sand. Don't overfertilize in an attempt to improve the soil or plant growth. Amending your soils by mixing in well-rotted compost or leaf mold will usually help make any soil better. In addition, don't over-mulch in an effort to retain water and prevent weeds. A light mulch is excellent; it prevents the soil from getting too hot and retains some moisture. A heavy mulch of more than 1 in. may actually form a moisture-repellant crust and prevent water from penetrating.

As long as the beauty and appeal that make wildflowers worth growing outweigh the tiresome efforts to nurture and control them, they will be candidates for the pleasure garden in some capacity. Some have but brief moments of floral beauty that last only a day or two. Others bloom for weeks. Some add great structural beauty to the garden, are fragrant, or can be used as long-lasting cut flowers; others just gladden our hearts when we see them in bloom for the first time, no matter their appearance or longevity. Because we rely on plants for our very life sustenance, wildflowers remind us, perhaps more than any other type of organism, of our human connection to nature.

INVESTIGATE THE FAMILY RELATIONSHIPS OF WILDFLOWERS

All plant species are aligned into families, whose Latin names are derived from a familiar member of the group, whose genus name is combined with the suffix "-aceae" to form the family name. Membership in a family is based on the floral structure of the species, because these reproductive flower parts (number and arrangement) reflect what botanists consider the evolutionary development of the species through eons of time.

Knowing a plant's family might help you with cultural practices even more than knowing the name of a plant, as knowledge of relationships can give you clues about culture and other traits. For example, most members of the buttercup family (Ranunculaceae) are spring-flowering woodland species, rarely edible, and sometimes poisonous, while members of the aster family (Asteraceae) are generally harmless, fragrant, sun-loving, and later flowering. Members of the lily family (Liliaceae and its modern segregates) tend to be bulbous, difficult to propagate from seeds, and like moisture and shade, while mallow relatives (Malvaceae) are robust perennials, readily spreading by seeds, and like moisture and sun. Of course, there are exceptions.

The largest family, Asteraceae, has a peculiar floral construction that is evident in looking at any of the flowers. In the aster family, two kinds of flowers are grouped together into a larger head, which looks like a complete flower itself. The small disk flowers are in the center of the head, and larger ray flowers, which look like petals, are arranged around the outer edge of the head. The ray flowers may be flat, as in coreopsis, or they may be tubular, as in Stokes' aster. The flowerlike heads may then be arranged in spikes, branched plumes, flat-topped structures, or one-sided wands. Look up your favorite plants to see what families they are in, and see if you recognize their relatives. Check to see which may be related to our edible flora, and which may have unusual traits or folklore. You will be surprised and amused by these associations.

Aconitum uncinatum

Actaea pachypoda, flower

Actaea pachypoda, fruit

Aconitum uncinatum ★★★

Eastern monkshood

Ranunculaceae (buttercup family)

HABITAT & RANGE moist, less-acidic woods, mostly mountains, but scattered across the Upper South **ZONES** 5–7 **SOIL** moist, well drained **LIGHT** part shade to part sun.
DESCRIPTION Deciduous, sprawling semivine, 2–6 ft. tall, from a tuberous rhizome, very slowly forming spreading colonies. Leaves alternate, widely spaced, deeply lobed like a geranium leaf, 2–3 in. wide. Flowers showy, deep blue-violet, upper petal cupped like a monk's cloak, 1 in. long, blooming in sequential clusters on the tips of the brittle vinelike stems, long blooming in late summer. All parts, especially tubers, are very poisonous. Fruits are 1-in. elongate pods maturing in late autumn.
PROPAGATION stem-tip cuttings in summer, division of newly formed knotty tubers **LANDSCAPE USES** specimens, mixed in with other plants upon which they can find support **EASE OF CULTIVATION** moderate **AVAILABILITY** rare.
NOTES Eastern monkshood is one of my favorite wildflowers because the flowers are so dark purple, so unusual, and so rarely grown. This charming garden plant is worth anticipating because of its rich autumn bloom. It is viewed as a cool-night-loving plant found only in the mountains (which is mostly true), but I grow a beautiful, robust heat-tolerant strain I call "Emily Allen" after a famous wildflower lady who grew it for decades in her Winston-Salem, North Carolina, garden, and it should become more widely enjoyed, at least in the cooler parts of Zone 8. Monkshood can be grown in loamy soil, or even clay-loam, and should be kept moist. You will love its purple flowers in close combination with those of orange jewelweed (*Impatiens capensis*).

Actaea pachypoda ★★

Doll's-eyes

Ranunculaceae (buttercup family)

HABITAT & RANGE rich woods throughout the Southeast, except absent in the Atlantic Coastal Plain **ZONES** 3–8 **SOIL** moist, well drained **LIGHT** part shade to part sun.
DESCRIPTION Deciduous, clump forming, 18–24 in. tall, from a knotty rhizome. Leaves usually one from the base, 1 ft. or more wide, much divided, the 2–3 in. leaflets three-lobed and toothed. Flowers lack petals, but have showy white stamens, clustered at tip of stalk, blooming in spring. Fruits are very showy, white, 1/4 in. berries, each with a dark eye (like a porcelain doll's eye), on a thick red stalk, in late summer, readily eaten by birds.
PROPAGATION seeds, cold stratified for three months **LANDSCAPE USES** specimens, better as masses in woodland gardens **EASE OF CULTIVATION** easy **AVAILABILITY** frequent.
NOTES Doll's-eyes is a striking plant in fruit and is worth the wait for them to mature. The broad foliage is attractive and holds up well, with a typical compound leaf seen on many

Actaea racemosa

woodland plants such as black cohosh (*Actaea racemosa*) and false goat's-beard (*Astilbe biternata*). It is a worthy plant for the woodland garden, but has only a brief sparkle in spring and then in late summer.

Actaea pachypoda 'Misty Blue' has attractive bluish green foliage.

Actaea racemosa ★★★

Black cohosh

Syn. *Cimicifuga racemosa*
Ranunculaceae (buttercup family)

HABITAT & RANGE moist to dry, rocky or rich woods, mostly throughout the Southeast, except absent in Mississippi and Florida **ZONES** 4–8 **SOIL** moist, well drained **LIGHT** part shade to part sun.

DESCRIPTION Deciduous, clump forming, slowly spreading from a knotty rhizome, with foliage to 3 ft. tall and flowering stalk 4–8 ft. tall. Leaves very large, to 3 ft. wide, three- to four-times compound with numerous three-lobed and toothed leaflets 2–6 in. long. Flowers lack petals, but have showy white stamens, are numerous, densely clustered along the upper 1–2 ft. of stalk, blooming in early summer. Fruits are little dry pods, not particularly showy, maturing in summer.

PROPAGATION seeds, warm stratified for three months, cold stratified for three months; division not recommended **LANDSCAPE USES** specimens, masses in informal woodland gardens **EASE OF CULTIVATION** easy **AVAILABILITY** frequent.

NOTES Black cohosh is a stunning garden plant, coming into flower after most other spring flowers are past, giving a wonderful boost to the open shade garden. Native bees love it. It provides bold architectural form with its very broad leaves and tall flowering stalks, even after flowers have faded. Specimens slowly spread to form large patches, which make wonderful scenes in the garden, so let them increase. Black cohosh is a victim of a recent name change, finding itself transferred from the long-known genus *Cimicifuga* to *Actaea*, a change which can be appreciated based on leaf and floral characters, if not immediately accepted.

Ageratina altissima

Ageratina altissima ★★★

White snakeroot

Syn. *Eupatorium rugosum*
Asteraceae (aster family)

HABITAT & RANGE moist roadsides and woodland edges, mostly throughout the Southeast **ZONES** 4–8 **SOIL** moist, well drained **LIGHT** part sun to sun.

DESCRIPTION Deciduous, clump forming, spreading, 2–4 ft. tall, stems strong branching and of medium texture. Leaves opposite, triangular-rounded, sharp-toothed, 2–4 in. long. Flowers showy, white, in tight clusters, blooming in late summer. Seeds have tufts of bristles for dispersal.

PROPAGATION seeds easy; nonflowering tip cuttings are easy; division of clump **LANDSCAPE USES** specimens, mixed borders **EASE OF CULTIVATION** easy **AVAILABILITY** frequent.

NOTES White snakeroot is a wonderful species that covers itself with long-lasting white flowers. In the mountains where it's very common, it's a sign of late summer and can be abundant along roadsides.

The cultivar 'Chocolate' with dark purple foliage and snow-white flowers is striking. The foliage is darker in full sun, but may need afternoon protection further south.

Allium cernuum

Allium cuthbertii

Allium cernuum ★★

Nodding onion

Amaryllidaceae (amaryllis family), formerly in Liliaceae (lily family)

HABITAT & RANGE open woodlands and around rock outcrops, mostly scattered throughout the lower mountains and Piedmont of North Carolina and Tennessee, into northeastern Alabama **ZONES** 5–7 **SOIL** moist, well drained to dryish **LIGHT** mostly sun

DESCRIPTION Deciduous, spreading from bulbs, to 1 ft. tall. Leaves arising from the base, grasslike. Flowers showy, pink to white, ½ in. wide, numerous to few in spherical starlike clusters (umbels), nodding, blooming in summer. Fruits are dry pods with shiny black seeds, maturing in summer.

PROPAGATION seeds, cold stratified for three months; division of bulbs **LANDSCAPE USES** masses in meadows **EASE OF CULTIVATION** easy **AVAILABILITY** frequent.

NOTES This attractive onion is a more northern prairie species. It is barely tolerant of hot summers south of Zone 7, but this may help keep it in check as it can spread by its proliferating bulbs. Don't overwater, just let it die down in the meadow. It can produce striking flowers that bees love, and the dried seed heads can be left for ornament.

A less common species that I really like is Cuthbert's onion (*A. cuthbertii*) from dry rocky outcrops. It has big erect balls of ⅝ in. pinkish flowers with striking green ovaries that look great in bloom with orange butterfly-weed (*Asclepias tuberosa*) in early summer. It is easy to grow and may naturalize in a sunny rock garden. It should absolutely be grown more as it is charming, heat tolerant, and is perhaps our most ornamental native species of *Allium*.

Amsonia tabernaemontana 

Common bluestar

Apocynaceae (dogbane family)

HABITAT & RANGE moist woods and stream edges throughout the Southeast **ZONES** 3–9 **SOIL** moist, well drained **LIGHT** part shade to sun.

DESCRIPTION Deciduous, clump forming, slowly enlarging, 2–3 ft. tall, with leaves uniquely spiraled along the stem. Leaves alternate, 3–4 in. long, tapering, exuding milky sap when bruised, turning yellow in autumn. Flowers showy, light blue, star-shaped, ½ in. wide, in loose clusters, blooming in early spring. Fruits are 3-in. long, thin pods, usually paired, maturing in summer.

PROPAGATION seeds, cold stratified for three months; 4-in. stem cuttings taken as flowers fade; division of clump in early spring before growth starts **LANDSCAPE USES** specimens, borders, naturalized in open woodlands **EASE OF CULTIVATION** easy **AVAILABILITY** common.

NOTES Common bluestar is a tough, beautiful plant, robust in the garden, sometimes more leaf than flower, especially in too much shade, but it looks good all summer. It likes full sun, but appreciates some afternoon shade in the Deep South.

The excellent cultivar 'Blue Ice' and the floriferous variety *montana* are smaller. The sandhills bluestar (*A. ciliata*), from sandy soil in the coastal plain, is more delicate with very narrow, hairy leaves and smaller flowers. It is similar to the well-known just-west-of-our-range threadleaf bluestar (*A. hubrechtii*). Both species have prominent straw-yellow fall color.

Amsonia tabernaemontana

Anemone quinquefolia ★★★

Wood anemone

Ranunculaceae (buttercup family)

HABITAT & RANGE rich woods throughout the Upper Southeast **ZONES** 5–8 **SOIL** moist, well drained **LIGHT** shade to part shade.

DESCRIPTION Deciduous, colony forming, slowly spreading, from a slender rhizome, 2–5 in. tall. Leaves a single set of three, each with three leaflets that are lobed and toothed. Flower showy, solitary, white, with four to six sepals, petals lacking, 1 in. wide, blooming in spring. Fruits are tiny pods, not showy, maturing in summer.

PROPAGATION division of clump **LANDSCAPE USES** naturalized in woodland gardens **EASE OF CULTIVATION** easy **AVAILABILITY** frequent.

NOTES Wood anemone is a delicate early spring wildflower producing charming colonies. It will not choke anything out and fits in with other similar-sized wildflowers. The back of the sepal may be rose-red, so observe the sepals before the flowers open. After the brief bloom, the foliage is a delicate groundcover.

Anemone quinquefolia

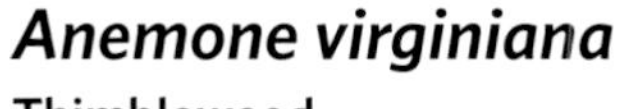

Anemone virginiana

Thimbleweed

Ranunculaceae (buttercup family)

HABITAT & RANGE rich to dryish woods throughout most of the Southeast, except absent in Florida **ZONES** 5–8 **SOIL** moist, well drained **LIGHT** part shade to part sun.

DESCRIPTION Deciduous, slowly enlarging, coarse-textured clump 2–4 ft. tall. Basal leaves clustered, to 10 in. long, blades divided into three leaflets that are toothed. Another set of three leaves halfway up the stem, each about 4 in. long and divided into three toothed leaflets. Flower showy, solitary, white, with four to six sepals, petals lacking, 1–1½ in. wide, each on its own tall stalk, up to nine per plant, blooming in summer. Fruits soon mature into interesting fluffs of cottonlike seeds.

PROPAGATION seeds, cleaned and cold stratified for three months; division of clump when not in active growth **LANDSCAPE USES** specimens, naturalized in open woodland gardens **EASE OF CULTIVATION** easy **AVAILABILITY** frequent.

Anemone virginiana

NOTES Thimbleweed is much taller and coarser than wood anemone (*A. quinquefolia*). The sturdy flowers with many stamens show a lot of detail. They are fun to find in bloom in early summer after most other woodland plants are finished. The fruiting heads are attractive, and the plant provides good form and texture.

Aquilegia canadensis ★★★★

Eastern Columbine

Ranunculaceae (buttercup family)

HABITAT & RANGE forests, woodlands, and rock outcrops throughout the Southeast **ZONES** 4–8 **SOIL** moist, well drained **LIGHT** part shade to part sun.

DESCRIPTION Short-lived perennial, semievergreen, clump forming, 1–3 ft. tall. Leaves arise from the base of the stems, each leaf divided three times, the ultimate leaflets ovate with three lobes and/or large teeth. Flower very showy, red tubular with a yellow center, dangling, 1–1½ in. long, blooming in late winter to early spring. Fruits are 1 in. dry pods, splitting in early summer to release shiny black seeds.

PROPAGATION seeds, without pretreatment, sown when mature **LANDSCAPE USES** specimens, massed in woodland gardens or semishady borders **EASE OF CULTIVATION** easy **AVAILABILITY** common.

NOTES Eastern columbine, a classic favorite for people and early hummingbirds, is among the first wildflowers to bloom and continues until late spring. Plants self-sow readily so you will always have a few; they are actually short-lived perennials, each persisting for one to two years. Let them grow where they will; you can remove the tangled stems after flowering to keep them from setting seeds, and new leaves will emerge to grow and overwinter. The one drawback is their susceptibility to disfiguring leaf miners. Remove infected leaves regularly to reduce the damage.

Several charming dwarf selections to about 8 in. tall seem to resemble the species closely: 'Canyon Vista', 'Little Lanterns', and 'Nana'. They make good massing and edging plants, may hold up better through the summer as foliage plants, but may not bloom as long nor reseed as regularly.

Arisaema triphyllum 

Common Jack-in-the-pulpit

Araceae (arum family)

HABITAT & RANGE rich woods and floodplains throughout the Southeast **ZONES** 4–10 **SOIL** moist, well drained to somewhat wet **LIGHT** part shade

DESCRIPTION Deciduous, solitary plant or colony forming, 1–3 ft. tall, from a large disc-shaped tuber (or corm). Leaves alternate, compound, 6–24 in. long, with three leaflets, male plants have one compound leaf, female plants have two. "Flowers" are large, showy green to purple structures consisting of a leaflike spathe wrapped around and folded over a spike (spadix, the "jack") of the actual tiny unisexual flowers down inside, blooming in late spring. Fruits are red berries, in a cluster, maturing in autumn.

PROPAGATION seeds, cleaned from the red pulp, cold stratified for three months; division of tubers from colony (but not division of single tubers) **LANDSCAPE USES** specimens naturalized in woodland gardens **EASE OF CULTIVATION** easy **AVAILABILITY** common.

NOTES Jack-in-the-pulpit's as common and well known a wildflower as there is. Some folks confuse it with a pitcherplant or ladys'-slipper orchid, but the flowering structure (spathe and spadix) that has a hood is designed to keep out rainwater and attract tiny flies that carry pollen from the tiny male flowers to the female—Jack-in-the-pulpits don't eat the flies as do pitcherplants. You will need plants of both sexes nearby to get the red berries. Oddly, this species can change sex. The young plant will first bloom as

Aquilegia canadensis

Arisaema triphyllum

a single-leaved male, then as it grows larger will become female and produce two leaves, presumably to photosynthesize more to feed the developing seeds. If stress (shade, drought) comes, the plant may revert to being a one-leaved male. Adequate water and some bright light will allow larger plants to develop. Some plants are all green, others deep purple with white stripes. Every woodland garden should have a Jack somewhere about, at least for the kids to see.

The related green-dragon (*A. dracontium*) is a curiosity. It has a single leaf up to 3 ft. tall, with a horseshoe-shaped midrib that radiates several leaflets. The flowering spathe (with both sexes) is much less conspicuous than that of the common Jack, but a long yellow appendage sticks way out. The large fruit cluster can be stunning. I love these plants for their bold architectural forms, and they can grow big with adequate water in silt-loam woodland soil. As with Jack (often called Indian turnip), don't eat the tubers as they contain poisonous calcium oxalate crystals.

Arnoglossum atriplicifolium ★★★

Pale Indian-plantain

Syn. *Cacalia atriplicifolia*
Asteraceae (aster family)

HABITAT & RANGE moist forests and woodland edges throughout the Upper Southeast **ZONES** 5–8 **SOIL** moist, well drained **LIGHT** part sun to sun.

DESCRIPTION Deciduous, clump forming, to 6–10 ft. tall. Leaves alternate, simple, triangular, coarsely toothed, white underneath, 3–6 in. long. Flowers showy as a group, in small clusters, creamy white, arranged in flat-topped inflorescences at the top of the plant, blooming in late summer. Fruits are inconspicuous dry seeds with fluffy parachutes, maturing in autumn.

PROPAGATION seeds, cold stratified for three months **LANDSCAPE USES** specimens naturalized in open woodland gardens, specimens in meadows **EASE OF CULTIVATION** moderate **AVAILABILITY** infrequent.

NOTES Pale Indian-plantain is a fabulous plant, bold in structure and color with its unusual white triangular leaves. It is one of our tallest herbaceous plants and really stands out in meadow or woodland. This prairie species likes to be in sun, or at least woods edge. It attracts interesting and useful bees, wasps, and beetles.

Arnoglossum atriplicifolium

Aruncus dioicus ★★

Goat's-beard

Rosaceae (rose family)

HABITAT & RANGE moist forests and roadbanks, mostly throughout the Piedmont and mountains of the Southeast **ZONES** 5–8 **SOIL** moist, well drained **LIGHT** part shade to part sun.

DESCRIPTION Deciduous, clump forming, 3–6 ft. tall, from a woody rootstock. Leaves very large, alternate, three- to four-times compound, to 3 ft. long and wide, the ultimate segments ovate-tapering with teeth. Flowers small, white, ⅛ in. wide, densely arranged in branched inflorescences (panicles), male and female flowers on separate plants, the males (each with twenty stamens) showier *en masse*, blooming in early summer. Fruits are tiny pods, densely arranged, maturing in autumn.

PROPAGATION seeds, without pretreatment; division of clump not recommended **LANDSCAPE USES** specimens in

Aruncus dioicus

Asarum canadense

Asclepias tuberosa with spicebush swallowtail

half-shady borders or open woodland gardens **EASE OF CULTIVATION** moderate **AVAILABILITY** frequent.
NOTES Goat's-beard makes a wonderful upright clump of large proportions, so give it room. The flower plumes are fragrant and attract many small bees. Wild plants can be found on every damp north-facing roadbank throughout the southern Appalachian Mountains and beyond. For maximum size and fullness, keep plants moist and protect them from hot afternoon sun. Goat's-beard can be confused with false goat's-beard (*Astilbe biternata*) and black cohosh (*Actaea racemosa*) when nonflowering, but differs in having nonhairy, unlobed leaflets and twenty stamens per flower.

Asarum canadense ★★

Common wild ginger

Aristolochiaceae (birthwort family)

HABITAT & RANGE moist woods throughout most of the Upper Southeast **ZONES** 5–8 **SOIL** moist, well drained **LIGHT** shade to part sun.
DESCRIPTION Deciduous, colony-forming groundcover, from a creeping rhizome. Leaves 3–6 in. tall, with ginger odor when crushed, a single pair per stem tip, each leaf rounded kidney-shaped, fuzzy, 3–5 in. wide. Flower unusual, a single brownish ½ in. fat tube with three long diverging sepal lobes, borne on the ground in the notch between the leaves, interesting but not readily visible from above, blooming in spring. Fruit's an enlarged berrylike pod with several seeds, maturing in early summer.
PROPAGATION seeds, cold stratified for three months; division of clump **LANDSCAPE USES** groundcover in woodland gardens **EASE OF CULTIVATION** easy **AVAILABILITY** common.
NOTES Wild ginger makes an attractive dense groundcover with soft texture, and it covers well, growing around (and over) anything in its way. In that sense, it's too aggressive for associating with delicate wildflowers. It readily seeds around, and you will have little satellite colonies popping up everywhere, as the seeds are dispersed by ants. Fortunately, the seedlings are easy to dig and remove.

Asclepias tuberosa ★★★★

Butterfly-weed

Apocynaceae (dogbane family), formerly in Asclepiadaceae (milkweed family)

HABITAT & RANGE open woods, meadows, and roadsides throughout the Southeast **ZONES** 5–10 **SOIL** moist to dry, very well drained **LIGHT** part sun to sun.
DESCRIPTION Deciduous, clump forming, from a deep, fleshy taproot, 1–2 ft. tall, often much-branched, stem and leaves exuding a watery sap. Leaves alternate, simple, narrowly elongate, 2–3 in. long, turning yellow in autumn. Flowers very showy, orange (to yellow and rarely red), ½ in. long, in dense flat-topped clusters, opening sequentially over a long period in summer. Fruits are long hornlike pods to 4 in., maturing in summer in sequence, releasing exciting masses of iconic fluffy parachuted seeds that kids should see and appreciate.
PROPAGATION seeds, fresh or dry-stored in cold; open just-ripe seedpods in a paper bag so you can crush off the buoyant hairy parachutes and separate the seeds before opening the bag outdoors to expel the fluff; plant in deep pots to allow taproots to develop; stem cuttings and root sections will root; difficult to dig from the wild **LANDSCAPE USES** specimens in borders and meadows **EASE OF CULTIVATION** easy **AVAILABILITY** common.
NOTES Orange-flowered butterfly-weed is a classic roadside plant and a best-loved wildflower, so don't let the "weed" designation deter your attention. It is a fabulous garden plant, adapting to well-drained sites and blooming over a long period. If cut back and watered during droughts, it will grow back and bloom some more. Don't overwater. It attracts many insects, especially butterflies. It is especially good, as are all milkweeds, for the monarch butterfly in late summer to lay its eggs and rear a brood of caterpillars, which will pupate and emerge to fly south for the winter—all in a matter of three to four weeks. A miracle of nature to watch before your very eyes.

A seed strain called 'Gay Butterflies' produces good plants in an array of bright colors.

Asclepias incarnata

Astilbe biternata

All milkweeds attract annoying yellow-orange aphids in late summer. Spray them off with a strong stream from the garden hose, soapy water, or mild pesticide. Likewise, all milkweeds except for *A. tuberosa* have milky sap. It is best to avoid extensive contact with the sap on sensitive skin.

Other species of milkweed are readily available and may bring pleasure and challenge. Purple milkweed (*A. purpurascens*) is a northern species extending into Tennessee and North Carolina. It has gorgeous balls of rich purple flowers and forms a spreading colony.

Swamp milkweed (*A. incarnata*) is a very popular species with pink flowers, but there are rose-pink ('Cinderella') and pure white ('Ice Ballet') selections.

For something surprisingly different, try the easy-to-grow green milkweed (*A. viridis*) with strange cuplike green and pink flowers.

Common milkweed (*A. syriaca*) has sweet-smelling pink flowers but is a very coarse and aggressive spreader and not recommended for the garden; keep it in the meadow.

Astilbe biternata ★★★

False goat's-beard

Saxifragaceae (saxifrage family)

HABITAT & RANGE moist cove hardwoods and north-facing roadbanks throughout the southern Appalachian Mountains **ZONES** 4–8 **SOIL** moist, well drained **LIGHT** part shade to part sun.

DESCRIPTION Deciduous, clump forming, from an enlarged tuberous crown, flower stalks to 5 ft. tall. Leaves very large, alternate, mostly basal, three- to four-times compound, to 3 ft. long and wide, the ultimate segments ovate-tapering, toothed, rough hairy, the tip segments also three-lobed, unlike goat's-beard (*Aruncus*) with which it's often confused. Flowers small, white, 1/8 in. wide, densely arranged in branched inflorescences (panicles), male and female flowers on separate plants, the males (each with ten stamens) showier *en masse*, blooming in early summer (after goat's-beard). Fruits are tiny pods, densely arranged, maturing in autumn.

PROPAGATION seeds, without pretreatment; division of clump not recommended **LANDSCAPE USES** specimens in half-shady borders and in open woodland gardens **EASE OF CULTIVATION** easy **AVAILABILITY** rare.

NOTES False goat's-beard is a big and beautiful wildflower, but it's practically unknown and underutilized. Perhaps it's overshadowed by its look-alike, goat's-beard, or has been pushed aside by attention to the colorful Asian *Astilbe* hybrids that are so popular in cooler climates north of Zone 8. The tall formal white plumes of *A. biternata* are a welcome sight in early summer in the shade garden. Its coarse leaves offer strong structural appeal. Many little insects like the flowers. Let it form a big clump in the background. You will come to admire its graceful form.

Baptisia alba ★★★★

White wild indigo

Fabaceae (legume family)

HABITAT & RANGE open woods and meadows throughout the Southeast but not Virginia **ZONES** 5–9 **SOIL** moist to dryish, well drained to seasonally damp **LIGHT** sun.

DESCRIPTION Deciduous, clump forming, 3–4 ft. tall, from a tough crown. Leaves alternate, with three leaflets ovate and toothless, abundant on the stems such that older plants form dense rounded herbaceous "bushes." Flowers showy, white, pealike, 3/4 in. long, blooming in late spring on stalks rising just above the foliage. Fruits are inflated pea-pods with numerous small seeds, maturing in late summer and turning black.

PROPAGATION seeds collected fresh when pods darken will germinate readily, seeds dry-stored in refrigerator for twelve weeks should be nicked or soaked overnight before being sown; rooting of robust leaf segments in summer; division of the crown in late winter **LANDSCAPE USES** specimens in borders and meadows **EASE OF CULTIVATION** easy **AVAILABILITY** common.

NOTES White wild indigo is a grand wildflower, forming attractive, huge leafy mounds that occupy space and look good all summer in the sunny border. It increases in size

Baptisia alba

Baptisia australis

every year, so give it plenty of room. A well-grown specimen can outlive your cars and furniture if established in a sunny, well-drained site. The new shoots look like emerging asparagus spears and are interesting in their own right. The formal-looking yet charming flowers are showstoppers when a large clump is in bloom.

Mass plantings of baptisia can become monotonous after flowering, so it's probably better to intersperse them with other large plants that can offer summer color. Extra water during droughts keeps the plants looking good, but they are certainly drought tolerant.

Boltonia asteroides

There are many species and hybrids to choose from, some are tall and some are shorter, in a range of colors. False indigo (*B. australis*) is a tall, beautiful blue. The leaves are darker than those of white wild indigo.

The hybrid between *B. alba* and *B. australis* is 'Purple Smoke', discovered as a chance seedling at the North Carolina Botanical Garden about 1990. It is a fabulous plant of magnificent stature to 5 ft., producing numerous robust smoky-purple flowers with gray-green leaves.

Of equal beauty is the tall *B.* 'Carolina Moonlight', a hybrid between the bright yellow *B. sphaerocarpa* and *B. alba*. The hybrid has light yellow flowers, produced in profusion.

For an electric turn-on, try 'Screaming Yellow', a selection of yellow wild indigo (*B. sphaerocarpa*), a prairie species, with outlandish yellow color on a tall robust plant.

Boltonia asteroides ★★★

Eastern doll's-daisy

Asteraceae (aster family)

HABITAT & RANGE marshes and ditches, mostly in the coastal plain of South Carolina, then skipping into Alabama and Mississippi **ZONES** 4–9 **SOIL** moist, well drained to wet **LIGHT** part shade to sun.

DESCRIPTION Deciduous with evergreen basal leaves, clump forming, slowly enlarging, 3–5 ft. tall. Leaves alternate, narrow, 1–2 in. long. Flowers showy, daisylike, white with yellow center, 1 in. across, in multibranched profusion, blooming in mid to late summer. Fruits are rarely produced.

PROPAGATION stem cuttings root readily; division of clump in autumn **LANDSCAPE USES** specimens in borders, naturalized

Camassia scilloides

Campanula americana

in open wetlands **EASE OF CULTIVATION** easy **AVAILABILITY** common.

NOTES Eastern doll's-daisy is a very worthwhile plant for late summer. Long blooming and charming, it slowly enlarges its clump and will eventually need dividing, which is easy to accomplish. The common cultivar 'Snowbank' is highly recommended over the species. It is a bit smaller and neater, with bluish green leaves.

Camassia scilloides ★★★

Wild hyacinth

Agavaceae (agave family), formerly in Liliaceae (lily family)

HABITAT & RANGE moist woods and floodplains, mostly west of the Appalachian Mountains **ZONES** 5–8 **SOIL** moist, well drained to seasonally wet **LIGHT** part shade to part sun.

DESCRIPTION Deciduous, clump forming, from a bulb, 1–2 ft. tall. Leaves grasslike, from the base, somewhat floppy. Flowers showy, light blue with yellow stamens, 1–1½ in. wide, arranged on unbranched stalks, blooming briefly in spring. Fruits are small pods with flattened seeds, maturing in early summer.

PROPAGATION seeds, left to come up on their own; division of bulbs **LANDSCAPE USES** specimens in mixed borders, naturalized in moist woodlands or floodplains **EASE OF CULTIVATION** easy **AVAILABILITY** frequent.

NOTES While not as striking as *Camassia* species in the American West, our wild hyacinth has a charm all its own. I look forward to the early appearance of the pale blue flowers each year, as a single plant of mine has done reliably for over thirty years, producing a few offspring over that time. The best use for wild hyacinth is in mass plantings to give a breathtaking effect in moist spring woodland.

Campanula americana ★★

American bellflower

Syn. *Campanulastrum americanum*
Campanulaceae (bellflower family)

HABITAT & RANGE moist woods, rocky roadsides, and shady streambanks, mostly in the Piedmont and mountains throughout the Southeast, and across Tennessee **ZONES** 5–8 **SOIL** moist, well drained **LIGHT** part sun to sun.

DESCRIPTION Deciduous, biennial with evergreen overwintering leaves, clump forming, 3–5 ft. tall. Leaves alternate, simple, lanceolate to long tapering, toothed, 2–6 in. long. Flowers showy, consisting of five pale blue petals that form an open 1-in. wide star, on a somewhat leafy stem, blooming in summer. Fruits are dry pods with minute seeds, maturing sequentially in late summer.

PROPAGATION seeds, cold stratified for three months **LANDSCAPE USES** specimens in half-sunny borders or woodland gardens **EASE OF CULTIVATION** moderate **AVAILABILITY** frequent.

NOTES The flowers of American bellflower are simply charming, with details worth admiring, and they will remind you of beautiful encounters on mountain roadside seeps. The plant is tall and may need staking outside the cooler mountains. It is very worth trying, even though it only blooms once. Such scarcity of bloom makes the plant even more desirable.

Cardamine diphylla ★★★

Common toothwort

Brassicaceae (mustard family)

HABITAT & RANGE rich woods, mostly in the Piedmont and mountains of the Southeast, and across Tennessee **ZONES** 4–8 **SOIL** moist, well drained **LIGHT** shade to part sun.
DESCRIPTION Slowly spreading clump, from a thick rhizome, 4–8 in. tall, summer dormant in the Southeast. Leaves wintergreen, two, opposite, on the rhizome, thick textured, compound with three leaflets with rounded teeth, dark green, about 4 in. long. Flowers showy, white, four-petaled, ¾ in. long, several on stalk, blooming in spring. Fruits are slender pods, maturing soon after blooming.
PROPAGATION division of rhizomes **LANDSCAPE USES** evergreen groundcover in moist woodland gardens **EASE OF CULTIVATION** easy **AVAILABILITY** frequent.
NOTES The thick dark green leaves of common toothwort provide an excellent wintergreen groundcover. Plant enough so it fills in quickly. Remember, this species normally goes dormant in summer, so surround it with other delicate deciduous wildflowers, like bleeding-heart or windflower, which can take over as toothwort goes to sleep. Cultivars 'Eco Moonlight' and 'Eco Cut Leaf' are good selections.

I also especially like the cutleaf toothwort (*C. concatenata*, syn. *C. laciniata*) with its pinkish white flowers in early spring. Its deeply divided and toothed leaves are borne on the delicate flowering stems, which die down in summer and grow back from creeping fleshy rhizomes.

Chamaelirium luteum 

Fairy-wand, devil's-bit

Heloniadaceae (swamp-pink family), formerly in Liliaceae (lily family)

HABITAT & RANGE rich woods, mostly throughout the Southeast **ZONES** 6–9 **SOIL** moist, well drained **LIGHT** part shade to part sun.
DESCRIPTION Deciduous with semievergreen basal leaves, 2–4 ft. tall, from a tuberous rootstock. Leaves alternate, simple, forming a basal rosette, oblong, widest toward the tip, to 8 in. long, getting much smaller up the flowering stem. Flowers showy, white fading to faint yellowish, small, six-parted, tightly arranged on an unbranched stalk that elongates and tapers to the point of gently arching over (hence the common name), male and female flowers on separate plants, male flowering stalks longer than female stalks, blooming in spring. Fruits are small pods with tiny seeds, maturing in autumn.
PROPAGATION seeds, cold stratified for three months, sow fresh as seeds don't store well **LANDSCAPE USES** colonies in moist woodland gardens **EASE OF CULTIVATION** moderate **AVAILABILITY** infrequent.
NOTES Fairy-wand is a most charming wildflower, seemingly delicate from its tapering flowering stem. It is a bit difficult to establish, perhaps requiring extra water in hot summers. A small population that includes both males and females is necessary for the production of seed. I have grown numerous seedlings to have them succumb in summer. They are worth every effort to grow, however; and a marvel to behold when grown well.

Chelone glabra 

White turtlehead

Plantaginaceae (plantain family), formerly in Scrophulariaceae (figwort family)

HABITAT & RANGE streambanks, wet ditches, and low woods, mostly throughout the Upper Southeast **ZONES** 3–8 **SOIL** moist, well drained to seasonally wet **LIGHT** part sun to sun.
DESCRIPTION Deciduous, expanding clump, 2–4 ft. tall. Leaves opposite, simple, lanceolate, thick textured, toothed, 4–7 in. long. Flowers showy, white, tubular, 1½ in. long, tightly arranged in a four-ranked cluster at the tip of unbranched leafy stems, barely open at tip with small petal "lips" around mouth, blooming in late summer. Fruits are pods with tiny seeds, maturing in autumn.
PROPAGATION seeds, cold stratified for two months; stem cuttings root easily; division of clump **LANDSCAPE USES** specimens in moist, sunny borders, naturalized in open, moist woodlands **EASE OF CULTIVATION** easy **AVAILABILITY** frequent.
NOTES White turtlehead is a neat plant; its flowers are shaped like a turtle's head, stretching out of the shell, with a mouth even. Inside the tube, where normally only bees can see, are four fertile stamens and one sterile. That fifth stamen can help identify the species: it's green in *C. glabra*, white in *C. obliqua*, and rose-tipped in *C. lyonii*. The latter two have pinkish purple flowers. All three species behave about the same—long blooming, slowly spreading clumps in open, moist habitats. *Chelone lyonii* 'Hot Lips' has bright pink flowers. *Chelone glabra* 'Black Ace' has dark purple leaves. As you can imagine, since *chelone* means "tortoise" in Greek, stories of turtles and mythology abound.

Cardamine diphylla

Chamaelirium luteum

Chelone glabra

Chrysogonum virginianum ★★★★

Green-and-gold, gold-star

Asteraceae (aster family)

HABITAT & RANGE rich, moist to dry woods throughout the Southeast, mostly east of the Blue Ridge mountains **ZONES** 6–8 **SOIL** moist, well drained to dryish **LIGHT** shade to part sun.

DESCRIPTION Evergreen, clumping to creeping, colony forming, 2–4 in. tall, plantlets can form on runners. Leaves opposite, ovate, hairy, 1–3 in. long. Flowers showy, yellow, daisylike heads about 1 in. across, on short stems rising barely above the leaves, blooming heavily in early spring and sporadically later. Fruits are single nutlets, maturing in early summer.

PROPAGATION seeds, cold stratified for three months; division of runners and clumps **LANDSCAPE USES** groundcover on shady banks and woodland settings **EASE OF CULTIVATION** easy **AVAILABILITY** common.

NOTES A gem of a garden plant, green-and-gold holds through heat and cold, and dry. You could probably find at least one flower in your colony nearly any day of the growing season. However, the plant does not like to be overwatered. It is a cheerful yellow to greet you *en masse* on a rough bank or path edge. I am very fond of this species and encourage you to find a place for it in your garden.

Chrysogonum virginianum

Claytonia virginica

Collinsonia verticillata

Clematis ochroleuca

It has two growth forms: variety *virginianum* is more clump forming but still forms a nice mass. 'Allen Bush' is an excellent long-blooming vigorous selection that grows up to 8 in. tall.

Variety *australe*, the more southern variety, sends out runners to form a colony more quickly. Its selection, 'Norman Singer', forms a compact mat about 4–6 in. tall. 'Eco Lacquered Spider' is weak blooming with unbelievable runners up to 3 ft. long. It covers an area quickly with glossy leaves, but then goes on to cover the path and then the property next door, and may become something of a pest. It would be a fun plant for a kid to try to tame.

Claytonia virginica ★★

Spring-beauty

Portulacaceae (portulaca family)

HABITAT & RANGE rich forests and low woods, mostly throughout the Southeast, except absent in Florida **ZONES** 3–8 **SOIL** moist, well drained **LIGHT** shade to part sun.
DESCRIPTION Deciduous, tuberous single plant, to 3–4 in. tall, forming extensive groundcover. Leaves alternate, with narrowly lanceolate to grasslike blades to 3 in. on long petioles arising from the deep marble-sized tubers (or corms), dying down by late spring. Flowers showy, pinkish white with deep pink veins, opening two or three at a time in sequentially flowering clusters and continuing for weeks starting in late winter. Fruits are small pods, maturing soon after flowering.
PROPAGATION seeds, self-sown; separation of clustering tubers if necessary and replanted as leaves die down **LANDSCAPE USES** naturalized in clumps or as groundcover in woodland gardens **EASE OF CULTIVATION** easy **AVAILABILITY** frequent.
NOTES Spring-beauty is as delicate a wildflower as you can find, but can completely colonize a moist to dryish woodland site by starting with three plants and letting the seeds

Conoclinium coelestinum

self-sow. The real charm of these earliest spring flowers is in the gentle pink veins on the petals.

Carolina spring-beauty (*C. caroliniana*), a species of the higher mountains, has similar flowers and much wider leaves, but is not as heat tolerant and would not be happy south of Zone 7.

Clematis ochroleuca ★★

Curlyheads

Ranunculaceae (buttercup family)

HABITAT & RANGE dry woodlands and borders, usually in less-acidic soil, limited mostly to the Piedmont from Virginia to northeastern Georgia **ZONES** 6–8 **SOIL** moist to dryish, well drained **LIGHT** part shade to part sun.

DESCRIPTION Deciduous, clump forming, 12–18 in. tall. Leaves opposite, simple, broadly ovate triangular, very white-hairy, 3–6 in. long. Flower showy, greenish yellow with a hint of lavender, single, nodding, with five or six petal-like sepals curled back, covered with white hair, ¾ in. long, blooming in late spring. Fruits are ball-like clusters of nutlike achenes with feathery tails 2 in. long, maturing in early autumn.

PROPAGATION seeds, cold stratified for three months; avoid disturbing clumps **LANDSCAPE USES** specimens, better when massed **EASE OF CULTIVATION** easy **AVAILABILITY** infrequent.

NOTES Unlike most clematis which are vines with showy white, pink, or purple flowers, curlyheads is an erect herbaceous species with single stems arising straight up, topped with a solitary, quiet flower. It does not grab your attention until it's in fruit, when the head of flattened seeds fluffs up like a permanent wave. It is long-lived and becomes more robust with age. This modest wildflower with unassuming traits needs some marketing pizazz to bring it to the attention of those who would like to see it more often "dancing with the stars."

Collinsonia verticillata ★★★

Whorled horsebalm, early stoneroot

Lamiaceae (mint family)

HABITAT & RANGE moist woodlands, usually in less-acidic soil, widely scattered in the low mountains and Piedmont from Virginia to Alabama **ZONES** 6–8 **SOIL** moist, well drained **LIGHT** part shade to part sun.

DESCRIPTION Deciduous, clump forming, slowly spreading, from tuberous rootstock, 1–2 ft. tall. Leaves opposite, simple, two pairs so closely spaced as to appear whorled on the stem, toothed, broadly ovate 2–8 in. long. Flowers showy, pink-lavender, 1 in. long, with four stamens sticking out of open mouth, arranged on an unbranched stem covered with sticky hairs, blooming in mid to late spring. Fruits are tiny nutlets, maturing in summer.

PROPAGATION seeds, cold stratified for three months; careful division of tuber after flowering **LANDSCAPE USES** specimens for choice spots in woodland gardens **EASE OF CULTIVATION** easy **AVAILABILITY** rare.

NOTES Whorled horsebalm is rarely seen in cultivation, and yet it's one of our most delicate and beautiful wildflowers. The four seemingly whorled leaves offer a bold texture for the garden after spring bloom. This elusive wildflower should be propagated and grown more; note that two different specimens are required for cross-pollination to get seeds.

Relatives that don't look much like it are Northern horsebalm (*C. canadensis*), Southern horsebalm (*C. anisata*), and late stoneroot (*C. tuberosa*). They have yellow flowers that are very similar, but bloom in late summer on taller, much-branched plants with many coarse opposite leaves, and some anise or lemon smell in the foliage. They make interesting airy woodland plants.

Conoclinium coelestinum ★★

Hardy ageratum, mistflower

Syn. *Eupatorium coelestinum*
Asteraceae (aster family)

HABITAT & RANGE moist to wet ditches and disturbed areas throughout the Southeast **ZONES** 6–10 **SOIL** moist, well drained to seasonally wet **LIGHT** part sun to sun.

DESCRIPTION Deciduous, colony forming, 1–3 ft. tall, spreading by long underground rhizomes. Leaves opposite, simple, broadly ovate, rounded-toothed, 2–4 in. long. Flowers showy, in tiny clusters of flat-topped aggregations, blue-purple, blooming in late summer through frost. Fruits are tiny nutlets with fluff, maturing in autumn

PROPAGATION seeds, cold stratified for three months; division of clump **LANDSCAPE USES** managed specimens in borders, as groundcover **EASE OF CULTIVATION** easy **AVAILABILITY** frequent.

NOTES Hardy ageratum is a welcome plant for color in the garden in late summer; it's a veritable butterfly magnet and looks attractive when in full bloom. However, it's a rampant spreader (by seeds and runners) and needs to be controlled in the garden. In the wild, it can form solid, spreading stands, but is limited by soil moisture and competition. I let them grow in my garden and just pull them up early on (easily) when they get in the way. They are wonderful to have around when most everything has faded, or in tough dry places where nothing else will grow.

Coreopsis major ★★

Woodland coreopsis

Asteraceae (aster family)

HABITAT & RANGE moist to dry woodlands, mostly throughout the Southeast **ZONES** 6–8 **SOIL** moist, well drained **LIGHT** part shade to part sun.

DESCRIPTION Deciduous, clump forming, slowly spreading, 1–3 ft. tall. Leaves opposite, simple but so deeply divided as to appear compound with three lanceolate leaflets, 2–3 in. long. Flowers showy, large yellow sunflower-like heads 2–3 in. across on leafy stems, the ray petals not notched or toothed, disk flowers yellow, blooming in summer. Seeds are dark nutlets, without bristles, maturing in late summer.

PROPAGATION seeds, cold stratified for three months; stem cuttings in summer; division of clump **LANDSCAPE USES** specimens in borders or open woodlands and meadows **EASE OF CULTIVATION** easy **AVAILABILITY** frequent.

NOTES *Coreopsis* is a large group of yellow-flowering species from a variety of habitats from shade to sun. They make great garden plants. Many can be recognized by the conspicuously three to five notched petal tips. The woodland coreopsis is easy to grow, produces large cheerful yellow flowers over a long period, and is adaptable to the dryish sunny woodland garden.

Lobed coreopsis (*C. auriculata*) is a tall species with opposite ovate leaves that have a pair of lobes at the base of the blade. It is best represented in the garden by its delightful cultivar 'Nana', a dwarf selection that forms an excellent semievergreen groundcover 2 in. tall, blooming heavily in spring with many bright yellow-orange ray flowers (with five teeth) about 6 in. tall. One of my favorites.

Swamp coreopsis (*C. gladiata*) is fairly new to cultivation. It is clump forming with basal leaves to 2–8 in. long and 2 in. wide. The charming flowers have yellow rays with three teeth and purple disk flowers.

Longstalk coreopsis (*C. lanceolata*) is a common clump-forming perennial with stiffly erect stems and long leaves. Often crossed with the similar large-flowered coreopsis (*C. grandiflora*) to make many garden hybrids.

Broadleaf coreopsis (*C. latifolia*) is the yet undiscovered sleeper in the group. Here is an easy-to-grow native species considered rare in the southern Blue Ridge Mountains, but which naturally grows and flowers in the shady woods in mid to late summer and forms large clumps. What a great opportunity to develop a shade-loving summer-blooming wildflower–it just needs some refinement to make it larger-flowered and more compact. Please, someone work on it.

Tall coreopsis (*C. tripteris*) grows to 8 ft. or more, forming a most striking large spreading clump with many gray-green three-parted leaves and huge clusters of 2–3 in. flowers in midsummer. For that special roomy spot in the sunny garden; deadhead after flowering.

Threadleaf coreopsis (*C. verticillata*) is widely known and grown, mostly for its garden selections. It grows 1–2 ft. tall and forms a mound with threadlike leaves and yellow flowers. Good cultivars in various yellow shades include 'Golden Gain', 'Golden Showers', 'Moonbeam', and 'Zagreb'.

Cuthbertia rosea ★★★

Common roseling

Syn. *Callisia rosea, Tradescantia rosea*
Commelinaceae (dayflower family)

HABITAT & RANGE sandhills and dry woodlands, throughout the Piedmont and coastal plain from North Carolina to Alabama **ZONES** 7–9 **SOIL** moist, well drained to dryish **LIGHT** part shade to part sun.

DESCRIPTION Deciduous clump forming, to 1 ft. tall. Leaves narrow, erect arching. Flowers very showy, ¾ in. wide, with three bright pink petals and six yellow stamens with hairs, open sequentially from a cluster of buds on a few erect stems, blooming from spring through summer. Fruits are small pods maturing shortly after flowering.

PROPAGATION seeds, cold stratified for one month; division of clump **LANDSCAPE USES** specimens or masses in borders, beds, or dryish woodlands **EASE OF CULTIVATION** easy **AVAILABILITY** frequent.

NOTES This is one of the newer members of the dayflower clan to come into cultivation. Despite the fact that we don't know what exactly to call it, it makes a great garden plant, much more delicate than spiderwort (*Tradescantia* species). The flowers are delicate pink, have stamen hairs, and the

Coreopsis major

Coreopsis auriculata

Coreopsis gladiata

Coreopsis verticillata

leaves are much more delicate and grasslike, but the plant tolerates intense heat and drought and can bloom from spring through the entire summer with a little supplemental water during droughts. A real winner by any name.

The similar but smaller grassleaf roseling (*Cuthbertia graminea*), so charming and characteristic of the sandhills, is very difficult to establish in cultivation.

Cuthbertia rosea

Cypripedium parviflorum ★★★

Yellow lady's-slipper

Syn. *Cypripedium pubescens, C. calceolus*
Orchidaceae (orchid family)

HABITAT & RANGE very rich woods, mostly in the Piedmont and mountains of the Southeast, and across Tennessee **ZONES** 4–8 **SOIL** moist, well drained, extra organic matter **LIGHT** part shade to part sun.

DESCRIPTION Deciduous, clump forming, 1–2 ft. tall, from a mass of fleshy roots. Leaves alternate, simple, broadly ovate, 4–6 in. long, strongly ribbed. Flowers very showy, yellow, solitary (rarely two) on the tip of a leafy stem, two long-twisted petals and a third forming an inflated pouch (the "slipper"), long-lasting, blooming in mid to late spring. Fruit an enlarged pod with thousands of dustlike seeds, maturing in autumn.

PROPAGATION careful division of clump if you must **LANDSCAPE USES** choice specimens in moist, open woodland gardens **EASE OF CULTIVATION** moderate **AVAILABILITY** rare.

NOTES Wow! What wildflower gardener doesn't want to grow the elusive yellow lady's-slipper orchid, or even just to see one in the wild or in someone's garden? This is the rarest of wildflowers because is it difficult to propagate from the minute seeds that require a fungus-root relationship in just the right soil. They are becoming more available commercially, at a high price. An experienced gardening friend of mine once gave me a yellow lady's-slipper and told me, "Now, plant it only under a beech tree." I did, and was successful; and later realized that beech is the last tree to leaf out in the spring, so the orchid got a few extra precious days of bright sunlight before the canopy closed in. It is possible to grow reasonably well in less-acidic, moist soil rich in organic matter.

Of greater difficulty and thus perhaps evoking greater desire than yellow lady's-slipper is pink lady's-slipper (*C. acaule*), also known as moccasin-flower. This orchid with two basal leaves only is far more common, occurring in drier, open woods throughout the Upper South, often in large colonies in light shade where the soil is intensely acid and very well drained, a very different habitat from that of yellow lady's-slipper. Without understanding these requirements, it's notoriously difficult to transplant this species and keep it alive. The thin fleshy roots of pink lady's-slipper spread on top of the mineral (even clay) soil, just under a layer of leaf duff (not in a cubic foot of soil as old-fashioned beliefs suggest). To transplant when a rescue is in order, select an identical new location, pluck a young seedling

Cypripedium parviflorum

Cypripedium acaule

Delphinium tricorne

from the duff just as the flowers are fading (this is when new roots begin to grow), spreading the seedling's old roots gently on top of the soil in new location, then covering with the same duff to less than 1 in. deep. Water well once, then no more. It is critical not to break new roots when transplanting. This technique was perfected by nurseryman Don Jacobs of Decatur, Georgia.

Delphinium tricorne ★★★

Dwarf larkspur

Ranunculaceae (buttercup family)

HABITAT & RANGE rich woods, mostly throughout the Upper Southeast **ZONES** 4–7 **SOIL** moist, well drained **LIGHT** shade to part sun.

DESCRIPTION Deciduous plant, usually single-stemmed, 12–18 in. tall, from a knotty tuber. Leaves alternate, simple, deeply dissected and deeply toothed, 2–3 in. long. Flowers showy, lavender to purple (rarely white), 1 in. long with five spreading "sepals" and a long tubular nectar spur extended behind, opening in succession along an upright stem, blooming in spring. Fruits are dry pods shedding seeds shortly after flowering, the whole plant disappearing by late spring.

PROPAGATION seeds, allowed to self-sow **LANDSCAPE USES** masses in open woodland gardens **EASE OF CULTIVATION** moderate **AVAILABILITY** frequent.

NOTES Dwarf larkspur is a beautiful addition to the spring garden with dainty flowers of a unique color of blue-purple (often called "larkspur blue"). This species generally grows in the mountains and avoids the heat by dying down shortly after late-spring seed dispersal. Plant it in drifts, and they should cross-pollinate to give you an ever-enlarging population in moist organic soil. I have had one solitary plant bloom annually in my garden for over thirty years; it never has a companion to mate with, and produces no seeds, alas.

Dicentra eximia 

Bleeding-heart

Fumariaceae (fumitory family)

HABITAT & RANGE rich, less-acidic woodlands, mostly in the higher mountains of the Southeast **ZONES** 4–8 **SOIL** moist, well drained **LIGHT** part shade to part sun.

DESCRIPTION Deciduous, short-lived perennial, clump forming, from an enlarged crown, 6–12 in. tall. Leaves arising from the base, deeply divided, appearing fernlike, to 1 ft. long, very brittle. Flowers very showy, light to dark pink, with two inflated petals coming together to form a heart-shaped floral tube which hangs down, 1 in. long, in branched clusters on leafless stalks, blooming in spring and continuing through the summer as conditions allow. Fruits are 1-in. pods with black seeds, maturing quickly as flowers fade.

PROPAGATION seeds, collected when ripe and sown directly (never stored), or allowed to self-sow; careful division of young brittle crowns in spring or fall **LANDSCAPE USES** specimens, massed or naturalized as groundcover in woodland gardens **EASE OF CULTIVATION** easy **AVAILABILITY** common.

NOTES Bleeding-heart is the cutest wildflower, evoking

Dicentra eximia

Dicentra cucullaria

wonder as to how the strange flower is put together. You can see how the ancients invoked the Doctrine of Signatures to justify the use of this plant as (an ineffective) treatment for heart ailments. It is a wonderful, delicate, long-blooming, easily spreading wildflower. Seeds are dispersed by ants and can come up in many different places. I keep my bleeding-hearts regularly watered, and they grow and bloom nicely through the summer even in very hot weather.

There are several robust and attractive selections, even white ones, mostly derived from hybrids with Japanese bleeding-heart (*D. formosa*), such as the widely grown 'Luxuriant'. While this one is indeed wonderful in its exuberant growth and flowering, I find the straight species performs better in the warm Southeast. Perhaps it's because seedlings self-select in the garden for heat tolerance, while the hybrids are clones that prefer a specific set of conditions. The hybrids rarely set seed.

One of my favorite spring wildflowers is Dutchman's-breeches (*D. cucullaria*). Its nifty white flowers have spreading petal lobes, like pants hung out to dry. It has the same delicate foliage as bleeding-heart, but has underground bulblets that spread the plant to form a larger mass. The plants bloom early and quickly disappear for the summer. They barely tolerate conditions south of Zone 7, but there are some southern-derived clones.

I have never found the equally intriguing cohort, squirrel-corn (*D. canadensis*), to do well in the Southeast outside of its cool mountain haunts.

Dodecatheon meadia ★★★★

Shooting-star

Syn. *Primula meadia*
Primulaceae (primrose family)

HABITAT & RANGE rich, moist, less-acidic woods and streambanks, mostly throughout the Upper Southeast **ZONES** 4–8 **SOIL** moist, well drained **LIGHT** part shade to part sun.
DESCRIPTION Deciduous, clump forming, from fibrous roots, 1–2 ft. tall. Leaves clustered strictly at the base, elongate-rounded, smooth, 4–10 in. long, short-lived. Flowers showy, white, hanging down, five petals flared back, exposing yellow stamens that come to a point, 1 in. long, several arranged gracefully together on the top of a tall leafless stalk, blooming for a long time in spring. Fruits are rounded pods on erect stalks opening through a small hole just at the tips to release numerous seeds in early summer.
PROPAGATION seeds, cold stratified for three months **LANDSCAPE USES** specimens, massed or grouped in borders, beds, or informal woodlands **EASE OF CULTIVATION** easy **AVAILABILITY** frequent.
NOTES Yes, shooting-star is in the primrose family and may be just another primrose whose petals have been blow-dried too hard, but I like recognizing it as a separate genus to give it distinctiveness. Many western species are pink; ours is white and stately, and may be the most elegant plant in your spring garden. It also is long-lived. Watch the bumblebees visit, hang upside down on the tip, and *buzz-bzzz-bzzt* at different frequencies to vibrate the pollen out of the stamens. Don't feel bad when the plants die down completely in early summer; they do that to avoid hot weather, like swans migrating north.

Dodecatheon meadia

Echinacea purpurea ★★★

Purple coneflower

Asteraceae (aster family)

HABITAT & RANGE open woodlands and dry roadsides scattered throughout the Southeast, less common in the more eastern sectors **ZONES** 3–8 **SOIL** moist, well drained **LIGHT** sun.
DESCRIPTION Deciduous, often short-lived perennial, strongly erect, clump forming, from thick roots, 2–4 ft. tall, sometimes irritatingly rough-textured. Leaves alternate, narrow to elliptical, 4–6 in. long. Flowers very showy, in large sunflower-like heads, pink-purple rays held horizontally to slightly drooping, central portion tough and solid, becoming conelike in maturity, with sharp spine-like pegs, blooming in summer. Fruits are like flattened thistle seeds, relished by goldfinches, maturing soon after flowering.
PROPAGATION seeds, without pretreatment **LANDSCAPE USES** specimens in sunny borders, naturalized in well-drained meadows **EASE OF CULTIVATION** easy **AVAILABILITY** common.

NOTES Coneflowers are a beautiful mainstay of the sunny border. Their striking clumps of bold flowers are very attractive. Many colorful and fragrant hybrids and selections are available, but few have been long lasting or strong growing in our Southeastern gardens, perhaps because of too much rainfall. After all, *Echinacea purpurea* is a prairie species and is used to less than 25 in. of rain a year. Plant coneflowers in well-drained soils and hope for the best. There is some controversy as to how far east this species is truly native; many reports are perhaps of escaped plants from age-old medicinal gardens.

As for cultivars, 'Kim's Knee High' has gorgeous purple flowers on a shorter plant. The following were top-ranked in an evaluation at Mt. Cuba Center, which is in Zone 7: a coneflower bearing the trade name Pixie Meadowbrite and the cultivars 'Pica Bella', 'Elton Knight', 'Fatal Attraction', and 'Vintage Wine'. The other new colorful coneflower cultivars have not done as well in the long run in the Southeast. Smooth purple coneflower (*E. laevigata*) is an eastern cousin of purple coneflower and may be the most long-lived of the group. The narrow ray flowers droop strongly. Pale purple coneflower (*E. pallida*), also with drooping ray flowers, is a good choice with better heat and drought tolerance.

The Tennessee cone flower (*E. tennesseensis*) cultivar 'Rocky Top' is distinctive in having bold pink ray flowers that often flare upwards. It is a rare wild relative of pale purple coneflower from the Nashville area, and I like it the best.

Echinacea purpurea

Erigeron pulchellus ★★★

Robin's-plantain

Asteraceae (aster family)

HABITAT & RANGE rich wooded slopes, trail margins, and open roadbanks throughout most of the Southeast, except absent in Florida **ZONES** 5–8 **SOIL** moist, well drained **LIGHT** part sun to sun.

DESCRIPTION Deciduous, basal rosette, spreading by stolons, colony-forming groundcover, ½–1½ ft. tall. Leaves basal, ovate with rounded teeth, hairy, 3–6 in. long. Flowers showy, daisylike, soft pink with yellow center, heads 1–1½ in. wide clustered atop mostly leafless stems, blooming in spring. Fruits are nutlets with fluff, maturing in early summer.

PROPAGATION seeds, cold stratified for three months; division of colonies by separating stoloniferous plantlets **LANDSCAPE USES** groundcover, especially on slopes and edges **EASE OF CULTIVATION** easy **AVAILABILITY** frequent.

NOTES Robin's-plantain is a choice plant to occupy space for itself alone, where it creates a dense colony that hugs the ground like shelf paper, the leafy rosettes touching as the patch spreads. It loves edges: roadbanks, along trails, rocky slopes, even flat ground where it will not be covered by heavy leaf mulch. The early spring flowers are cheerful and a patch of them blooming at once is remarkable. They are fragrant, too.

Erigeron pulchellus

Erythronium umbilicatum

Erythronium umbilicatum ★★★

Dimpled trout lily, dogtooth-violet

Liliaceae (lily family)

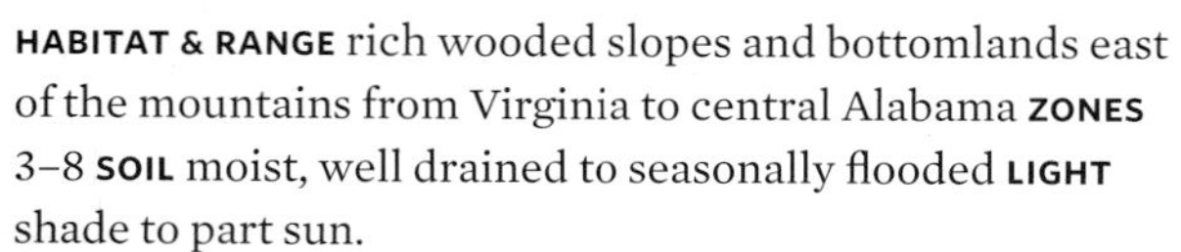

HABITAT & RANGE rich wooded slopes and bottomlands east of the mountains from Virginia to central Alabama **ZONES** 3–8 **SOIL** moist, well drained to seasonally flooded **LIGHT** shade to part sun.

DESCRIPTION Deciduous, bulbous plant forming an extensive colony by underground stolons, 4–6 in. tall. Leaves one per bulb, two per bulb when flowering, narrowly elongate, 3–6 in., usually mottled in greens and dark purples, disappearing by late spring. Flowers showy, one per stem per bulb, 1 in. across, with six yellow “petals” (tepals), often blooming either *en masse* or hardly at all, in late winter or very early spring. Fruits are rounded pods on the long stalks, maturing soon after flowering and reclining on the ground.

PROPAGATION seeds, self-sown; division of colonies by separating the deeply buried bulbs and stolons **LANDSCAPE USES** patches and masses in woodland gardens **EASE OF CULTIVATION** easy **AVAILABILITY** frequent.

NOTES This well-known and charming wildflower has beautiful yellow flowers in late winter, displays its characteristic speckled leaves, and forms solid groundcover (often one clone) in rich woods. The flowers are shy and hang down except when there is bright sun. It is a mystery why they don’t always bloom profusely given that there are so many plants in one place.

Dimpled trout lily is one of two common yellow-flowered species in the East. It is the more southern species, occurring in the mountains and to the east from Virginia to Alabama. The northern species, American trout lily (*E. americanum*), comes south mostly west of the mountains. The two species are superficially identical: American trout lily has yellow stamens and a seedpod held off the ground with a flattened tip, whereas dimpled trout lily has brown-purple stamens and a seedpod that rests on the ground with an indentation in its tip. Both trout lilies form similar extensive patches, though the southern one is more heat tolerant. It would be interesting to compare the two in your garden, side by side, and to inquire at the garden club, “Which trout lily do you have?”

Eupatorium perfoliatum

Eupatorium perfoliatum ★★★

Boneset

Asteraceae (aster family)

HABITAT & RANGE marshes, swamps, bogs, and wet pastures throughout most of the Southeast **ZONES** 3–9 **SOIL** moist, well drained to damp or seasonally wet **LIGHT** part sun to sun.

DESCRIPTION Deciduous, one to several stems from rootstock, 1–3 ft. tall. Leaves opposite, most pairs connected around the stem (connate or perfoliate), elliptic-long tapering, toothed, hairy, 4–6 in. long. Flowers showy, small white heads arranged into a broad, flat-topped inflorescence, blooming in late summer. Fruits are nutlets with fluff, maturing in fall.

PROPAGATION seeds, cold stratified for three months **LANDSCAPE USES** stately specimens in borders and beds **EASE OF CULTIVATION** easy **AVAILABILITY** frequent.

NOTES Boneset is a choice plant for the garden, neat and clean. Its rough texture is fun to feel. It loves a moist, sunny spot. The fused leaves directed the ancient use to heal broken bones based on the Doctrine of Signatures. There are other eupatoriums, but this one is the most ornamental.

Euphorbia corollata ★★

Flowering spurge

Euphorbiaceae (spurge family)

HABITAT & RANGE woodland borders, roadbanks, and old fields throughout most of the Southeast, except absent in Florida **ZONES** 3–9 **SOIL** dryish, well drained **LIGHT** part sun to sun.

DESCRIPTION Deciduous, one to several stems from a deep rootstock, 1–2 ft. tall, exuding an irritating milky sap. Leaves alternate, simple, variable, 1–3 in. long. Flowers showy, small cuplike structures (cyathia) with white marginal petal-like bracts, usually produced in profusion on branching stems over a long period all summer. Fruits are small green pods with three seeds, maturing during and just after the flowering period.

PROPAGATION seeds, without pretreatment **LANDSCAPE USES** curiosity in dry open or shady areas **EASE OF CULTIVATION** easy **AVAILABILITY** infrequent.

NOTES Flowering spurge is not for everyone. I mention it because it's so different, and yet it's widespread and persistent as a woodland specimen, flowering well even in part sun. You might have it in a scruffy disturbed area, and it seems to have some appeal in its long-blooming behavior. The minute floral structure is interesting; while the plant evokes a notion of coming from some other place (like a desert, where most of its relatives live), it makes you stop and want to look at it. Large patches along a sunny roadbank are actually noticeable from a car.

Eurybia divaricata ★★

White wood aster

Syn. *Aster divaricatus*
Asteraceae (aster family)

HABITAT & RANGE moist to dry forests, mostly throughout the Upper Southeast **ZONES** 4–8 **SOIL** moist to dryish, well drained **LIGHT** part shade to part sun.

DESCRIPTION Evergreen, colony forming by coarse woody creeping rhizomes and sprawling dark stems, 6–24 in. tall. Leaves alternate, simple, broadly heart-shaped with large teeth, 3–6 in. long. Flowers showy, daisylike, white with a yellow center, 1 in. wide, produced in broad flat-topped clusters atop black zigzag stems, blooming late summer into autumn. Fruits are nutlets with fluff, maturing in autumn.

PROPAGATION division of leafy growth from colonies **LANDSCAPE USES** groundcover in woodland gardens **EASE OF CULTIVATION** easy **AVAILABILITY** frequent.

Euphorbia corollata

Eurybia divaricata

NOTES White wood aster is a beautiful late-summer plant even though it's not as colorful as some of its relatives. It spreads tenaciously in any soil, and thus is wonderful on dry rocky slopes or difficult situations where you can't (or don't want to) grow anything else. You can't manage it by pulling it up, the stems break off, so watch where you plant it. The more you cut it back, the more it grows back and blooms, especially in shade.

'Eastern Star', 'Raiche', and 'Silver Spray' are excellent cultivars with clean white flowers that are a bit shorter.

Big-leaved aster (*E. macrophyllus*) is larger than white wood aster and coarser in all respects. It is drought tolerant and blooms in shade, often with slightly more lavender flowers.

The several species of *Eurybia* used to be in the genus *Aster*, but have been segregated out along with *Symphyotrichum* and other genera. I can't give you a simple reason why, but it's not a new idea, only an unsettling one that we have to accept. There are no longer any true native species of redefined genus *Aster* in the Southeast.

Eutrochium fistulosum

Eutrochium dubium

Eutrochium fistulosum ★★★★

Joe-pye weed

Syn. *Eupatorium fistulosum*
Asteraceae (aster family)

HABITAT & RANGE moist forests, marshes, and ditches, mostly throughout the Southeast **ZONES** 6–9 **SOIL** moist, well drained to seasonally wet **LIGHT** part sun to sun.
DESCRIPTION Perennial, clump forming, from a tough, slowly enlarging crown, 2–10 ft. tall. Leaves whorled, four to seven in each ring, regularly spaced all along the stem, 4–10 in. long, toothed. Flowers showy, fragrant, pink, individually small but clustered together into large elongate-rounded masses atop the tall stems, blooming in late summer. Fruits are nutlets with fluff, maturing in autumn.
PROPAGATION seeds, without pretreatment, or allowed to self-sow; division of clump **LANDSCAPE USES** specimens in borders, naturalized in wetlands or in open woodland gardens **EASE OF CULTIVATION** easy **AVAILABILITY** common.
NOTES Joe-pye weed (supposedly named after an eighteenth-century medicine man) is a giant of a tall wildflower. It is conspicuous in late summer in moist to wet habitats where it rises above all other herbaceous vegetation. It also is a butterfly magnet; butterflies can cover the big flowering head like dust-bunnies on a feather duster, and they are difficult to shake off, so drawn to the sweet nectar are they. As striking as Joe-pye weed is (no garden should be without one), it may be too tall for your plot. To remedy this situation, however, you can do one of two things: cut the stem down halfway by mid-June and it will regrow shorter and bushier, or plant one of the shorter cultivars such as 'Atropurpurea', which may reach only 7 ft. tall.

Three-nerved Joe-pye-weed (*E. dubium*) has only three or four leaves per whorl, heavily spotted stems, and more floral units that are more delicate. Its cultivar 'Little Joe' is definitely smaller, only 2–3 ft. That is too small in my opinion, but it might just fit your space and be a bonanza for you.

Spotted Joe-pye-weed (*E. maculatum*) has purple-marked stems, and the darker-flowered inflorescence is more flat-topped. Its cultivar 'Gateway' is 4–5 ft. tall.

Gaillardia pulchella ★★

Beach blanket-flower

Asteraceae (aster family)

HABITAT & RANGE sandy fields, roadsides, and dunes along the coast and widely scattered across the Southeast **ZONES** 6–10 **SOIL** dryish, well drained **LIGHT** sun.

DESCRIPTION Deciduous, short-lived perennial, forming a single well-branched semiwoody "bush" about 1 ft. tall. Leaves alternate, hairy, more or less lobed or dissected, 3 in. long. Flowers very showy, red and yellow, like bachelor's-button, with heads about 2 in. wide on erect stalks, blooming all summer until frost. Fruits are dry nonfluffy nutlets, maturing throughout summer.

PROPAGATION seeds, without pretreatment, or allowed to self-sow **LANDSCAPE USES** specimens in well-drained borders or rock gardens **EASE OF CULTIVATION** moderate **AVAILABILITY** frequent.

NOTES Beach blanket-flower is so colorful, it's as if it's already a selected cultivar. It grows in sandy soil and is especially prominent on beach dunes and sandy roadside waste areas. You will probably start them as seeds. They will grow better if you add extra coarse sand to the soil, and don't overwater. Heck, you could probably grow them in pure sand, and they must have full sun. Many similar selections behave as colorful annuals, but some are hybrids of more western species. Seeing these at home may remind you of your beach trips, and you can scatter some sand on the walk to give you that gritty feel.

Gaillardia pulchella

Galax aphylla ★★

Galax

Diapensiaceae (diapensia family)

HABITAT & RANGE rocky woodlands, mostly throughout the Upper Southeast **ZONES** 5–8 **SOIL** moist to dryish, acidic, well drained **LIGHT** shade to part sun.

DESCRIPTION Evergreen, densely clump forming, from a slow-growing creeping crown, 4–6 in. tall. Leaves alternate, closely spaced on short semiwoody stems, round with sharp teeth or wavy edges, dark green, turning red in autumn and winter in the sun. Flowers showy, very small, pure white, individually ¼ in. long, densely crowded on flowering stalk 12–18 in. tall, blooming in midsummer. Fruits are tiny pods with many brown seeds, maturing in autumn.

PROPAGATION seeds, germinate on damp surface without pretreatment; division of clump **LANDSCAPE USES** specimens or masses in woodland gardens **EASE OF CULTIVATION** difficult **AVAILABILITY** infrequent.

Galax leaves in fall

NOTES Galax is a unique species, so common in the mountains and beyond in very acidic, thin, well-drained soil, growing with members of the heath family, such as trailing arbutus and mountain-laurel. The shiny round leaves form carpets under evergreen rhododendrons, and have been gathered for ages as Christmas greens and dried for florist's foliage. We once brought home a freshly collected bouquet of galax leaves, put them in water on the table, and the next morning the whole house smelled like the mountain forests—you know, that fresh musty smell that is hard to figure out where it comes from when you're walking through a damp woods. Galax is tricky to establish outside the mountains, but keep the soil acidic and don't overwater. It is worth trying because if it takes, you will have a great feeling of success.

Gaultheria procumbens

Gaura biennis

Gaultheria procumbens ★★

Teaberry, wintergreen

Ericaceae (heath family)

HABITAT & RANGE rock outcrops and rocky woods with mountain laurel throughout North Carolina to central Tennessee, and hardly further south **ZONES** 4–8 **SOIL** moist, well drained **LIGHT** shade to part shade.

DESCRIPTION Evergreen shrublet, clump forming, slowly spreading, 2–6 in. tall, with wintergreen odor. Leaves alternate, simple, thick, ovate, a few crowded at tip of stem, 1–2 in. long. Flower showy, white, several, borne under the leaves, tubular, ¼–½ in. long, blooming in summer. Fruit is a red berry, maturing in autumn and persisting into winter and longer.

PROPAGATION seeds, cold stratified for three months; division of clump **LANDSCAPE USES** specimens as clumps in raised beds, as groundcover in shady gardens **EASE OF CULTIVATION** moderate **AVAILABILITY** frequent.

NOTES Teaberry is a delightful plant to find in the woods, to crush the leaf and smell the wintergreen fragrance, and to nibble the pithy berry. It forms carpets mixing with trailing arbutus and galax. Although it's a true woody plant (a subshrub), I am treating it as a wildflower because of its small stature and slow growth. The little white flower is barely noticeable, the red berry much more so. Keep acidic and well drained.

Gaura biennis ★★★

Biennial gaura, beeblossom

Syn. *Oenothera gaura*

Onagraceae (evening-primrose family)

HABITAT & RANGE streambanks, meadows, and roadsides, coming southward in the mountains, staying west of that line into southern Alabama **ZONES** 5–8 **SOIL** moist, well drained **LIGHT** sun.

DESCRIPTION Deciduous, biennial or short-lived perennial, clump forming, much-branched from tough roots, 1–6 ft. tall. Leaves alternate, elliptic, to 4 in. long. Flowers showy, white, open, 1 in. long, with four petals, many dangling along branched spikes, opening sequentially, blooming all summer. Fruit is a small pod, ¼ in. long, maturing shortly after flowering.

PROPAGATION seeds, self-sown and germinating best in warm temperatures fluctuating between 60 and 90°F; transplant self-sown seedlings when young **LANDSCAPE USES** specimens in sunny borders **EASE OF CULTIVATION** easy **AVAILABILITY** frequent, as seeds

NOTES Gaura is a very delicate and wispy plant, its long slender branches moving in the breeze, the loosely tethered flowers playing like butterflies. This species has received scant attention, being used mostly as fast-growing annuals in wildflower mixes for quick meadows. Its relative from the southwest barely reaches our area, but whirling butterflies (*G. lindheimeri*) sparked a rash of fabulous selections in the early 2000s to produce a surprisingly successful array of wonderful garden plants. Its whimsical cultivars are much more compact and vigorous, heat tolerant, with larger flowers in various shades of white to deep pink. Some of these are 'Blushing Butterflies', 'Dauphin', 'Whirling Butterflies', and the most famous, 'Siskiyou Pink'. In all cases, don't overwater or overfertilize, and cut back after flowering to rejuvenate.

Gentiana saponaria

Gentiana autumnalis

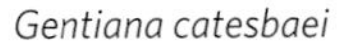

Gentiana catesbaei

Gentiana saponaria ★★

Soapwort gentian

Gentianaceae (gentian family)

HABITAT & RANGE bogs, marshes, and low woods scattered throughout the Southeast, except absent in Florida **ZONES** 6–8 **SOIL** moist, well drained to damp **LIGHT** part shade to part sun.

DESCRIPTION Deciduous, clump forming, 1–2 ft. tall. Leaves opposite, simple, broadly elongate, 3–5 in. long. Flowers showy, rich dark to light blue, tubular, 2 in. long, closed at the mouth, blooming in late summer. Fruits are many-seeded pods, maturing in autumn.

PROPAGATION seeds, stratified for three months; division of clump **LANDSCAPE USES** specimens as clumps in woodland gardens or moist rock gardens **EASE OF CULTIVATION** moderate **AVAILABILITY** infrequent.

NOTES Gentians are among those charismatic wildflowers that everyone loves, but that rarely grow well in cultivation. The closed or bottle gentians are especially interesting because it takes a large bumblebee to force her way into the closed petal tube for effective cross-pollination. Soapwort gentian is one of the easiest to grow.

Catesby's gentian (*Gentiana catesbaei*) is very similar to soapwort gentian. The other purple bottle gentians, *G. andrewsii* and *G. clausa*, are very worth trying in moist, shady woodlands, forming clumps when happy. Their tightly closed flowers also offer bees a challenge.

More challenging yet is the enigmatic pinebarrens gentian (*G. autumnalis*), blooming in September to December in longleaf pine flatwoods (wiregrass is a typical understory plant in this ecosystem) with the most beautiful 2-in. flowers. Its opposite leaves are very narrow, widest near the tips, and twisted like an airplane propeller. Hard to see when not in bloom and difficult to grow; keep it acidic and well drained.

Geranium maculatum

Wild geranium

Geraniaceae (geranium family)

HABITAT & RANGE rich woods, mostly throughout the Upper Southeast **ZONES** 3–8 **SOIL** moist, well drained **LIGHT** part shad to part sun

DESCRIPTION Deciduous, clump forming, slowly enlarging, from a thick elongate rootstock, 1–2 ft. tall. Leaves alternate at the base, opposite halfway up the stem where it forks,

Geranium maculatum

Gillenia trifoliata

round, deeply divided and toothed, slightly hairy. Flowers showy, pink-blue-lavender, 1–1½ in. wide, borne on tall stalks above the leaves, opening sequentially, lasting two days each, blooming in spring. Fruits are quick to mature into 1-in. long pods with five seeds, mature seeds separate at the base, then coil up like a spring and are flung off in long-dispersal.

PROPAGATION seeds, collected just before ripe, then dried and cold stratified for three months; division of rhizomes in a colony **LANDSCAPE USES** clumps in woodland gardens **EASE OF CULTIVATION** easy **AVAILABILITY** common.

NOTES Wild geranium is easy to grow and produces numerous cheerful blue flowers in springtime abundance. The dark leaves add great texture to the spring garden. This tough short-lived plant usually dies down for the summer. Plants can have various shades of flower color, and a white form is available. The cultivar 'Espresso' is more compact.

Gillenia trifoliata ★★

Indian-physic, bowman's-root

Syn. *Porteranthus trifoliatus*
Rosaceae (rose family)

HABITAT & RANGE moist to dry forests edges and roadbanks, mostly in the mountains and immediate adjacent areas **ZONES** 4–8 **SOIL** moist, well drained **LIGHT** part sun to sun.

DESCRIPTION Deciduous, clump forming, slowly enlarging, 1–3 ft. tall. Leaves alternate, compound with three toothed leaflets 2–3 in. long, practically lacking a petiole. Flowers showy, white, tubular-flaring, 1 in. long, in loose clusters, sepal tube reddish, blooming in late spring. Fruits are small dry pods, maturing in late summer.

PROPAGATION seeds, cold stratified for three months; division of larger clumps **LANDSCAPE USES** specimens in sunny borders or beds **EASE OF CULTIVATION** moderate **AVAILABILITY** frequent.

NOTES Indian-physic produces dense clumps, reminiscent of a dwarf spiraea and is common on well-drained roadsides in the mountains. Its bloom time is brief, but the flowers are striking *en masse*, and the foliage makes a good-looking mound throughout the summer. Plants like cooler weather and will need afternoon shade south of Zone 7. A charming pink-flowered form with reddish new growth is called 'Pink Profusion'.

A similar species, Midwestern Indian-physic (*G. stipulata*) has more distinctly toothed leaves that give the plant a finer texture, and it's more heat tolerant. I like it better in Zone 8.

Helenium autumnale ★★★

Dogtooth-daisy

Asteraceae (aster family)

HABITAT & RANGE pastures, meadows, and ditches, mostly throughout the Southeast **ZONES** 3–8 **SOIL** moist, well drained **LIGHT** sun.

DESCRIPTION Deciduous, clump forming, several stems arise from a crown, 2–5 ft. tall. Leaves alternate, narrow, toothed, 3–6 in. long, their bases running down the stems as "wings." Flowers showy, yellow, daisylike, about 1 in. across, ray petals three-toothed, blooming in late summer. Fruits are nutlets without fluff, maturing in autumn.

PROPAGATION seeds; cold stratified for three months; division of larger clumps **LANDSCAPE USES** specimens in borders and moist meadows **EASE OF CULTIVATION** easy **AVAILABILITY** common.

Helenium autumnale

NOTES Dogtooth-daisy is a great plant for showy long-lasting flowers in late summer. The flowers are well known, characterized by the yellow petals with three teeth. Plants become tall and may need staking if overfertilized and overwatered. Many new cultivars are shorter and stockier, with new colors of reds and bronzes. Try 'Butterpat' with golden yellow flowers; 'Helena Gold' and 'Helena Red' are heat tolerant; 'Moerheim Beauty' has darker red flowers.

Purple-headed sneezeweed (*H. flexuosum*) is an overlooked summer-blooming native that is well worth trying. It grows 2–3 ft. tall with winged stems, and is distinctive in having the central disk flowers dark purple.

Helianthus angustifolius ★★★

Swamp sunflower, narrowleaf sunflower

Asteraceae (aster family)

HABITAT & RANGE savannahs, marshes, ditches, and wetlands, mostly throughout the Southeast, rarer in mountains **ZONES** 5–9 **SOIL** moist, well drained to seasonally wet **LIGHT** sun. **DESCRIPTION** Deciduous, spreading from short rhizomes to make an ever-enlarging dense colony, 3–8 ft. tall. Leaves opposite, simple, narrow, 3–10 in. long. Flowers very showy, bright yellow, the heads large to 3 in. wide, produced in profusion on the tops of the stems, blooming in late summer. Seeds are nutlets without fluff, maturing in autumn. **PROPAGATION** seeds, cold stratified for three months; division of clump **LANDSCAPE USES** specimens of strong clumping form for back of border, naturalized in meadows and ditches **EASE OF CULTIVATION** easy **AVAILABILITY** common. **NOTES** Sunflowers are signature plants for North America. They are very important for wildlife and humans as food. They are usually robust, clump-forming and spreading plants, blooming in late summer with their cheerful big yellow flowers with dark centers. The well-known species grown for ornament, seed, and oil is an annual, *Helianthus anuus*. They all may best be grown in dry or moist meadows and rough areas outside the formal garden.

Swamp sunflower is a beauty, so characteristic of wet ditches in early fall with clouds of flowers. Be aware of its urge to spread even more vigorously under garden conditions and keep it lean on fertilizer and water. You will marvel at its profusion.

The old-fashioned Jerusalem artichoke (*H. tuberousus*) is a coarse hairy plant and vigorous spreader, producing edible tubers and gigantic yellow flower in sun or half-day sun.

Hairy sunflower (*H. resinosus*, syn. *H. tomentosus*) is a strong clump-former with very hairy stems, large rough leaves, and pale yellow flowers. It grows and flowers well in

Helenium flexuosum

Helianthus angustifolius

Heliopsis helianthoides

light shade and as such is a good choice for the woodland garden. Though prone to seed around, it's easily controlled by removing unwanted seedlings.

Naked-stem sunflower (*H. occidentalis*) is not too aggressive. Tall thin rough-hairy stems have large basal leaves and open clusters of medium-sized yellow flowers for sun.

Sunflowers that form slowly enlarging colonies include small-headed sunflower (*H. microcephalus*) for the sunny meadow, and forest sunflower (*H. decapetalus*) in light shade.

But be careful: thirty-five years ago I planted rough-leaf sunflower (*H. strumosus*), which has long rhizomes, and I am still trying to get rid of it running through my garden. One good thing is that if you remove the new stems in spring, they will not grow back that same year.

Heliopsis helianthoides ★★★

Ox-eye, false sunflower

Asteraceae (aster family)

HABITAT & RANGE forests and woodland borders throughout the Southeast but rare in the coastal plain **ZONES** 3–9 **SOIL** moist, well drained **LIGHT** sun.

DESCRIPTION Deciduous, clump forming, 4–6 ft. tall, somewhat sandpaper-rough. Leaves opposite, ovate-triangular, 4–8 in. long. Flowers very showy, large, yellow like sunflowers but with the center becoming conical in age and the petals remaining after they fade, 2–3 in. wide, blooming in summer. Fruits are nutlets without fluff, maturing in autumn.

PROPAGATION seeds, cold stratified for three months; division of clump **LANDSCAPE USES** specimens in borders and meadows **EASE OF CULTIVATION** easy **AVAILABILITY** common.

NOTES Ox-eye is very much like a sunflower, only more refined and more readily adaptable to every sunny garden. The plants are shorter and much more clump forming, the flowers a little smaller and longer blooming. A good 2- to 3-ft. cultivar is 'Summer Sun'.

Hepatica americana ★★★

Round-lobed liverleaf, round-lobed hepatica

Syn. *Anemone americana*

Ranunculaceae (buttercup family)

HABITAT & RANGE rich to dryish woods, mostly throughout the Upper South **ZONES** 3–8 **SOIL** moist, well drained **LIGHT** shade to part shade.

DESCRIPTION Evergreen, delicate clump-former, from fibrous roots, 3–6 in. tall. Leaves in a basal tuft, simple, distinctly three-lobed, rounded, mottled green and purple, 2 in. wide. Flowers showy, white to blue-lavender to pink, ¾ in. wide, with five to eight sepals (no petals), single on a slender erect stalk, blooming late winter into spring. Fruits are small nutlets, maturing in spring.

PROPAGATION seeds, collected when ripe and planted, will germinate the following spring; division of larger clumps **LANDSCAPE USES** massed or scattered in woodland garden, rockeries **EASE OF CULTIVATION** easy **AVAILABILITY** frequent.

NOTES Round-lobed liverleaf is not so much a harbinger of

Hepatica americana

Heuchera americana

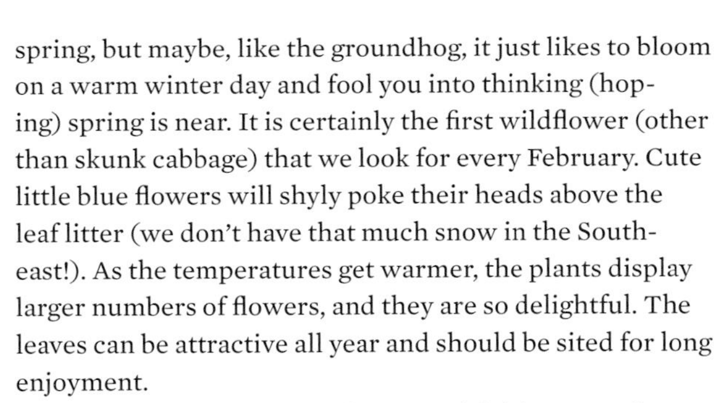

spring, but maybe, like the groundhog, it just likes to bloom on a warm winter day and fool you into thinking (hoping) spring is near. It is certainly the first wildflower (other than skunk cabbage) that we look for every February. Cute little blue flowers will shyly poke their heads above the leaf litter (we don't have that much snow in the Southeast!). As the temperatures get warmer, the plants display larger numbers of flowers, and they are so delightful. The leaves can be attractive all year and should be sited for long enjoyment.

The sharp-lobed liverleaf (*H. acutiloba*) is more of a mountain species, slightly larger and more floriferous, and less evergreen. It has not performed as well warmer than Zone 7. It prefers less acid soil.

Some botanists have placed these species into the genus *Anemone*, while others lump them both into the European species *Hepatica nobilis*. Sigh! At least we all know what "liverleaf" means.

Heuchera americana ★★★

American alumroot

Saxifragaceae (saxifrage family)

HABITAT & RANGE rocky forests and outcrops throughout most of the Upper Southeast **ZONES** 3–8 **SOIL** moist to dryish, well drained **LIGHT** part shade to part sun.
DESCRIPTION Evergreen with a basal rosette of leaves, clump forming, from a branching crown, 12–18 in. tall. Leaves alternate, simple, on long petioles, round with rounded lobes and teeth, mottled green, silver, and purple, blade to 6 in. across. Flowers not particularly showy, small, greenish, numerous on tall branching stalks to 3 ft. long, blooming from spring into summer. Fruits are tiny pods with numerous small seeds, maturing soon after flowering.
PROPAGATION seeds, sown immediately without pretreatment; division of clump not recommended **LANDSCAPE USES** specimens in partly shaded borders or beds, in rocky woods and rock walls **EASE OF CULTIVATION** easy **AVAILABILITY** common.
NOTES American alumroot is one of our toughest wildflowers, with beautiful long-lasting leaves. It has staying power on well-drained sites; I have had one clump for over thirty years. The plants can tolerate strong sunlight. The small flowers offer little—I usually remove mine to encourage more leaves. Giving plants extra water and fertilizer may shorten their life span. The many new beautiful hybrids rarely last more than one or two years in the warm Southeast, but you may keep searching for one that works for you. The only selection I have found with staying power equal to the wild type is 'Dale's Strain', which is seed-grown and hardly differs from the best-marked leaves of the wild.

Hairy alumroot (*H. villosa*) is typical of mountainous seepage ledges and is difficult to grow. *Heuchera villosa* var. *macrorhiza* (syn. *H. macrorhiza*), from west and north of the Blue Ridge, is fuzzy and can look great in the garden, but may not be so heat tolerant. The two cultivars 'Purpurea' and 'Autumn Bride' are wonderful for a while, but slowly melt in the summer heat and humidity.

Hexastylis naniflora

Hexastylis speciosa

Hexastylis arifolia ★★★

Heartleaf-ginger, piggies ginger

Syn. *Asarum arifolium*
Aristolochiaceae (birthwort family)

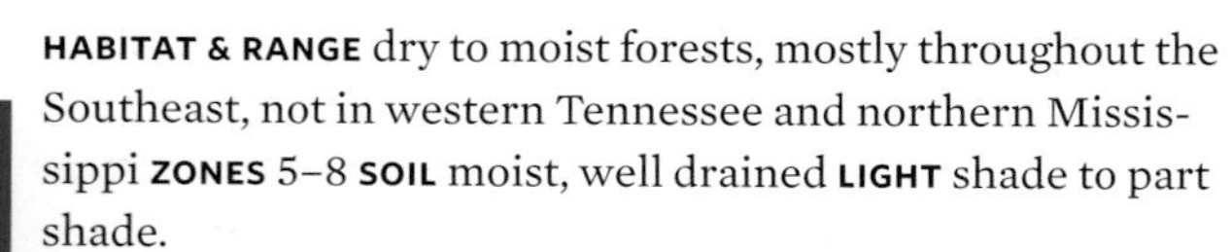

HABITAT & RANGE dry to moist forests, mostly throughout the Southeast, not in western Tennessee and northern Mississippi **ZONES** 5–8 **SOIL** moist, well drained **LIGHT** shade to part shade.

DESCRIPTION Evergreen, clump forming, from a branching crown, 6–12 in. tall, with deep-penetrating, fleshy brittle roots. Leaves alternate or basal, simple, leathery, arising from the crown, elongate-heart shaped, mottled with green-silver between the veins. Flowers interesting, clustered (hidden) under the leaves at the base of the plant, tan or brown, tubular, 1 in. long with slightly flaring mouth, blooming in spring. Fruits are rarely seen, seeds are dispersed by ants in early summer.

PROPAGATION seeds, self-sown; division of large clumps sets them back a year, but they will recover **LANDSCAPE USES** clumps in woodland gardens **EASE OF CULTIVATION** moderate **AVAILABILITY** infrequent.

NOTES Heartleaf gingers produce our most beautiful evergreen foliage wildflowers for the woodland garden, forming robust leafy clumps that are a marvel to behold (creeping forms are known). To see the hidden flowers, you peer beneath the leaves and say, "there they are." They are all slow to establish, but become better with each passing year. It is best not to disturb a fine specimen (because people will want a piece of it).

There are some twelve species and varieties of *Hexastylis*, all are beautiful. They are often listed in the genus *Asarum*; I prefer to keep them all in *Hexastylis*. Dwarf-flower heartleaf (*H. naniflora*), an endangered species but readily available from tissue culture, has smaller round leaves (2–3 in. wide) with striking green-silver markings *along* the veins.

The most robust species in the genus is showy heartleaf (*H. speciosa*) from central Alabama. It has larger leaves and unbelievable 2 in. open flowers with dark chocolate and light butterscotch markings. It is worth peering at.

Harper's heartleaf (*H. shuttleworthii* var. *harperi*) is a creeping form with small leaves, available as the exceptional distinctively mottled cultivar 'Callaway'.

Hibiscus coccineus ★★★★

Scarlet hibiscus

Malvaceae (mallow family)

HABITAT & RANGE marshes, swamps, and swales, mostly along the Gulf Coast and Peninsular Florida, also appearing native at one location in the Carolinas (perhaps escaped) **ZONES** 6–9 **SOIL** moist, well drained to seasonally wet **LIGHT** part sun to sun.

DESCRIPTION Deciduous, strong clump-forming crown with three to five or more stems, from massive fleshy roots, 4–8 ft. tall, stems can be quite red. Leaves alternate, simple but so deeply divided into three to five narrow, toothed segments as to appear almost compound, 6–8 in. long. Flowers showy, scarlet, very large, 4–6 in. across, blooming from midsummer to autumn. Fruits are 1-in. dry brown pods that split to release many round seeds, maturing sequentially.

Hibiscus coccineus

Hibiscus moscheutos

PROPAGATION seeds, germinate easily, but cold stratification for one month or scarification may improve germination rate **LANDSCAPE USES** specimens in sunny borders, massed in natural wetlands **EASE OF CULTIVATION** easy **AVAILABILITY** common.

NOTES No wildflower is more striking than scarlet or cardinal hibiscus. The giant red flowers are showstoppers and have been favorite garden plants for generations. There is nothing quite like it in the world, and it has been used with several other species to make wonderful garden hybrids (see below). It becomes very large and wide; cutting it down by half in June will keep it shorter with more flowering branches. A new flower opens each day. Keeping the fruits picked off will prolong flowering (true of many perennials). Butterflies love it, and you will too.

Among several new albino forms, the white Texas star hibiscus 'Summer Snow' has striking pure white flowers and no red in the stems. It is less cold tolerant, especially in a pot.

Hibiscus moscheutos ★★★

Rose mallow, marsh mallow

Malvaceae (mallow family)

HABITAT & RANGE marshes, swamps, ditches, and wetlands throughout the entire Southeast **ZONES** 5–9 **SOIL** moist, well drained to wet **LIGHT** part sun to sun.

DESCRIPTION Deciduous, strong clump-forming crown with three to five or more stems from massive fleshy roots, 3–6 ft. tall. Leaves alternate, simple, three- to five-lobed and toothed, 6–8 in. long. Flowers showy, white to pink, very large, 4–6 in. across, blooming from midsummer to autumn. Fruits are 1-in. dry brown pods that split to release many round seeds, maturing sequentially.

PROPAGATION seeds, without pretreatment; stem cuttings **LANDSCAPE USES** specimens in borders, naturalized in wetlands **EASE OF CULTIVATION** easy **AVAILABILITY** common.

NOTES Our native rose mallow has been a favorite garden perennial for a long time. The giant flowers and handsome foliage present an attractive show, but the plants do take up space and they become larger with age. Cutting them down by half on June 1 will shorten their stature and give more flowering branches. Removing the seedpods will prolong blooming. Do site them to give abundant sunshine or they will weaken over time. They are heavy feeders. The many new cultivars and hybrids offer much refinement and variety over the wild types.

Velvet mallow (*H. grandiflorus*) has unique silvery green leaves and elegant rich pink flowers on very tall plants.

Hybridization with native hibiscus (including *H. coccineus*, *H. moscheutos*, and *H. laevis*) goes way back, producing a broad array of hardy hibiscus to choose from, with more new ones to come. The old-fashioned hard-to-beat 'Lady Baltimore' (pink) and 'Lord Baltimore' (red) have large bright flowers with rich green broad-lobed leaves. 'Kopper King' has beautiful pink flowers and dark foliage.

I am not as fond of the giant-flowered Disco series of hybrids in the native garden, but they are certainly striking and worthy of growing.

Hydrastis canadensis ★★★

Goldenseal

Ranunculaceae (buttercup family)

HABITAT & RANGE rich, less-acidic woods, uncommon throughout the central portion of the Upper Southeast **ZONES** 3–8 **SOIL** moist, well drained **LIGHT** shade to part sun.

DESCRIPTION Deciduous, clump forming, from a stocky crown, slowly spreading, 1 ft. tall, roots and crowns are bright yellow inside. Leaves borne two per flowering stalk, round, deeply lobed and toothed, expanding after flowering to become 6–8 in. wide. Flower intriguing, one per stalk, showy by virtue of the stocky white stamens, without sepals or petals, short-lived, blooming in early spring. Fruit a single conspicuous red berry with black seeds, maturing in midsummer, loved by birds.

PROPAGATION seeds, cold stratified for three months; division of clump in late winter **LANDSCAPE USES** specimens allowed to form a colony in woodland gardens **EASE OF CULTIVATION** easy **AVAILABILITY** frequent.

NOTES No spring wildflower captures my attention like goldenseal. The strange little flowers are alluring; they work well with the expanding leaves to produce a uniquely appealing combination. The bright yellow roots indicate medicinal quality (true or imagined). This northern species is quite heat tolerant. It can hold its own in strong sunlight if kept watered, but shade is its preferred home. While it's not flashy, I guess you would call it a "charmer."

Hydrastis canadensis

Hydrophyllum virginianum ★★

Eastern waterleaf

Boraginaceae (borage family), formerly in Hydrophyllaceae (waterleaf family)

HABITAT & RANGE rich, moist woods, just coming south into the Piedmont and mountains of North Carolina and adjacent Tennessee **ZONES** 3–8 **SOIL** moist, well drained **LIGHT** shade to part sun.

DESCRIPTION Evergreen, clump forming, slowly spreading, 1 ft. tall. Leaves alternate (basal), compound or deeply lobed and toothed, very brittle, mottled, 3–6 in. long. Flowers showy, white-pink-purple, 1 in. long including conspicuous stamens, formed at the tip of an uncoiling branched stalk, blooming in spring. Fruits are little pods, maturing in summer.

PROPAGATION division of spreading clumps; seeds, allowed to self-sow **LANDSCAPE USES** clumps in woodland gardens **EASE OF CULTIVATION** easy **AVAILABILITY** frequent.

NOTES Eastern waterleaf is a delicate wildflower of several uses. The light mottling on the leaves is supposed to evoke the notion of being stained by water for the common name. It also grows in very moist habitats. The tender new leaves can be eaten in spring salads. The flowers are fleeting but charming. It makes a delicate evergreen groundcover, but is very brittle so you can't walk on it. It helps fill in with many wildflower situations. It may be confused with fernleaf phacelia (*Phacelia bipinnatifida*) when not in bloom.

Hydrophyllum virginianum

Hymenocallis occidentalis

Hymenocallis occidentalis ★★★

Woodland spider-lily

Syn. *Hymenocallis carolinianus*
Amaryllidaceae (amaryllis family)

HABITAT & RANGE floodplain forests and upland slopes and flats, mostly in South Carolina and westward throughout the Southeast **ZONES** 7–10 **SOIL** moist, well drained to seasonally wet **LIGHT** part shade to part sun.
DESCRIPTION Deciduous, clump forming, from bulbs, 1–2 ft. tall. Leaves alternate, arising in a basal rosette from the bulb, narrow straplike, about 1 ft. long. Flowers very showy, white, delicate, spiderlike, 3 in. wide with narrow sepals and petals, six stamens connected by a unique cup-forming white membrane, two to six flowers in a cluster atop a stocky hollow stalk, blooming over a long period from spring to summer. Fruits are green egg-shaped fleshy berries, maturing in summer.
PROPAGATION seeds, planted as soon as the green berries are ripe; division of the clump of bulbs **LANDSCAPE USES** specimens in borders, or in open, sunny rocky or boggy woodland settings **EASE OF CULTIVATION** moderate **AVAILABILITY** infrequent.
NOTES This is perhaps the most eye-catching woodland wildflower when in bloom. The ghostly white flowers appear suddenly one morning, last a day each, and then, whoosh, disappear. The clump of semierect leaves is gracefully attractive and may simply remind you of what you missed and were dying to see. Worth growing even if you see it in its glory only once.

Hypoxis hirsuta ★★

Yellow star-grass

Hypoxidaceae (star-grass family), formerly in Amaryllidaceae (amaryllis family)

HABITAT & RANGE woodlands, roadbanks, savannas, and meadows throughout the Southeast, except absent in Florida **ZONES** 3–9 **SOIL** moist, well drained **LIGHT** part shade to part sun.
DESCRIPTION Deciduous, colony forming, from a bulblike corm, 3–6 in. tall. Leaves grasslike, several tightly clumping from the corm, up to 1 ft. long and ½ in. wide. Flowers showy, bright yellow, with six tepals (sepals and petals looking alike), ½ in. wide, several on sparsely branched stalks, shorter than the leaves, blooming over a long period from spring to summer. Fruits are inconspicuous pods, maturing in summer.

Hypoxis hirsuta

PROPAGATION seeds, self-sown; division of clump **LANDSCAPE USES** delicate plants form colonies on thin soil in open woods and meadows **EASE OF CULTIVATION** easy **AVAILABILITY** infrequent.
NOTES I would not even mention these catchy little yellow flowers except that they have a way of forming their own modest colonies in thin grassy spots, being there in bloom almost all summer, transplanting well, and providing something a little fun to talk about and to look for.

Impatiens capensis ★★

Orange jewelweed

Balsaminaceae (touch-me-not family)

HABITAT & RANGE rich, moist woods and floodplains, mostly throughout the Southeast **ZONES** 4–11 **SOIL** moist, well drained **LIGHT** shade to part sun.
DESCRIPTION Annual, single fleshy stem containing slimy sap, from fibrous roots, 2–5 ft. tall. Leaves alternate, simple, ovate with shallow wavy teeth, 2–3 in. long. Flowers showy, orange with dark orange spots, tubular with a curled nectar spur, mouth flaring with two large lobes, blooming all summer. Fruits are plump pods that burst open forcefully when touched to disperse black seeds, maturing continuously.
PROPAGATION seeds, self-sown **LANDSCAPE USES** scattered in woodland gardens **EASE OF CULTIVATION** easy **AVAILABILITY** infrequent, collect your own seeds
NOTES This is the only true annual to be mentioned here with the perennial woodland wildflowers, and some folks would consider it just a weed, even though it has so many

Impatiens capensis

Iris cristata

benefits. But, it's not really a weed because weeds require, by definition, disturbed soil. Impatiens come up year after year in rich woodland leaf litter and help fill the summer woods with cute orange flowers. Also, they are fun to watch as bees and hummingbirds often encounter each other on frequent visits. Furthermore, the slimy juice in the stems is the age-old remedy for poison-ivy itch in the woods. What kid (in all of us) has not freaked out at pinching the ripe, turgid seedpods and having them suddenly split like a hand-buzzer to our delight? Yes, these can become too common in garden situations, especially if you irrigate in dry summers, but they are easy to remove as seedlings, and I'd rather have a few too many than none at all. Besides, the fresh orange flowers look great paired with purple monkshood in the early autumn, and I just love to see that!

Yellow jewelweed (*I. pallida*) is a frequent compatriot, though not nearly as prolific in the garden, even proving somewhat difficult to keep going outside its cooler mountain home.

Iris cristata ★★★

Dwarf crested iris

Iridaceae (iris family)

HABITAT & RANGE rich, moist woods and roadbanks, mostly throughout the Upper Southeast **ZONES** 3–8 **SOIL** moist, well drained **LIGHT** shade to part sun.

DESCRIPTION Deciduous tufts or fans of leaves arise from a short-creeping shallow rhizome to form a large colony, 3–6 in. tall. Leaves alternate, simple, closely overlapping at base, flattened vertically, 3–6 in. tall and up to 1 in. wide. Flowers showy, blue-lavender with white, yellow, and purple markings, shorter than the leaves, 2–3 in. tall, blooming briefly in spring. Fruits are small pods, inconspicuous, maturing in summer.

PROPAGATION division of clump by leaf tufts **LANDSCAPE USES** groundcover in woodland gardens **EASE OF CULTIVATION** easy **AVAILABILITY** common.

NOTES Dwarf crested iris is a true charmer, spreading in characteristic patches in lush leaf humus. The plants brighten the spring woodland with their flowers for a brief time, and then they sort of lay back and just vegetate. Spread them around and let them be understory wildflowers for taller shrubs and ferns.

Several selections are available. 'Eco Bluebird' is more vigorous with darker flowers. 'Edgar Anderson' is lighter blue. 'Powder Blue Giant' is very vigorous and noticeably larger. 'Shenandoah Sky' has darker blue flowers. 'Tennessee White' is a pure albino.

Dwarf iris (*I. verna*) is a favorite of mine, with blue-purple flowers and a bright yellow blotch. It has a mountain form that makes distinct clumps, and a coastal plain form that is more spreading. It likes more sun and drier soil than dwarf crested iris.

Jeffersonia diphylla 

Twinleaf

Berberidaceae (barberry family)

HABITAT & RANGE rich, less-acidic woods, reaching into the Southeast in middle and eastern Tennessee, tiptoeing into northern Alabama and northwestern Georgia **ZONES** 4–8 **SOIL** moist, well drained **LIGHT** shade to part shade.

DESCRIPTION Deciduous, clump forming, from fibrous roots, 1–2 ft. tall. Basal leaves arise from the crown and are compound with two leaflets ("twinleaf") shaped like butterfly wings, 3–4 in. long. Flower showy, white, 1 in. wide, solitary on a leafless stalk, very short-lived, blooming in spring. Fruit is an interesting 1 in. tan pod cracking open on one

Jeffersonia diphylla, flower

Jeffersonia diphylla, fruit

side at the top like a mouth to release several brown seeds, maturing in early summer.

PROPAGATION seeds, collected as soon as ripe and stratified for three months; division of larger crowns **LANDSCAPE USES** specimens in shady borders, beds, or woodland gardens **EASE OF CULTIVATION** moderate **AVAILABILITY** rare.

NOTES Twinleaf is one of the classiest wildflowers you can grow. There is one species in America (another in Asia). Like so many things in life, the most desirable are often fleeting. Here both the flowers and the seeds are elusive. The single flower begins to open as the leaves unfurl, lasts a day or two (if it does not rain), and then is gone. A fruit will slowly develop to form a unique receptacle like a tall cookie jar. It barely cracks open to reveal the seeds which will shake out and be gone if you don't get them . . . well, actually just before the pod opens, as it turns yellow. Every day, every hour, counts in this plot to get those seeds. But, even if you miss them, the leaves are the things anyway; they last all summer and add a wonderful unique shape to the garden palette. While the plant is not considered heat tolerant, I have seen robust plants in cooler parts of Zone 8. Rarely available commercially and always in demand, twinleaf should be sought out and taken a chance on whenever you can.

Kosteletzkya pentacarpos ★★★★

Seashore mallow

Syn. *Kosteletzkya virginica*
Malvaceae (mallow family)

HABITAT & RANGE brackish to freshwater tidal marshes, hugging the coast line from New Jersey to Texas, inland in Peninsular Florida **ZONES** 6–10 **SOIL** moist, well drained **LIGHT** sun.

Kosteletzkya pentacarpos

Liatris spicata

Lilium michauxii

Lilium superbum with pipevine swallowtail

DESCRIPTION Deciduous, clump forming, 3–4 ft. tall. Leaves alternate, simple, ovate, three-lobed and toothed, rough-hairy, 2–4 in. long. Flowers very showy, large, bright pink with yellow stamen tube, 2 in. wide, blooming in late summer. Fruits are flattened pods that split into five black seeds, maturing in late autumn.
PROPAGATION seeds, cold stratified for three months; don't disturb established plants **LANDSCAPE USES** specimens in borders or beds **EASE OF CULTIVATION** easy **AVAILABILITY** infrequent.
NOTES Seashore mallow is a magical plant. All who see it are mesmerized and stop to look. There is a special charm in this stately plant, so symmetrical and perfect in form. The leaves and flowers complement each other. It is late to emerge in spring. Butterflies love it. It adds greatly to the late summer garden color (which does not have much else that looks good), and should be much better known and widely used.

Liatris spicata ★★★

Gayfeather, blazing-star

Asteraceae (aster family)

HABITAT & RANGE roadside, bogs, seeps, and pine savannahs throughout the Southeast **ZONES** 3–9 **SOIL** moist, well drained **LIGHT** sun.
DESCRIPTION Deciduous, clump forming, from a crown with tuberous swellings, 2–3 ft. tall. Leaves alternate, narrow grasslike, 3–6 in. long, thick along the lower part of the stem, gradually reducing upwards. Flowers showy, purple (to white), in small heads clustered tightly along the unbranched sturdy stem, opening from top to bottom (a bit unusual). Fruits are nutlets with fluff, maturing in autumn.
PROPAGATION seeds, may self-sow, cold stratified for three months; division of clump **LANDSCAPE USES** specimens in borders and beds, naturalized in moist meadows and bog gardens **EASE OF CULTIVATION** easy **AVAILABILITY** frequent.
NOTES Gayfeather is a well-known garden plant, and its stiffly upright flower stalks are appealing and good for cutting. The bright flowers are attractive, and the plant is very adaptable. I can't imagine a perennial border without a liatris of some sort. There are several selections, the shorter 'Kobold' is the most well known.

Spreading liatris (*L. squarrulosa*) naturally grows in rather dry open woods with basal gray-green leaves, sporting large individual flower heads about 1 in. wide.

Other liatris are available, many from western prairie regions. Grow them in lean, well-drained sunny locations and hope our higher rainfall does not make them too lush.

Lilium michauxii 

Carolina lily

Liliaceae (lily family)

HABITAT & RANGE dry woods and mountain slopes, mostly throughout the Southeast, except Tennessee outside the high mountains, and very rare in Florida **ZONES** 5–8 **SOIL** moist, well drained **LIGHT** part shade to part sun.
DESCRIPTION Deciduous single stem from large scaly bulb 1–2 feet tall; leaves whorled in sets along the stem, becoming

Lilium grayi

alternate above, somewhat fleshy, narrow 3–6 in. long, widest towards the tip; flowers very showy, six tepals recurved back, orange with brown spots, 2–3 in. wide, nodding, one to three from the top of the stem, slightly fragrant, blooming in early summer; fruit an elongate 2-in. pod, with flattened brown seeds stacked like Pringles potato chips, maturing in late summer

PROPAGATION seeds, cold stratify three months, may take seven years to bloom; divide scales from bulb, keep in warm moist soil (even a plastic bag) until buds develop **LANDSCAPE USES** specimens in border or patches in woodland garden **EASE OF CULTIVATION** moderate **AVAILABILITY** rare.

NOTES Lilies are considered the most sophisticated and regal of wildflowers, and yet are among the most difficult to grow. Carolina lily is more heat tolerant than most other lilies and is the state wildflower of North Carolina. It should accept any well-drained garden soil in morning sun. It is said to be the only fragrant lily in eastern North America, and the butterflies know it.

Turk's-cap lily (*L. superbum*) is certainly a favorite, quite prominent in summer along the Blue Ridge Parkway, often growing 10 ft. tall with dozens of flowers (and dozens of butterflies). South of Zone 7 it will be shorter and less floriferous, but can be grown well and should be staked.

Gray's lily (*L. grayi*) is so beautiful and unusual in that it's the only humming bird-pollinated native lily. It is not easy to grow, but it has been tissue cultured and could be available. It grows at high elevations and is often called Roan lily after Roan Mountain on the Tennessee–North Carolina border. Grow it in Zones 6 and 7 in very well drained soil.

Lobelia cardinalis

Lobelia cardinalis ★★★★

Cardinal flower

Campanulaceae (bellflower family)

HABITAT & RANGE marshes, streambanks, wet meadows, and low woods throughout the Southeast **ZONES** 2–9 **SOIL** moist, well drained to wet **LIGHT** part shade to sun.

DESCRIPTION Short-lived perennial with evergreen basal leaves, clump forming, 1–8 ft. tall (usually shorter). Leaves alternate, ovate-elongate, toothed, 2–8 in. long, basal ones larger. Flowers very showy, bright red, long-tubular, two-lipped with three lower petals and two flaring above, 2 in. long, closely spaced on upright branching stems, blooming sequentially from midsummer. Fruits are pods with tiny seeds, maturing in late summer as flowers fade.

PROPAGATION seeds, without pretreatment, often self-sowing in the garden; division of clump in early spring; stem cuttings before flowering **LANDSCAPE USES** specimens in borders and beds, colonies in open woods or damp spots, naturalized in wet meadows or pond margins **EASE OF CULTIVATION** easy **AVAILABILITY** common.

NOTES Cardinal flower is so widely recognized and grown, it could be our national flower. It is very adaptable to growing conditions from shady to sunny and from wet to dryish. It is so beautiful when in bloom that you can sit and look at it for hours; if so, you will see a hummingbird visit every five

Lobelia siphilitica

Lupinus perennis

minutes (maybe the same one). It grows readily from fresh seed and may bloom the following year.

Great blue lobelia (*L. siphilitica*) has striking blue flowers that are less tightly tubular (bee pollinated) on robust perennial plants. Very nice in the garden. There are many hybrids with cardinal flower. My favorite is 'Ruby Slippers', but 'Purple Towers' is also good.

Lupinus perennis ★★

Sundial lupine

Fabaceae (legume family)

HABITAT & RANGE dry, sandy woodlands and roadsides, mostly throughout the coastal plain and sandhills region of the Southeast **ZONES** 4–8 **SOIL** dryish, very well drained **LIGHT** part sun to sun.

DESCRIPTION Deciduous, colony forming, from a long underground rhizome, 1–2 ft. tall. Leaves alternate, basal ones compound with a whorl of ten or more leaflets, not hairy. Flowers showy, pealike, blue-purple, ¾ in. long, on a dense spike, blooming in spring. Fruits are dry bean pods, 1–2 in. long, maturing in summer.

PROPAGATION seeds, scarified, or allowed to self-sow; difficult to divide **LANDSCAPE USES** naturalized colonies in sandy, open ground **EASE OF CULTIVATION** moderate **AVAILABILITY** infrequent.

NOTES Sundial lupine is a magical plant: the leaves are unique and the flowers are charming on the spikes. The bad habit is the unstoppable spreading, so plant this one by itself. It is reportedly difficult to establish, but I have seen it grown well in the Southeast. Try it in almost pure sand and plenty of sun with some afternoon shade, sowing the scarified seeds *in situ*. Be sure your seeds come from the Southeast, so the plants will be more heat tolerant (it grows also in the far north). Summer heat is the weakness of garden lupines. Our native is your only hope.

The sandhills lupines (*L. diffusus* and *L. villosus*) might be a challenge for you to try in similar pure sand. They make exquisite clumps of silvery gray–fuzzy leaves with spikes of blue or pink flowers, respectively. They are weird in having noncompound leaves (single leaflets), and being evergreen, often behaving as biennials.

Lysimachia quadrifolia ★★

Whorled loosestrife

Primulaceae (primrose family)

HABITAT & RANGE moist to dry woodlands and roadsides throughout the Upper Southeast **ZONES** 5–8 **SOIL** moist, well drained to dryish **LIGHT** part sun to sun.

DESCRIPTION Deciduous, colony forming, spreading from a long-creeping underground rhizome, 2–3 ft. tall. Leaves

Lysimachia quadrifolia

Lysimachia ciliata

in whorls of three to six, lanceolate, 3–5 in. long. Flowers showy, yellow, 1/2 in. wide, each on a delicate stalk arising from a leaf base, blooming through midsummer. Fruits are inconspicuous pods, maturing in autumn.
PROPAGATION division of rhizomes from the colony **LANDSCAPE USES** naturalized colonies in open woodlands, edges, and meadows **EASE OF CULTIVATION** easy **AVAILABILITY** frequent.

Maianthemum racemosa, foliage and flowers

NOTES Whorled loosestrife produces delicate flowers that are a joy to see, and the structural presence of the whorled-leaved stems bring a special feature to the garden. This plant does spread, but does not choke out other plants and can fit in nicely with coreopsis, wood asters, liatris, goldenrods, and other summer flowers of the open, dryish woods and meadows.

Fringed loosestrife (*L. ciliata*) is also a colony-forming species, but of moist habitats. Use it as a vigorous groundcover with 3/4 in. yellow flowers in more sunny damp places. It can even grow in full sun in a perennial border with proper "removal management." The purple-leaved cultivar 'Firecracker' is exceptional as a sun-loving groundcover, the bright yellow flowers rising above the deep purple mounding leaves.

Maianthemum racemosa ★★★

Solomon's-plume, false Solomon's-seal

Syn. *Smilacina racemosa*
Ruscaceae (ruscus family), formerly in Liliaceae (lily family)

HABITAT & RANGE woodlands of all sorts throughout most of the Southeast, except absent in the Gulf Coast **ZONES** 4–8 **SOIL** moist, well drained **LIGHT** shade to part sun.
DESCRIPTION Deciduous, tight colony forming, slowly spreading from a thick rhizome, 1–2 ft. tall. Leaves alternate, arranged along the stem in a flattened plane with one row on each side of the stem, ovate-elongate, 3–5 in. long. Flowers showy, white, starlike, small 1/4 in. wide, arranged in a formal branched inflorescence at the tip of the leafy stem, blooming in spring. Fruits are showy round berries, turning solid or speckled red in late summer, eaten by birds.
PROPAGATION division of thick rhizomes **LANDSCAPE USES** clumps in woodland gardens **EASE OF CULTIVATION** easy **AVAILABILITY** frequent.

Maianthemum racemosa, fruits

Marshallia obovata

Medeola virginiana

NOTES When I go on wildflower field trips with others, I notice that we always have to differentiate between "true" Solomon's-seal (*Polygonatum commutatum*) and "false" Solomon's-seal—it's a classic endeavor. To call this lookalike "Solomon's-plume" is an effort to "be less negative." The tiny flowers in plumes (compared to the larger bell-shaped flowers that hang under the leaves of true Solomon's-seal) provide a special show in spring and are displayed conspicuously. They are followed by the unique array of small red berries in late summer as the leaves are yellowing and dying. Few other woodland wildflowers are as pretty in flower and in fruit, doll's-eye (*Actaea pachypoda*) being one example. I am as disheartened as you are to accept the name change (from *Smilacina*), but it makes more sense than some name changes we have noted.

Canada mayflower (*M. canadense*) is a charming little wildflower with tiny white spring flowers and petite bright red autumn berries—a miniature (4 in. tall) much-simplified version of Solomon's-plume. It forms extensive groundcover in acidic mountain heath-dominated woods, and is difficult to grow well outside the cooler mountains. Try it in well-drained acidic soil.

Marshallia obovata ★★★

Spoon-leaved Barbara's-buttons

Asteraceae (aster family)

HABITAT & RANGE woodlands and borders, mostly in the Piedmont and coastal plain from the Carolinas through central Alabama **ZONES** 6–8 **SOIL** moist, well drained **LIGHT** part shade to part sun.

DESCRIPTION Deciduous, clump forming, 6–12 in. tall. Leaves basal, simple, elongate, widest near the tip, 3–5 in. long. Flowers showy, pinkish white, a single head atop each barely leafy stem, blooming in spring. Fruits are nutlets without fluff, maturing soon after flowering.

PROPAGATION seeds, easy to germinate; division of clump

LANDSCAPE USES minor specimens in shady borders or beds

EASE OF CULTIVATION easy **AVAILABILITY** infrequent.

NOTES Although this species is rare in cultivation, that situation could change since the plant is easy to propagate. But don't just grow one, plant a colony. Better yet, try to naturalize it in open woodland garden or open meadow edge. The floral details are exquisite, and there is a delicate fragrance.

Grassleaf Barbara's-buttons (*M. graminifolia*) is just becoming more known (see description in bog chapter). It is a plant of water-saturated pitcherplant bogs but should be growable in a moist, sunny border. It has more numerous 1 in. pink heads and strong sweet fragrance.

Medeola virginiana ★★

Indian cucumber-root

Liliaceae (lily family)

HABITAT & RANGE rich, moist forests, mostly throughout the Southeast **ZONES** 4–8 **SOIL** moist, well drained **LIGHT** part shade to part sun.

DESCRIPTION Deciduous, colony forming, from a white creeping rhizome, 1–2 ft. tall. Leaves whorled in two sets, one with six to ten leaves at stem middle, each ovate-elongate 4–6 in. long, the second whorl at stem tip with three or four smaller leaves, turning yellow in late summer. Flowers interesting, yellowish, ¼ in. wide, with three conspicuous red stigmas, recurved on slender stalks among the upper set of leaves, blooming in spring. Fruits are ⅜ in. black berries, erect among the upper leaves that have now developed scarlet-red blotches to offset the black berries, presumably to attract birds, maturing in late summer.

PROPAGATION seeds, cold stratified for three months; division of rhizomes **LANDSCAPE USES** colonies in open woodland garden **EASE OF CULTIVATION** easy **AVAILABILITY** infrequent.

NOTES Indian cucumber-root is a unique species and has no close relatives. It is very common and easily identified by the one set of whorled leaves before it blooms. In the garden, it provides show-stopping late summer leaf and fruit color changes in its own way.

Mertensia virginica ★★★★

Virginia bluebells

Boraginaceae (borage family)

HABITAT & RANGE low woods and floodplains from Virginia to central Alabama, skirting most of the Carolinas and Georgia **ZONES** 3–8 **SOIL** moist, well drained to seasonally wet **LIGHT** shade to part shade.

DESCRIPTION Deciduous, clump forming, from a thickened taproot, 1–2 ft. tall. Leaves alternate, simple, ovate and wavy, bluish green, 4–8 in. long, kind of floppy, disappearing by the end of spring. Flowers showy, pink in bud, opening blue, 1–1½ in. long, tubular with trumpet-like flaring mouth, opening from tight coil at stem tip, somewhat dangling, blooming in earliest spring. Fruits are nutlets, maturing in late spring before the whole plant disappears.

PROPAGATION seeds self-sown; division of branches of taproot with buds **LANDSCAPE USES** clumps in shady beds, naturalized in woodland gardens **EASE OF CULTIVATION** easy **AVAILABILITY** infrequent.

NOTES One of my all-time favorites, Virginia bluebells

Mertensia virginica

emerges early and shows flower bud color as the leaves develop. Both the cluster of flowers and the stance of the plant are very appealing. But this wildflower plays the most illustrious disappearing trick for a plant its size, and the whole thing turns to compost in a very short time—you would not even know it had been there at all. Thus, you have to plan and have some other plants coming along to take its place, a nifty trick in itself. Mix Virginia bluebells with modest-sized deciduous ferns that start slow and get larger with time (for example maidenhair, silvery glade, lady fern), or other late wildflowers such as Jack-in-the-pulpit (*Arisaema triphyllum*), turtleheads (*Chelone* spp.), alumroot (*Heuchera* spp.), yellow wood poppy (*Stylophorum diphyllum*), or my favorite companion, goldenseal (*Hydrastis canadensis*). I still would not be without these long-lived perennial bluebells, and early absence makes the heart grow fonder. Don't worry; they will be back next spring.

Mitchella repens 

Partridge-berry

Rubiaceae (madder family)

HABITAT & RANGE rich woods and streambanks throughout the Southeast **ZONES** 3–9 **SOIL** moist, acidic, well drained **LIGHT** shade to part sun.

DESCRIPTION Evergreen, mat-forming colony, from a creeping superficial stem, about 1 in. tall. Leaves opposite, entire, ovoid to round, leathery ½–1 in. long. Flowers showy, white, in pairs at a leaf, ½ in. long, with four spreading petals, male and female flowers on separate plants, blooming in late spring to early summer, the two flowers fused at the base to a pair of ovaries. Fruits are bright red "Siamese-twin"

berries 3⁄8 in. wide, maturing in summer, and if not eaten by birds, they persist until next season.
PROPAGATION seeds, cold stratified for three months; division of the mat **LANDSCAPE USES** low evergreen groundcover **EASE OF CULTIVATION** easy **AVAILABILITY** frequent.
NOTES Partridge-berry is a wonderful plant with a paradoxical character. It has been called a subshrub with minute woody growth, or just a tougher-than-usual herbaceous plant. It grows in very acidic soil under the perpetual shade of hemlock trees where few other plants can thrive (it does not want to be covered up by heavy leaf litter), and yet I have seen it form a dense mat in full sun in ordinary sandy soil. I have also seen it in dry woods and in wet woods. Male and female flowers are on separate plants, and yet seemingly compatible clones fail to set seed when interplanted. The bright red berries should be attractive to birds, and yet most of them are never eaten and look perfect a year later. So, what do we make of these mysteries? One explanation is that the species has a very broad range, from Canada to central Florida, in all kinds of woods and soil types, and they are adapted to local conditions. A clone from Michigan might not be sexually compatible with a clone from South Carolina. This is a lesson on the importance of knowing where your plants come from and to what specific condition they might be adapted. The solution: try your best to get local material, but be happy that they survive and grow as well as they do.

Mitchella repens

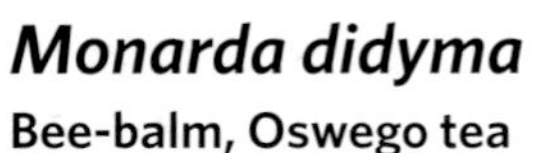

Monarda didyma ★★★

Bee-balm, Oswego tea

Lamiaceae (mint family)

HABITAT & RANGE woods, thickets, roadside seeps, and ditches, limited in the mountains and Piedmont of western Virginia and the Carolinas **ZONES** 3–8 **SOIL** moist, well drained to seasonally wet **LIGHT** part sun to sun.
DESCRIPTION Evergreen basal leaves, colony forming, from a strong-running rhizome, 2–4 ft. tall stems are square in cross section, plant has minty smell. Leaves opposite, ovate-triangular, toothed, 2–4 in. long. Flowers very showy, red, tubular and two-lipped, 2 in. long, tightly clustered into single heads atop leafy stems, blooming in summer. Fruits are small nutlets, maturing in late summer.
PROPAGATION division of the rhizomes of a colony **LANDSCAPE USES** specimens in sunny borders and beds, naturalized in moist meadows, ditches, wetlands **EASE OF CULTIVATION** easy **AVAILABILITY** common.
NOTES Bee-balm may be as famous with humans for its tea as for its flowers. Hummingbirds and bees certainly like the

Monarda didyma

bright red nonfragrant flowers. Because it's so aggressive, this plant can't be mixed in the border without considerable work with periodic division and discarding. We still grow it like that and it makes a wonderful summer show with other tall perennials. Let them fight it out; just don't overwater.

There are many selections and hybrids; some are an improvement on mildew resistance and drought tolerance. Some are hardly an improvement on the red color of the wild type, but they do add variety. You will have to try them to find the ones best for your garden. 'Jacob Cline' is the best red with mildew resistance.

Wild bergamot (*M. fistulosa*) is a widespread prairie species with lovely pink flowers and soft hairy stems. Best as a hybrid in the Southeast.

Spotted horsemint (*M. punctata*) is a crazy-looking species from very dry habitats. I love it. The white gaping flowers with red spots sit above bright pink bracts in several whorls on the stems, reminding me of pink nonpareils in a candy store. It is an annual; try it in pure sand and full sun.

Lemon beebalm (*M. citriodora*) is a prairie species that has scattered occurrences in the Southeast. It is striking, with rich pink flowers dotted with purple in several tight whorls. It is best as a meadow plant with no excess water, and is often part of the meadow-in-a-can mixtures. It blooms all summer and mixes with various yellow meadow "daisies" and orange butterfly-weed (*Asclepias tuberosa*).

Oenothera fruticosa ★★★

Southern sundrops

Onagraceae (evening-primrose family)

HABITAT & RANGE dry woods, roadsides, and meadows throughout the Southeast **ZONES** 4–9 **SOIL** moist, well drained **LIGHT** sun.

DESCRIPTION Deciduous, clump forming, 1–2 ft. tall. Leaves alternate, simple, elliptical, 2–3 in. long. Flowers showy, bright yellow, with four petals, about 2 in. wide, lasting one day, produced in a cluster atop branching stems, drooping buds can be red, blooming in spring. Fruits are 1-in. pods, maturing in summer.

PROPAGATION seeds, without pretreatment; stem cuttings after flowering; division of larger clumps **LANDSCAPE USES** specimens in borders, beds, rock gardens **EASE OF CULTIVATION** easy **AVAILABILITY** common.

NOTES Southern sundrops is a great meadow plant, good for massing and interplanting with other species (such as gayfeathers, nodding onion, beardtongue, and Barbara's-buttons) for mixed-meadow effect. Its bright yellow flowers are cheerful. Grow it in a well-drained, sunny location. There are several selections, including 'Fireworks'.

Oenothera fruticosa

An almost identical species is Appalachian sundrops (*O. tetragona*), which could be available and is also good. The well-known spreading pink evening primrose (*O. speciosa*) is not native in the Southeast but widely naturalized from further west.

Pachysandra procumbens ★★★★

Alleghany-spurge

Buxaceae (boxwood family)

HABITAT & RANGE moist woodlands, mostly in middle Tennessee, scattered southward and to northwestern South Carolina **ZONES** 5–9 **SOIL** moist, well drained **LIGHT** shade to part shade.

DESCRIPTION Evergreen, colony forming, from a short-creeping rhizome, 6–8 in. tall. Leaves alternate, simple, ovate, toothed, 4–6 in. long, often green with silver mottling, bronzing in winter sun. Flowers interesting but often inconspicuous, each flower with four thick white stamens but no sepals or petals, several separate male and female flowers on the same short stem arising from the rhizome, blooming in late winter to very early spring. Fruits are inconspicuous small pods, maturing soon after flowering.

PROPAGATION division of clump, keeping at least one leaf on a piece of rhizome with some roots **LANDSCAPE USES** evergreen groundcover for woodland gardens **EASE OF CULTIVATION** easy **AVAILABILITY** common.

NOTES Alleghany-spurge is our finest evergreen groundcover and has a wonderful presence. It is tough, adaptable, and dense. Because it chokes out other plants, it can't be used in close combinations with other wildflowers, though I have seen trilliums come up in the middle of a patch. It

Pachysandra procumbens

Packera aurea

is heat and drought tolerant, and can take flooding. It can grow in dense shade, is especially good on uneven sites, and can be divided from the edges to keep the mass confined. The leaves are beautiful, especially the silver-mottled forms, and always seems to look great. This pachysandra should be in every shade garden.

A cultivar with consistently good silver markings is 'Silver Streak'.

Packera aurea ★★★

Golden ragwort

Syn. *Senecio aureus*
Asteraceae (aster family)

HABITAT & RANGE moist forests, bogs, bottomlands, and streambanks, mostly in the Piedmont and mountains across the Upper Southeast **ZONES** 4–9 **SOIL** moist, well drained to seasonally wet **LIGHT** part shade to part sun.

DESCRIPTION Evergreen basal leaves, clump forming, slowly enlarging from a short rhizome, 1–3 ft. tall. Leaves alternate, basal ones heart-shaped, ovate, toothed, stem leaves deeply dissected like a feather, 3–4 in. long. Flowers very showy, bright yellow, daisylike heads about ¾ in. wide, borne in branched flat-topped arrays on branched leafy stems, blooming in spring. Fruits are nutlets with white fluff, maturing in early summer.

PROPAGATION seeds, cold stratified for three months; division of clump **LANDSCAPE USES** evergreen groundcover in moist or boggy to average soils **EASE OF CULTIVATION** easy **AVAILABILITY** common.

NOTES Golden ragwort is one of the first spring wildflowers to make a great show as a tall exuberant plant. It can create an evergreen mass in virtually any soil and light situation. It does not mix well with other plants, but let it have the whole show. Its relatives are mostly yellow-flowered weeds of field and roadside, so here we have a species to strengthen the family image as a great woodland wildflower. It has been segregated, wisely so, with its closest relatives, from the huge worldwide genus *Senecio*.

Panax quinquefolius ★★

Ginseng

Araliaceae (ginseng family)

HABITAT & RANGE rich, moist woods throughout the Southeast, avoiding the Atlantic Coastal Plain and immediate Gulf Coast **ZONES** 5–8 **SOIL** moist, well drained **LIGHT** shade to part shade.

DESCRIPTION Deciduous, single stem, from a thickened branched taproot, 6–24 in. tall. Leaves whorled, three or four arising from the top of each stem, 6–12 in. long, compound with five ovate toothed leaflets palmately arranged (radiating), both leaves and leaflets held in a symmetrical circular pattern. Flowers tiny, inconspicuous, greenish white, many, on stalks in a tight ball (umbel) arising from the central point of leaf attachment at the top of the stem, blooming in early summer. Fruits are bright red berries

Panax quinquefolius

Parnassia asarifolia

3/8 in. wide, maturing in late summer.
PROPAGATION seeds, cold stratified for three months **LANDSCAPE USES** specimens or groupings in woodland gardens **EASE OF CULTIVATION** easy **AVAILABILITY** infrequent.
NOTES No wildflower is more famous than ginseng for its purported medicinal qualities, nor collected for so many centuries for human use. Still, it occurs in virtually every rich woods throughout the Southeast where rattlesnake fern (*Botrychium virginianum*) grows. Ginseng is difficult to spot (it mimics Virginia creeper and other low-growing wildflowers in its pattern), but it's fun to find. It is not a showy garden plant, but the leaf pattern is distinctive and the red berries add a great surprise to the late summer viewing experience (see if your wildflower friends can notice it, and check with them if it goes missing). Because ginseng is so famous, it's like having a celebrity in the garden.

Parnassia asarifolia 

Kidney-leaved grass-of-Parnassus

Parnassiaceae (grass-of-Parnassus family), formerly in Saxifragaceae (saxifrage family)

HABITAT & RANGE bogs, seeps, and acidic brooksides throughout the mountains into central Alabama **ZONES** 6–8 **SOIL** moist to wet **LIGHT** part sun to sun.
DESCRIPTION Deciduous, clump forming, slowly spreading, 6–8 in. tall. Leaves alternate, basal ones from the rhizomes but not tufted, simple, kidney-shaped, 2–3 in. wide, light green. Flowers showy, white with distinct green lines (veins), five petals, five stamens, and five sets of three-forked yellowish false stamens (staminodes), each flower on its own spindly leafless stalk arising in a mass above a patch of leaves, blooming in late summer. Fruits are oval pods, maturing in autumn.
PROPAGATION seeds, cold stratified for three months; division of clump **LANDSCAPE USES** specimens at edge of water gardens, bogs, or splash-zone of waterfalls **EASE OF CULTIVATION** difficult **AVAILABILITY** rare.
NOTES Grass-of-Parnassus is an intriguing wildflower because of the delicate beauty of the flowers and the unusual reniform (kidney-shaped) leaves. I could have put it with the aquatic or bog plants, but it's more a denizen of moist soil and wet seeps than standing water. It requires just the right moisture level to thrive.

Penstemon smallii 

Small's beardtongue, Blue Ridge beard-tongue

Plantaginaceae (plantain family), formerly in Scrophulariaceae (figwort family)

HABITAT & RANGE woodlands, cliffs, and roadbanks, mostly in the mountains **ZONES** 6–8 **SOIL** moist to dryish, well drained **LIGHT** part sun to sun.
DESCRIPTION Short-lived perennial with evergreen basal leaves, clump forming, 1–2 ft. tall. Leaves opposite, simple, ovate-triangular, toothed, hairy, 3–6 in. long. Flowers showy, pink-purple, tubular with two-lipped mouth, three lower lobes longer than the upper, 1 in. long, with fuzzy sterile stamen exerted through the mouth, abundant clusters on conspicuously leafy stem, blooming through spring. Fruits are 1/2-in. conical pods with many seeds, maturing in summer.
PROPAGATION seeds, without pretreatment; division of young clumps only as plants are short-lived (two years) **LANDSCAPE USES** specimens or groupings, naturalized in rocky

Penstemon digitalis

Phacelia bipinnatifida

Penstemon smallii

woodland gardens **EASE OF CULTIVATION** easy **AVAILABILITY** common.

NOTES Small's beard-tongue is the prettiest of the eastern pink-flowered species and a must for the open woodland garden. The lush foliage is handsome in itself, and the large pink flowers can be stunning. Being a short-lived perennial, it gives one good spring bloom and then perhaps one more spring from a tattered old plant. It's best to start new plants yearly or to encourage self-sowing by having several plants to cross-pollinate.

Southern beardtongue (*P. australis*) and Eastern beard-tongue (*P. laevigatus*) are common roadside plants with showy pink to lavender flowers, but with smaller leaves don't have the presence of Small's beardtongue. Grow them in moist to drier, sunny sites for best results.

Tall white beardtongue (*P. digitalis*) does excellent in the Southeast, with 2- to 3-ft. stems bearing 1 in. flaring white tubular flowers. The handsome basal leaves are also a plus, and they look good in winter. Expect at least two years of bloom from these before decline. The cultivar 'Husker Red' has excellent dark purplish evergreen foliage.

Phacelia bipinnatifida ★★★★

Fernleaf phacelia

Boraginaceae (borage family), formerly in Hydrophyllaceae (waterleaf family)

HABITAT & RANGE rich, less-acidic woods, rocky slopes, and creekbanks, mostly in mountains of central Tennessee into Alabama **ZONES** 6–8 **SOIL** moist, well drained **LIGHT** shade to part sun.

DESCRIPTION Biennial, branching plant, from a weak tap-root, 1–3 ft. tall, all parts clammy from minute gland-tipped hairs. Leaves alternate, compound with irregular divisions, lobes, and teeth, glandular-hairy, brittle, 4–8 in. long. Flowers showy, lavender-blue, round, with five petals, opening sequentially (one bloom per day) from an uncoiling elongating stem tip, blooming in late winter through spring. Fruits are round pods with four seeds, recurved on elongating stems, maturing in early summer as plants deteriorate.

PROPAGATION seeds, without pretreatment, sown when ripe or stored dry, germinate later that season and overwinter as seedings **LANDSCAPE USES** specimens in beds, groundcover in woodland gardens **EASE OF CULTIVATION** easy **AVAILABILITY** rare, as plants

NOTES Fernleaf phacelia is a remarkable plant with delicate beauty. It literally takes over the forest floor for two months each spring. As a biennial, its seeds germinate in summer and overwinter as lush leafy rosettes with mottled markings. In spring, they start to flower early and are most charming then, growing constantly larger and coarser with stems spreading to 3–4 ft. wide. Be sure to collect seeds before they shed so you can sow them where you like. You will need several seedlings close-by to cross-pollinate to get more seeds. You will also need other species to come along when these disappear by late spring. I would not be without it as it offers two seasons of interest.

Phlox carolina

Phlox divaricata

Phlox nivalis

Phlox carolina ★★★★

Carolina phlox

Polemoniaceae (Jacob's-ladder family)

HABITAT & RANGE open woods, roadsides, and ditches, mostly throughout the Upper Southeast **ZONES** 3–8 **SOIL** moist, well drained **LIGHT** sun.

DESCRIPTION Deciduous, clump forming, slowly spreading, 2–4 ft. tall. Leaves opposite, simple, narrow, 1–2 in. long. Flowers showy, fragrant, pink-magenta, tubular with flaring trumpet mouth, 1½ in. long, in dense clusters above the leafy stems, blooming in summer. Fruits are small pods, maturing in late summer.

PROPAGATION seeds, cold stratified for one month; division of clump **LANDSCAPE USES** specimens in borders and beds, naturalized in sunny meadows and open areas **EASE OF CULTIVATION** easy **AVAILABILITY** common.

NOTES With various meaningful common names of garden phlox, summer phlox, and tall phlox, the group of tall summer-flowering phloxes are mainstays of the perennial border. Various species are abundant along the Blue Ridge Parkway and sunny roadsides and moist ditches throughout the Upper South. While the wild phloxes are certainly good enough, there are many better garden selections from which to choose. Several species of tall phlox seem to form a complex in which it's sometimes difficult to separate the different ones, at least for gardeners. Here I am lumping *P. carolina, P. glaberrima, P. maculata,* and *P. paniculata*. Their differences may be real, but technically difficult to discern, and you may not get the correct one if you ask for a given species. You may find some outstanding forms in old gardens, like the unnamed pink one I still grow from my grandmother's garden. There are named selections for each species, and some may be hybrids. Powdery mildew and spider-mite resistance are factors to consider, along with length of flowering time and heat tolerance. Enjoy them without worrying too much about their proper names. A few proven suggestions follow.

'Morris Berd' is 1 ft. tall, deep pink. 'David' is the old standard white, but you might try the newer 'Delta Snow' (developed in Mississippi), or 'Danielle' with a long blooming time. 'Flower Power' is a sturdy white with more elongate inflorescences. 'Natascha' is excellent with star-pattern white and pink flowers. 'Robert Poore' is a great pink-magenta, and 'Shortwood' is rosy-pink.

Phlox divaricata

Woodland phlox

Polemoniaceae (Jacob's-ladder family)

HABITAT & RANGE moist, less-acidic forests, mostly west of the mountains, hardly in the Carolinas and Georgia **ZONES** 3–8 **SOIL** moist, well drained **LIGHT** shade to part sun.

DESCRIPTION Evergreen leafy basal shoots, colony forming, slowly spreading, 1–2 ft. tall. Leaves opposite, simple, narrow,

1–2 in. long. Flowers showy, fragrant, light blue, tubular with flaring trumpet mouth, 1½ in. long, in dense clusters, blooming in spring. Fruits are tiny pods, maturing in early summer. **PROPAGATION** rooting of sterile leafy basal shoots; division of clump **LANDSCAPE USES** groundcover in beds and woodland gardens **EASE OF CULTIVATION** easy **AVAILABILITY** common. **NOTES** Woodland phlox can form a beautiful mass of fragrant flowers in spring. The plant is easy to grow and divide. It can mix with other equally robust wildflowers, as its short-creeping stems don't form an impenetrable mat as Alleghany-spurge (*Pachysandra procumbens*) does, for example.

Several selections are available. 'Clouds of Perfume' is light blue and does well. 'Fuller's White' is pure white. 'London Grove Blue' has lilac-blue flowers.

Creeping phlox (*P. stolonifera*) is a fabulous garden plant with leafy stolons than run above the ground a foot or more each year to produce a patch of evergreen foliage. Few-leaved flowering stems arise in spring, bearing striking clusters of slightly larger flowers but are not as dense as woodland phlox. It occurs in the Piedmont and mountains. Selections 'Blue Ridge' and 'Pink Ridge' are 6–8 in. tall; 'Bruce's White' is one of the very best; 'Sherwood Purple' is very popular with mauve-purple flowers.

Phlox nivalis ★★★★

Trailing phlox

Polemoniaceae (Jacob's-ladder family)

HABITAT & RANGE rock outcrops, dry woodlands, and roadbanks, mostly in the Carolinas through Alabama and Peninsular Florida **ZONES** 6–9 **SOIL** moist to dryish, well drained **LIGHT** sun.

DESCRIPTION Evergreen, mat-forming mass. Leaves opposite, needlelike, ½ in. long. Flowers showy, pink, tubular, on leafy stems to 4 in. tall, blooming mainly in spring. Fruits are tiny pods, maturing soon after flowering.

PROPAGATION cuttings, or divisions of clumps **LANDSCAPE USES** dense groundcover in sunny rock gardens, hillsides, ditchbanks, rocky slopes **EASE OF CULTIVATION** easy **AVAILABILITY** common.

NOTES Trailing phlox, like the other mat-forming phloxes, is a fabulous plant, providing a tight groundcover and beautiful dense masses of spring flowers. It is adaptable to open situations too rough or steep to manage other plants, or just to provide a short cover for beauty and erosion control. I have even seen low-growing phloxes treated as a lawn, mixed with grasses and mowed periodically. Grow them all in sun, and don't overwater. Trailing phlox is heat-tolerant. Try cultivar 'Camla', a very vigorous cultivar with mauve flowers.

The other famous species that are widely grown are more northern and western. They sometimes have problems with heat and humidity, and would be better in the mountains. Check into the Southeastern phlox trials at Plant Delights Nursery or the University of Georgia. Ten-point phlox (*P. bifida*) has deeply notched petals and gets into central Tennessee. Try cultivar 'Betty Blake'.

Moss phlox (*P. subulata*) has shorter leaves and comes into the mountains of North Carolina and Tennessee. Try cultivar 'Candy Stripe'.

Physostegia virginiana ★★

Obedient plant

Syn. *Dracocephalum virginianum*
Lamiaceae (mint family)

HABITAT & RANGE bogs, seeps, and grassy meadows, scattered throughout the Southeast, perhaps escaped from cultivation **ZONES** 2–9 **SOIL** moist, well drained to seasonally wet **LIGHT** part sun to sun.

DESCRIPTION Deciduous, clump forming, vigorously spreading, 2–4 ft. tall, stems are square in cross section. Leaves opposite, simple, toothed, sometimes pointing downwards,

Physostegia virginiana

1–6 in. long. Flowers showy, pink-purple spotted with red, tubular with open mouth, 1 in. long, closely spaced in clusters on upper part of leafy stem, blooming in late summer. Fruits are small nutlets, maturing in autumn.
PROPAGATION division of clump **LANDSCAPE USES** bold clumps in borders and beds, naturalized in moist meadows or ditches **EASE OF CULTIVATION** easy **AVAILABILITY** frequent.
NOTES Obedient plant has been widely cultivated, and escapes into disturbed wetlands. It is still pretty, but sometimes it's too aggressive in the garden, where it lacks competition. The common name derives from the notion that the flowers will stay pointed in one direction if you push them that way. But, guess what? They don't stay. Don't tell the kids, and don't grow the plant where it can overtake others. 'Miss Manners' is one of several cultivars; it's white-flowered and a bit less aggressive.

Savanna obedient-plant (*P. purpurea*) is a coastal wetland species that is a little shorter, with less pointed leaf teeth and a little more charm.

Podophyllum peltatum ★★★

Mayapple

Berberidaceae (barberry family)

HABITAT & RANGE deciduous forest, low woods, and moist meadows and roadbanks throughout the Southeast, except absent in Florida **ZONES** 3–8 **SOIL** moist, well drained to seasonally wet **LIGHT** shade to part shade.
DESCRIPTION Deciduous, colony forming, with widely spaced 1- to 2-ft. tall stems arising from a robust long-creeping rhizome. Leaves are one per stem, or two with a single flower in the fork between them, round, deeply lobed and toothed, to 1 ft. across. Flowers showy but hidden beneath the paired leaves, large, 1½ in. wide, petals six, blooming in spring. Fruit is a single large fleshy egg-shaped berry 1 in. long, turning yellow when ripe (and only then can it be eaten by humans if you get them first), maturing in early summer as the plants die down.
PROPAGATION division of rhizomes **LANDSCAPE USES** colonies in moist to drying areas **EASE OF CULTIVATION** easy **AVAILABILITY** frequent.
NOTES Mayapple is well known, which explains why it's sought-after at our spring wildflower plant sales. It grows easily to form an attractive expanse of large leaves and is not fussy about soil conditions. Because it's invasive, it should not be grown with other wildflowers. It is best in a swamp or floodplain by itself, or even in a parklike forest with little understory; it requires no care and may die down in mid to late summer in dry conditions. It looks great when the leaves are just emerging in spring, with their new-green shine. It is fun to go find the flowers and fruits.

Polemonium reptans ★★★★

Jacob's-ladder

Polemoniaceae (Jacob's-ladder family)

HABITAT & RANGE rich, moist woods throughout the Upper Southeast west of the mountains **ZONES** 2–8 **SOIL** moist, well drained **LIGHT** shade to part shade.
DESCRIPTION Deciduous, clump forming, to 1 ft. tall. Leaves alternate, compound like a feather, with toothed leaflets, 4–8 in. long. Flowers showy, light blue, with five rounded petals, ½ in. wide, on abundant upright loose clusters, blooming all spring. Fruits are conspicuous, small rounded pods inside the persistent sepals, maturing in summer.

Podophyllum peltatum

Polemonium reptans

PROPAGATION seeds, sown immediately when ripe; division of larger clumps **LANDSCAPE USES** specimens in beds or woodland gardens **EASE OF CULTIVATION** easy **AVAILABILITY** frequent.

NOTES Jacob's-ladder is a long-lasting spring wildflower with beautiful blue flowers. It appears delicate, and it is, but I had a clump live for thirty years, and it grew a little larger every year. Jacob's-ladder is adaptable and fits in well with other wildflowers; thus, you can have several individuals that will cross-pollinate and make seedlings to self-sow about the garden.

Polygonatum biflorum ★★★

Large Solomon's-seal

Syn. *Polygonatum commutatum*
Ruscaceae (ruscus family), formerly Liliaceae (lily family)

HABITAT & RANGE moist to dry woods, mostly throughout the Southeast **ZONES** 3–8 **SOIL** moist, well drained **LIGHT** shade to part sun.

DESCRIPTION Deciduous, clump forming, from a thick rhizome, moderately spreading, 3–4 ft. tall. Leaves alternate, ovate-elongate, hairless, 4–6 in. long, floppy, distinctly veined. Flowers showy but hidden beneath the leaves, greenish white, tubular, ½ in. long, in clusters of three at each leaf node, blooming in late spring. Fruits are dark purple berries, maturing in autumn.

PROPAGATION seeds, cleaned from pulp and cold stratified for three months; division of rhizomes with a large tip bud in early autumn **LANDSCAPE USES** colonies in woodland gardens **EASE OF CULTIVATION** easy **AVAILABILITY** frequent.

NOTES Solomon's-seal quickly grows into a large colony. It provides excellent dynamic structural form because all the stems seem to arch in the same direction, and the leaves are droopy, playing off the light. The common name refers to the large round pitted scar left by the leaf on the top of the rhizome.

The largest form could be referred to as *P. biflorum* var. *commutatum* (syn. *P. canaliculatum*) and is found in moist forest and roadside seeps in the mountains, growing up to 8 ft. tall in large colonies with leaves to 8 in. long and flowers almost 1 in. long. These plants are impressive by any name and a wonder to behold.

The smaller hairy Solomon's-seal (*P. pubescens*) grows up to 2 ft. tall. Not bad for the garden, on a smaller scale.

Pycnanthemum incanum ★★★

Silverleaf mountain-mint

Syn. *Pycnanthemum pycnanthemoides*
Lamiaceae (mint family)

HABITAT & RANGE woodlands, thickets, pastures, and old fields, mostly throughout the Southeast, except absent in Tennessee and Florida **ZONES** 6–8 **SOIL** moist to dryish, well drained **LIGHT** part shade to sun.

DESCRIPTION Evergreen basal leafy shoots, spreading, vigorous colony forming, from a long-creeping rhizome, 2–6 ft. tall, stems are square in cross section. Leaves opposite, simple, ovate, hairy, toothed, 1–4 in. long, exceptionally minty. Flowers showy, lavender-white with purple spots, very

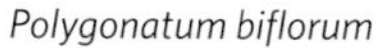

Polygonatum biflorum

tightly clustered on the tips of branched leafy stems, long blooming in midsummer. Fruits are tiny nutlets, maturing in autumn and persisting into winter.

PROPAGATION division of rhizomes **LANDSCAPE USES** patches in open woods or meadows **EASE OF CULTIVATION** easy **AVAILABILITY** frequent.

NOTES Silverleaf mountain-mint may be the best smelling mint you will find. The flowers are charming and delicate, and the fuzzy leaves are fun to hand someone to rub and smell. The plant is invasive in close quarters; I just pull out most of the extra plants each fall and let some grow back. In a meadow, it's superb, showing off its white-topped leaves and flower clusters like snow on mountain trees. The flowers attract the most unusual orange-black wasps and other insects, and beekeepers tell me it's a great honey plant in summer when others have faded. It is drought tolerant. It sounds like the kind of plant you might want to find a place for, but keep it managed. Good news: it's deer resistant.

Its less-hairy relative is short-toothed mountain-mint (*P. muticum*), which grows in damp places, but is less aggressive when grown dry, in sun or light shade.

Slender mountain-mint (*P. tenuifolium*) is widespread, with a much-branched flat-topped array of flowers and needlelike leaves. It has great texture for the garden.

Pycnanthemum tenuifolium

Pycnanthemum incanum

Rudbeckia fulgida ★★★★

Orange coneflower

Asteraceae (aster family)

HABITAT & RANGE meadows, pastures, old fields, and roadsides throughout the Southeast **ZONES** 3–8 **SOIL** moist, well drained **LIGHT** sun.

DESCRIPTION Evergreen basal leaves, clump forming but moderately spreading by underground rhizomes (or stolons), 2–4 ft. tall. Leaves alternate, simple, rough-hairy, toothed, 6–8 in. long. Flowers showy, each 2½ in. wide, solitary or up to three on erect leafy stems, ray flowers yellow (often with brown basal blotch), disc flowers dark purple, blooming in midsummer. Fruits are thin nutlets (like thistle seeds), maturing in summer, loved by goldfinches.

PROPAGATION seeds, without pretreatment; division of leafy stolons from colony in autumn **LANDSCAPE USES** robust clumps for borders or beds **EASE OF CULTIVATION** easy **AVAILABILITY** common.

NOTES Orange coneflower (*R. fulgida*), also casually called black-eyed Susan, is a well-known wildflower of roadside meadows and a classic for the sunny garden border. I don't know of a better wildflower that says *summer*. Grow it in full sun with minimal water to retard aggressiveness.

A well-known cultivar is the taller and more robust 'Goldsturm' (meaning "storm of gold" in German), but it may not be as heat tolerant in the Southeast. Try 'Early Bird Gold' or 'Little Goldstar', both are shorter.

Black-eyed Susan (*R. hirta*) is an annual that has been much selected and there are countless cultivars.

Sweet coneflower (*R. subtomentosa*) is an amazing plant in our perennial border, growing 4 ft. tall and blooming profusely all summer. A cultivar named 'Henry Eilers' has the petals rolled into a thin tube. Like it or not, it does well.

Cutleaf coneflower (*R. laciniata*) grows in shade or sun, moist or average. It becomes 6–8 ft. tall, with highly

Rudbeckia fulgida

Ruellia caroliniensis

Rudbeckia laciniata

Rudbeckia hirta

Salvia coccinea

dissected leaves and a wonderful plethora of large pale yellow flowers in late summer. Dead head after flowering to prevent a mass of self-sown seedlings. 'Herbstsonne', the autumn sun coneflower, is a very robust, long-blooming clumper to 8 ft. tall. It may not be as heat tolerant in the Southeast.

Three-lobed coneflower (*R. triloba*) is a less robust much-branched heat- and drought-tolerant plant that can take half shade. I love it in late summer with its many 1 in. yellow flowers with big brown centers. It is a choice species for the Southeast.

Ruellia caroliniensis ★★

Carolina wild-petunia

Acanthaceae (acanthus family)

HABITAT & RANGE dry to moist woodlands throughout the Southeast **ZONES** 6–9 **SOIL** moist, well drained **LIGHT** shade to part sun.
DESCRIPTION Deciduous, single stem, from a knotty root system, 6–12 in. tall. Leaves opposite, simple, ovate-tapered, slightly hairy, 2–4 in. long. Flowers showy, light blue, tubular with flaring petals, flowers in clusters at the leaf bases, blooming one at a time per leaf cluster all summer. Fruits are small pods, maturing continuously.
PROPAGATION seeds, cold stratified for three months **LANDSCAPE USES** groupings or self-scattered in open woodlands or thin grassy areas **EASE OF CULTIVATION** easy **AVAILABILITY** infrequent.
NOTES Carolina wild-petunia is a charming little plant, not hurting anyone. It seems to delight in the Piedmont open woodlands among grasses and on shady banks, blooming every day, always there for you to notice. But do you? Take time to smell the wild-petunias, and delight in these little plants that self-sow and take care of themselves. They are difficult to pull up (they break off), so leave them be. You just might have some already in your backyard woods. They are suitable for kids to pick a flower, but probably not suitable for taming in the bed or border.

Salvia coccinea ★★★

Scarlet sage

Lamiaceae (mint family)

HABITAT & RANGE sandy, open areas in the Gulf Coastal Plain and Peninsular Florida **ZONES** 8–10 **SOIL** moist to dryish, well drained **LIGHT** sun.
DESCRIPTION Annual, much-branched, 2–3 ft. tall. Leaves opposite, triangular-ovate, rounded-toothed, 1–2 in. long. Flowers very showy, bright scarlet red, tubular, two-lipped with lower lobes notched, 1 in. long, in tight whorls along erect spikes, blooming all summer. Fruits are nutlets, maturing all summer.
PROPAGATION seeds, without pretreatment; stem cuttings taken before flowering **LANDSCAPE USES** sunny borders and beds **EASE OF CULTIVATION** easy **AVAILABILITY** common.
NOTES Scarlet sage is one of our showiest and most long-blooming (until frost) wildflowers. It attracts hummingbirds and bees. Because it's an annual, you will need to start new plants each year or have self-sown seedlings coming along. Several good cultivars have been developed for the bedding plant trade. 'Lady in Red' is scarlet and 'Cherry Blossom' is white and salmon bicolored. The self-sown seedlings from these would revert to being more like the wild type (less compact and fewer flowers) after a few years, and would be lovely additions to a low-maintenance rock garden, flower bed, or meadow.

Lyre-leaved sage (*S. lyrata*) is a common colony-forming weed of roadsides, lawns, and waste places. (The word *weed* is not necessarily derogatory, as weeds are plants that thrive in disturbance and are often needed to help hold the soil. Sometimes the plants can be attractive.) The big purplish-marked basal leaves sit flat on the ground, sending up 1- to 2-ft. leafless branched stalks with charming 1 in. blue flowers. Sometimes we have to bend down to see the beauty in the weeds. Many people would pay money for a weed this nice. You can do so to acquire the dark-leaved 'Purple Knockout'.

Sanguinaria canadensis ★★★★

Bloodroot

Papaveraceae (poppy family)

HABITAT & RANGE rich to dry, rocky woods throughout the Southeast **ZONES** 3–8 **SOIL** moist, well drained **LIGHT** shade to part shade.
DESCRIPTION Deciduous, clump forming, from a very thick, stocky rhizome, 6–12 in. tall, rhizome and leaves with bright red sap. Leaves one (or two) per growing tip, broadly kidney-shaped, usually lobed, 3–6 in. wide. Flowers showy, white, one per growing tip, about 1–2 in. across, petals in multiples of four, usually eight to twelve, short-lived, blooming in late winter to early spring. Fruit is a 2-in. pod with small brown seeds dispersed by ants, maturing in late spring.
PROPAGATION seeds cold stratified for three months; division of loose clumps of rhizomes **LANDSCAPE USES** clumps, or allowed to naturalize and form groundcover in woodland gardens **EASE OF CULTIVATION** easy **AVAILABILITY** common.

Sanguinaria canadensis

Scutellaria incana

Sedum ternatum

NOTES Bloodroot is a prized wildflower, well known and easy to grow. This cheerful harbinger of spring in the Southeast is naturally found in virtually any moist woodland that harbors small herbaceous plants. It is a dazzling sight to see hundreds of square feet of bloodroots in bloom, like snow on the ground in March. The leaf is unique and attractive as summer foliage. Start with a few plants and let the seeds fall where they may (or where the ants take them).

Occasionally, double-flowered bloodroot is available. Known as *S. canadensis* 'Multiplex', it's wonderful and lasts longer than the single flowers. However, the plant is much slower to multiply and does not produce seed because all the sexual parts have become petals.

Scutellaria incana ★★★

Downy skullcap

Lamiaceae (mint family)

HABITAT & RANGE open woods and margins throughout the Southeast, sparse in the Atlantic Coastal Plain **ZONES** 4–9 **SOIL** moist, well drained **LIGHT** part sun to sun.

DESCRIPTION Deciduous, single stem or clumping, 1–3 ft. tall. Leaves opposite, elliptical, rounded-toothed, 1–3 in. long, blue-green, hairy. Flowers showy, lavender-blue, tubular, two-lipped but somewhat closed-mouthed, characterized by a distinct knob on the top of the flower sepal tube, in much-branched spikes, blooming in midsummer. Fruits are small nutlets, maturing in late summer.

PROPAGATION seeds, cold stratified for three months; stem cuttings taken before flowering; division of clump **LANDSCAPE USES** specimens in sunny borders, or in open woodland gardens **EASE OF CULTIVATION** easy **AVAILABILITY** frequent.

NOTES Skullcaps of all kinds have been around in the woods for years. They are cute and noticeable, but so brief flowering that they make little splash in the garden. Downy skullcap is different. It is much larger, holding its own with other perennials in the mixed border. In fact, it looks great rising above or with other equally textured specimens. In a shaded situation without companions, it would probably need staking and would not be seen as much of an addition. Since it grows in all provinces, you would be prudent to seek a heat-tolerant or cold-tolerant form for your locality. I would definitely try it.

Other species, including helmet skullcap (*S. integrifolia*), hairy skullcap (*S. elliptica*), and showy skullcap (*S. serrata*), may be available and are charming. Heartleaf skullcap (*S. ovata*) is easy and floriferous, but spreads readily.

Sedum ternatum ★★★

Mountain stonecrop

Crassulaceae (stonecrop family)

HABITAT & RANGE rich rocky woods, cliffs, and large boulders from the Piedmont of Virginia and the Carolinas across Tennessee into northern Alabama **ZONES** 4–9 **SOIL** moist to dryish, well drained **LIGHT** shade to part sun.
DESCRIPTION Evergreen, creeping, loosely mat forming, with sterile shoots and flowering shoots, 3–6 in. tall. Leaves alternate, simple, in whorls of three, fleshy, about 1 in. long. Flowers showy, white, about ½ in. wide, numerous in three-branched structures that uncoil as they bloom in late spring. Fruits are little pods, maturing in summer.
PROPAGATION division of sterile shoots **LANDSCAPE USES** groundcover for rocky ledges, walls, boulders, rock outcrops, and hillsides **EASE OF CULTIVATION** easy **AVAILABILITY** frequent.
NOTES Mountain stonecrop is one of the few sedums that loves shade, and it makes a great groundcover in thin soil where other plants would not thrive. Because it's so brittle and delicate, it's probably overlooked when designing a woodland garden, but it can be a great plant in the right niche, places unsuitable for Alleghany-spurge (*Pachysandra procumbens*), for example. It is cute to see growing in a mossy rock crevice.

Shortia galacifolia

Shortia galacifolia ★★★

Oconee bells

Diapensiaceae (diapensia family)

HABITAT & RANGE acidic rhododendron woods and streambanks in six counties in the western Carolinas **ZONES** 4–8 **SOIL** moist, humus rich, acidic, well drained **LIGHT** shade to part shade.
DESCRIPTION Evergreen, clump forming, slowly spreading by slender rhizomes, 4–6 in. tall, very delicately rooted. Leaves alternate, in a basal clump, blades ovate-rounded with teeth, on slender stalks (petioles), shiny, turning red-purple in winter, 3–6 in. long. Flowers showy, pinkish-white, toothed petals forming an open bowl, ¾ in. wide, solitary on individual stalks, several arising from the basal leaves, facing forward, blooming in late winter; fruits are fleshy pods, maturing in late spring
PROPAGATION seeds, sown immediately on moist medium; division of clump **LANDSCAPE USES** specimens in acidic litter under rhododendrons, azaleas, and hemlocks **EASE OF CULTIVATION** difficult **AVAILABILITY** infrequent.
NOTES Shortia, or Oconee bells, is one of the most famous wildflowers in the world. It was first discovered in 1788 and then "lost" for nearly a hundred years. It has a mystique due to its charming delicate flowers and restricted range (in six counties in western North and South Carolina, where half of its territory was lost to lake damming in 1970). It has close relatives in Japan. Its early blooming ensures attention, and it's hard to improve on the beauty of the compact cluster of shiny leaves (resembling *Galax*). It is difficult to acquire and establish (it likes moist but well-drained soil), yet it's one of the most sought-after wildflowers. Once established, it will spread and make a patch, but slowly. You must try it.

Silene virginica

Silene caroliniana

Silene virginica ★★★

Fire-pink

Caryophyllaceae (pink family)

HABITAT & RANGE rocky woodlands, outcrops, and cliffs throughout the Southeast, except absent in the Gulf Coast **ZONES** 4–9 **SOIL** moist to dryish, very well drained **LIGHT** part sun to sun.

DESCRIPTION Deciduous, short-lived perennial with basal leaves, clump forming, 1–2 ft. tall, leaf-stem nodes swollen, covered all over with minute sticky hairs. Leaves opposite, narrow, 4–8 in. long. Flowers showy, red, tubular with flaring, notched petals, in branched inflorescences, blooming spring into summer. Fruits are pods with many seeds, maturing soon after flowering.

PROPAGATION seeds, cold stratified for three months **LANDSCAPE USES** specimens in rock gardens or sunny beds, naturalized in open woodland gardens **EASE OF CULTIVATION** moderate **AVAILABILITY** frequent.

NOTES Fire-pink is always a favorite—it's so bright red and long-flowering. Hummingbirds and butterflies love it; the latter hook their legs in the petal notches. The plant is fussy about growing in a well-drained site. I have found it best to plant several individuals and let their seeds fall and come up where they want to. Usually, they choose the middle of a well-worn path or a raw rocky edge.

Royal catchfly (*S. regia*) is taller and just as red, likes well-drained open areas and meadows.

Wild pink (*S. caroliniana*) is a charming cushion-forming species, inhabiting sandy woods and rock outcrops. It is excellent in a well-drained rock garden or raised bed.

Starry campion (*S. stellata*) does not seem to fit with the others—it's not covered with sticky hairs and it does like moist woods and roadsides, especially in the mountains. The floral structure is similar, and it has the typical notched petals, to an extreme. Grow it in part sun and well drained, however.

Silene stellata

Silphium perfoliatum

Sisyrinchium angustifolium

Silphium perfoliatum ★★

Common cup-plant

Asteraceae (aster family)

HABITAT AND RANGE floodplain forest openings from North Carolina through Tennessee and into Alabama and Mississippi **ZONES** 3–9 **SOIL** moist, well drained to seasonally wet **LIGHT** part shade to sun.

DESCRIPTION Deciduous, very coarse perennial, from a taproot, slowly spreading, 4–6 ft. tall, stems square in cross section. Leaves opposite, middle and lowers pairs fused around the stem (connate), simple, toothed, smooth to rough and sticky, 4–12 in. long. Flowers showy, yellow, sunflower-like, with large heads 3–4 in. across, atop the leafy stem, blooming all summer. Fruits are ovoid nutlets without fluff, maturing in late summer.

PROPAGATION seeds, cold stratified for three months **LANDSCAPE USES** bold plants for sunny, moist meadows or open woodlands, especially damp places **EASE OF CULTIVATION** easy **AVAILABILITY** frequent.

NOTES Common cup-plant is very tall and very coarse, suitable only for larger spaces in borders or woodlands. It forms clumps and may spread a bit, and is not something you can just pull up casually. It is wonderful for its size and showiness, especially *en masse*. The cupped leaves hold water; not sure why.

A less-robust non-cupping relative is rosinweed (*S. asteriscus*, including *S. dentatum*). It reaches 4–5 ft. tall, is still coarse, and can have quite rough-textured (scratchy) round stems and leaves. It also spreads, but the big yellow flowers are wonderful in late summer in a meadow or prairie. I think they are too bold for the mixed border.

Sisyrinchium angustifolium ★★★

Blue-eyed grass

Iridaceae (iris family)

HABITAT & RANGE moist meadows and woodlands throughout the Southeast **ZONES** 4–10 **SOIL** moist to dryish, well drained **LIGHT** part sun to sun.

DESCRIPTION Deciduous, thin-stemmed, strictly clump forming, 6–18 in. tall, from fibrous roots. Leaves irregular in length, basal, grasslike, ⅛–¼ in. wide. Flowers showy, blue with yellow center, with six tepals, about ½ in. wide, two to ten coming out sequentially from a sheath atop a leafless flanged (winged) stalk, blooming spring into summer. Fruits are small pods, maturing soon after flowering.

PROPAGATION seeds, cold stratified for three months; division of clump **LANDSCAPE USES** groups or masses, path edging, naturalized in open meadows **EASE OF CULTIVATION** easy **AVAILABILITY** common.

NOTES Blue-eyed grass is a delicate plant, with hardly any leaves, the stems being wide (winged) and taking over the function of the leaves. The flowers are charming, but small and few per individual stem. By itself, it's not much. Used in numbers, it's terrific because of its strict clumping nature, profuse blooming, and ease of growth—and it looks like grass after it has bloomed. Be warned, however: I took my class out to a meadow to see the blue-eyed grass early one morning and none could be found. I thought cows had eaten it all. When I went back a few hours later, all were in bloom. The flowers don't open until near noon, Eastern Daylight Time.

Slender blue-eyed grass (*S. mucronatum*, also *S. atlanticum*) has leaves barely ⅙ in. wide.

Solidago caesia ★★

Wreath goldenrod, axillary goldenrod

Asteraceae (aster family)

HABITAT & RANGE moist forests throughout the Southeast **ZONES** 3–9 **SOIL** moist, well drained **LIGHT** part sun to sun.

DESCRIPTION Deciduous, clump forming, very slowly enlarging, 1–4 ft. tall, stems bluish green. Leaves alternate, simple, narrow-tapered, 2–6 in. long. Flowers, showy, yellow, each head with short "petals" (rays), arranged into small clusters at the base of upper leaves, blooming in late summer into autumn. Fruits are nutlets with fluff, maturing in autumn.

PROPAGATION seeds, cold stratified for three months; division of clump **LANDSCAPE USES** specimens in open woods, perhaps not formal enough for borders and beds **EASE OF CULTIVATION** easy **AVAILABILITY** frequent.

NOTES Wreath goldenrod is a reliable performer, flowering beautifully in late summer even when dry. The attractive flowers are in small clusters on arching stems. This is a good pollinator plant for autumn. It represents those goldenrods with wandlike flowering stems, often one-sided with tight clusters of flowers.

Rough-stemmed goldenrod (*S. rugosa*) also has small clusters of flowers with each leaf. An excellent cultivar is 'Fireworks'. A single plant grows rapidly to form a dense colony. It will need management if you grow it with other perennials, but it can occupy a finite space (such as an island or space between sidewalks) and look great in late summer.

Solidago sphacelata 'Golden Fleece' has long wands covered with flowers into autumn. It grows in part shade and is a spreader.

Solidago caesia

Solidago odora

Solidago odora 

Licorice goldenrod

Asteraceae (aster family)

HABITAT & RANGE dry woodlands throughout the Southeast **ZONES** 6–10 **SOIL** moist, well drained **LIGHT** part sun to sun.

DESCRIPTION Deciduous, clump forming, slowly enlarging, 2–3 ft. tall. Leaves alternate, simple, elongate, not toothed, 1–4 in. long. Flowers showy, yellow, heads arranged on branched spikes flaring upward and outward, blooming mid to late summer. Fruits are nutlets with fluff, maturing in autumn.

PROPAGATION seeds, cold stratified for three months; division of clump **LANDSCAPE USES** specimens in sunny borders, beds, and meadows, planted in open woodland gardens **EASE OF CULTIVATION** easy **AVAILABILITY** frequent.

NOTES Licorice goldenrod is a stately species, tightly clumped with large heads of bushy flowering spikes forming symmetrical plumes. In addition, it has somewhat bluish green foliage, and the crushed leaves smell faintly like anise. This species represents a large group of late summer goldenrods with full heads of yellow flowers; some species are vigorously spreading plants. Check the behavior of your species before you plant.

Canada goldenrod (*S. canadensis*) is a robust (up to 5 ft. tall) full-bodied species that forms discrete spreading

Spigelia marilandica

colonies in old fields and meadows. Use it that way rather than trying to make it fit in a border.

Stiff goldenrod (*S. rigida*) grows to 2 ft. tall with a broad, flat-topped or rounded "head' of flowers. Its leaves are short, broad, and hairy.

Seaside goldenrod (*S. sempervirens*) comes from coastal dunes and is salt-spray tolerant. It grows 4–6 ft. tall in a clump, with beautiful tall-branched plumes late into autumn.

Remember, goldenrods don't cause fall hay fever. They are victims of mistaken identity because they bloom and look conspicuous when the nonshowy culprit, giant ragweed (*Ambrosia trifida*), is going full blast. Goldenrods look beautiful and hold up well as cut flowers. When finished, cut them back to prevent self-sowing.

Spigelia marilandica ★★★★

Indian-pink

Loganiaceae (strychnine family)

HABITAT & RANGE moist to dryish woods, usually in less-acidic soil, mostly throughout the Southeast **ZONES** 4–8 **SOIL** moist, well drained **LIGHT** part shade to part sun.

DESCRIPTION Deciduous, strictly clump forming, ever enlarging, 1–2 ft. tall, from fibrous roots. Leaves opposite, simple, broadly triangular, 2–5 in. long. Flowers very showy, scarlet and yellow, tubular, 1½ in. long, on an arching stalk, opening one at a time, blooming in late spring and sporadically later. Fruit is a heart-shaped pod with two seeds, maturing sporadically in summer.

PROPAGATION seeds, cold stratified for three months; stem cuttings with two or three nodes, taken after flowering; division of clump **LANDSCAPE USES** specimens in borders or beds, naturalized in woodland gardens **EASE OF CULTIVATION** easy **AVAILABILITY** frequent.

NOTES Indian-pink is certainly one of our best and most striking wildflowers. A big clump in bloom is a showstopper. The plants start out small, but steadily become 1- to 2-ft. wide clumps that bloom spectacularly. The plants are adaptable to various light levels, but don't like hot sun. Keep moist for best growth, and cut back stems for regrowth and reblooming. Flowers must be cross-pollinated to get seeds, either by a hummingbird or by you. Seeds are hard to catch when ripe, so collect them just before and let them ripen in custody, or just let them self-sow.

Stellaria pubera

Stellaria pubera ★★

Giant chickweed

Caryophyllaceae (pink family)

HABITAT & RANGE moist woods, mostly throughout the Upper Southeast **ZONES** 6–8 **SOIL** moist, well drained **LIGHT** shade to part shade.

DESCRIPTION Deciduous, clump forming, 6–12 in. tall, forming robust leafy sterile summer shoots after flowering. Leaves opposite, oblong-triangular, ½–2 in. long. Flowers showy, white, ½ in. wide, with the five deeply divided petals appearing as ten, in loose clusters on informal delicate brittle stems, blooming in spring. Fruits are pods, maturing soon after flowering.

PROPAGATION seeds, cold stratified for three months **LANDSCAPE USES** specimens and colonies in woodland gardens **EASE OF CULTIVATION** easy **AVAILABILITY** infrequent.

NOTES Giant chickweed is a wonderful little charmer, early to flower. It grows as informal clumps and looks great on slopes with other delicate wildflowers. Individual plants are tough, lasting for many years. The spring-flowering growth fades quickly and is followed by stronger sterile shoots, a pattern seen in some other wildflowers, such as Jacob's-ladder and woodland phlox.

Tennessee starwort (*S. corei*) is very similar, though less floriferous, but spreads vigorously and makes a fresh-looking delicate groundcover that other wildflowers can tolerate.

Stenanthium gramineum ★★

Common featherbells

Melanthiaceae (bunchflower family), formerly in Liliaceae (lily family)

HABITAT & RANGE open woodlands and borders, and damp meadows scattered throughout the Southeast, mostly in the Piedmont and mountains **ZONES** 6–8 **SOIL** moist, well drained to seasonally wet **LIGHT** part shade to sun.

DESCRIPTION Deciduous, clump forming, from a slender bulb, 2–6 ft. tall. Leaves alternate, basal, grasslike to 2 ft. long and ½ in. wide. Flowers showy, white, small with six tepals to ½ in. wide, numerous in informal branching clusters at the top of leafy stems, blooming in late summer. Fruits are pods maturing in late summer to autumn.

PROPAGATION seeds, cold stratified for three months; division of clump **LANDSCAPE USES** specimens in moist, sunny borders or open woodland gardens **EASE OF CULTIVATION** moderate **AVAILABILITY** infrequent.

NOTES My first encounter with featherbells was as a young student on a field trip to the mountains of North Carolina, standing next to a colony of plants taller than me, while my professor took a picture (which I still have on my desk). They were spectacular (and fragrant), and I have always wanted to grow them. However, these robust mountain plants are not heat tolerant south of Zone 7. Fortunately, they are widespread and all we need is to propagate adaptable selections for these elegant plants to adorn more gardens.

Stenanthium gramineum

Stokesia laevis 

Stokes' aster

Asteraceae (aster family)

HABITAT & RANGE bogs and moist pinelands, mostly along Gulf Coast, perhaps escaped elsewhere **ZONES** 5–9 **SOIL** moist, well drained to seasonally wet **LIGHT** part sun to sun.

DESCRIPTION Evergreen basal leaves, clump forming, 1–2 ft. tall. Leaves wide spreading, narrowly elongate, thick, to 1 ft. long, the upper stem leaves smaller, with a few spiny teeth. Flowers showy, pink to bluish lavender, like bachelor's-buttons, 2–3 in. wide, with deeply divided petals, flowering in summer. Fruits are nutlets, maturing in late summer.

PROPAGATION seeds, cold stratified for three months; division of clump; sprouting of larger root pieces **LANDSCAPE USES** specimens or groupings in borders and beds **EASE OF CULTIVATION** easy **AVAILABILITY** common.

NOTES I first saw Stokes' aster standing in a wet pitcherplant

Stokesia laevis

bog in Alabama, grabbing a poison sumac branch for stability. The plant was beautiful. Its bloom is much anticipated in the summer garden because the heads are so large and the plants so robust (so much so that the flower stems often fall over if not staked). They can grow rather large and informal, occupying some space, even prominent in winter. Cut them back after flowering and they may bloom some more later. They are grand plants, a must-have for the garden.

Several selections are improvements in tidiness. 'Peachey's Pick' (or 'Peachey') is my favorite, with lavender flowers on strongly upright stems. 'Omega Skyrocket' is also sturdy and upright with pale blue flowers. 'Silver Moon' has white flowers. 'Blue Danube' has lavender-blue flowers 4 in. wide.

Stylophorum diphyllum ★★★★

Yellow wood poppy, celandine poppy

Papaveraceae (poppy family)

HABITAT & RANGE moist woods, usually in less-acidic soil, mostly in eastern Tennessee (and northward), touching a toe into Alabama and Georgia **ZONES** 4–8 **SOIL** moist, well drained **LIGHT** shade to part sun.

DESCRIPTION Deciduous, clump forming, from a tough tuberous crown-forming rhizome, 1–2 ft. tall, exuding a yellow sap when any part is broken. Leaves basal, divided like a feather with five deep lobes and rounded teeth, 6–8 in. long, light green, a pair of leaves on the flowering stem. Flowers showy, bright yellow, 1 in. or more across, with many stamens, several flowers loosely on a stalk, blooming in spring and sporadically later. Fruits are 1-in. pods with interesting long white hairs, maturing in late spring.

PROPAGATION seeds, sown immediately; division of clumps with "eyes" or buds **LANDSCAPE USES** specimens in informal shady beds or naturalized in woodland gardens **EASE OF CULTIVATION** easy **AVAILABILITY** frequent.

NOTES Yellow wood poppy is luscious when the large, showy, cheerful flowers begin to open in early spring. They bloom well for a longer time than many spring wildflowers and hold their leaves all summer. The seeds ripen bountifully and self-sow plentifully. You can move the extra plants around gleefully or give them away generously.

Symphyotrichum oblongifolium ★★★★

Aromatic aster

Syn. *Aster oblongifolius*
Asteraceae (aster family)

HABITAT & RANGE dryish woodlands and rock outcrops from Alabama northward into Tennessee **ZONES** 3–8 **SOIL** moist, well drained **LIGHT** sun.

DESCRIPTION Few evergreen basal leaves, much-branched, hard-stemmed, clump forming, ever-enlarging, 2–3 ft. tall. Leaves alternate, simple, oblong, 1 in. long. Flowers showy, rays blue-violet with yellow central disk flowers, 1–1½ in. across, fragrant, in several-flowered clusters, blooming in autumn. Fruits are nutlets with fluff, maturing in late autumn.

PROPAGATION seeds, cold stratified for three months;

Stylophorum diphyllum

Symphyotrichum oblongifolium

division of clump; stem cuttings taken before June **LANDSCAPE USES** specimens or groupings in sunny borders **EASE OF CULTIVATION** easy **AVAILABILITY** common.

NOTES Aromatic aster is one of the great fall asters, blooming late and long, providing fragrance and attraction for fall pollinators for weeks. Unfortunately, it's not strictly an aster anymore (but that does not affect its beauty). All of the traditional species in the genus *Aster* have been moved into other genera (see *Ampelaster*, *Eurybia*) for technical reasons, but the bees don't know it and they still come, and we still grow it and call it an aster, and it's still wonderful. As usual, selection has improved the species for the formal garden. 'Raydon's Favorite' is a great cultivar, covering several square feet of space in a perennial border with flowers. 'October Skies' is excellent, slightly smaller and tighter.

Blue wood aster (*S. cordifolium*) is an early autumn-bloomer for light shade, producing a groundcover effect of light blue flowers. 'Avondale' and 'Photograph' are excellent cultivars.

Heath aster (*S. ericoides*) barely enters the Southeast in Alabama and Mississippi, but has tiny leaves and small white flowers in profusion. 'Snow Flurry' is excellent, creating a groundcover or cascading effect, with semi-evergreen leaves.

Georgia aster (*S. georgianum*) has the most beautiful formal blue flowers you ever saw; single large heads sit atop wispy stems with small stem-clasping leaves, blooming in autumn.

Smooth blue aster (*S. laeve*) creates an upright plant covered with a profusion of violet-blue flowers with yellow centers. 'Bluebird' is a first-rate cultivar.

Calico aster (*S. lateriflorum*) 'Lady in Black' is a superb cultivar producing a show-stopping specimen broader than tall, with smoky-green leaves, covering itself with ½ in. white flowers in late summer into autumn. 'Lovely' is also top-notch with dark green leaves.

Symphyotrichum georgianum

Tephrosia virginiana

Tephrosia virginiana ★★

Goat's-rue

Fabaceae (legume family)

HABITAT & RANGE dry woods, sandy pinewoods, outcrops, and dry roadbanks throughout the Southeast **ZONES** 6–9 **SOIL** dryish, well drained **LIGHT** part sun to sun.

DESCRIPTION Deciduous, clump forming, slowly enlarging, 1–2 ft. tall. Leaves alternate, compound like a feather with many narrow-rounded leaflets, overall 3–4 in. long. Flowers showy, bright pink with a creamy white upper "banner," pea-shaped, ¾ in. long, several on a single upright stalk above the leaves, blooming in late spring to early summer. Fruit is a thin, dry bean pod 1–2 in. long, maturing in summer.

PROPAGATION seeds, cold stratified for three months and/or scarified **LANDSCAPE USES** specimens in well-drained rock gardens, dry meadows, woods edge **EASE OF CULTIVATION** difficult **AVAILABILITY** infrequent.

NOTES Goat's-rue is rarely seen in cultivation, but has an attractive and unusual flower color. It has a mounding habit and is fine-textured gray-green. All who see it marvel at its remarkable leaves. Plant it as a seedling, and don't try to move or divide a mature specimen.

Thalictrum thalictroides

Thalictrum thalictroides ★★★

Rue-anemone, windflower

Syn. *Anemonella thalictroides*
Ranunculaceae (buttercup family)

HABITAT & RANGE moist, rich woods throughout the Southeast **ZONES** 4–8 **SOIL** moist, well drained **LIGHT** shade to part shade.
DESCRIPTION Deciduous, delicate clump, from tuberous roots, 4–6 in. tall. Leaves basal, two- to three-times compound, the little blades (¾ in. long) with three shallow rounded teeth. Flowers showy, white (to pinkish) with five to ten petal-like sepals, ¾ in. wide, in clusters of one to five, blooming in spring. Fruits are small greenish nutlets, maturing in late spring when the plants disappear.
PROPAGATION seeds, self-sow readily (or cold stratified); division of clumps **LANDSCAPE USES** groupings or naturalized in woodland gardens **EASE OF CULTIVATION** easy **AVAILABILITY** frequent.
NOTES Rue-anemone is a charming spring wildflower, delicate in all ways, but proportionally attractive. It can be wispy or form thick clumps. It is an all-around favorite woodland plant. A cultivar with pink flowers is 'Pink Pearl'.

Relatives of rue-anemone are wind pollinated and lack showy sepals, but the leaves and fruits are very similar. Furthermore, they are separated into male and female plants. The males are showier with their whimsical mop-heads of dangling stamens, while= the females have medusa-like heads of pointed ovaries. The spring-blooming early meadow-rue (*T. dioicum*) is rather interesting as a delicate early woodland wildflower.

The later spring blooming tall meadow rue (*T. polygamum*) and purple meadow rue (*T. dasycarpum*) are much coarser, taller (up to 7 ft.), and longer lasting. I like their columbine-like foliage on robust clump-formers, in moist meadows, sunny borders or very open woods.

Thalictrum polygamum

Thermopsis villosa ★★★

Carolina false lupine

Syn. *Thermopsis caroliniana*
Fabaceae (legume family)

HABITAT & RANGE floodplains, disturbed areas, and roadsides in the southern Appalachian Mountains **ZONES** 4–8 **SOIL** moist, well drained **LIGHT** sun.
DESCRIPTION Deciduous, clump forming, 3–6 ft. tall. Leaves alternate, compound with three leaflets, elongate-rounded, 3–4 in. long. Flowers showy, bright yellow, pealike, in spikes up to 2 ft. long on a stem above the leaves, blooming in early

Thermopsis villosa

summer. Fruits are hairy bean pods 3 in. long, maturing in late summer.
PROPAGATION seeds, easy to germinate, scarify by rubbing on sandpaper **LANDSCAPE USES** specimens in sunny borders or moist meadows **EASE OF CULTIVATION** moderate **AVAILABILITY** frequent.
NOTES Carolina false lupine may be the closest you will get to easily growing a lupine in the Southeast. This strikingly handsome tall plant makes a great garden specimen, but because it only grows wild in the mountains, it will suffer from the lowland summer blues, and may need staking south of Zone 7. I loved it and found it to bloom reliably for years in Charlotte.

Tiarella wherryi ★★★★

Foamflower

Syn. *Tiarella cordifolia var. collina*
Saxifragaceae (saxifrage family)

HABITAT & RANGE moist forests from the Piedmont of the Carolinas westward to coastal Alabama, and into Tennessee **ZONES** 3–8 **SOIL** moist, well drained **LIGHT** shade to part sun.
DESCRIPTION Evergreen, short-lived perennial, clump forming, slowly enlarging, 6–12 in. tall. Leaves basal, forming a neat mound, ovoid, lobed and toothed, often marked with red veining, blade 2–4 in. wide. Flowers showy, white, starlike, ½ in. wide, on dense tall spikes arising from the center of the clump, blooming in spring. Fruits are thin, dry pods, maturing in late spring.
PROPAGATION seeds, cold stratified for three months **LANDSCAPE USES** specimens in shady borders or beds, naturalized in woodland gardens, edging **EASE OF CULTIVATION** easy **AVAILABILITY** common.
NOTES Foamflower is a most satisfying wildflower, producing very attractive simple little flowers from the neat leafy clumps. It is a must-have for any wildflower venue. In fact, I suggest it will impart the native plant mystique into your garden better than any other species—use them abundantly. It grows rapidly, and the leaves look good for a long time. Allow plants to self-sow; seedlings are easily moved around. Sometimes, in the right place, plants will live longer.

Many nifty if atypical-looking hybrids and cultivars have been created, providing new forms and colors. 'Dark Eyes' has burgundy leaf markings. 'Iron Butterfly', 'Neon Lights', and 'Oakleaf' have deeply lobed leaves and colorful markings. 'Spring Symphony' is probably the best all-around for toughness and beauty.

Heartleaf foamflower (*T. cordifolia*) is a spreading groundcover with runners producing a carpet of leaves and with abundant flowers arising from the mass of growth. Divide offsets after flowering to propagate. Again, many cultivars exist with the same tendency towards deeply cut leaves and colorful veining. I especially like 'Brandywine' and 'Elizabeth Oliver' as they are more like simple improvements on the wild type.

Tradescantia ohiensis ★★★

Smooth spiderwort

Commelinaceae (dayflower family)

HABITAT & RANGE woodlands, bottomlands, and open areas throughout the Southeast, less common in the mountains **ZONES** 5–8 **SOIL** moist, well drained **LIGHT** shade to part sun.
DESCRIPTION Deciduous, clump forming, slowly enlarging, from a tough rootstock with fleshy roots, 1–3 ft. tall, stems and leaves with faint whitish waxy coating (glaucous). Leaves alternate, simple, grasslike, to 1 in. wide and 16 in. long. Flowers showy, blue-lavender, with three rounded petals and very hairy stamens, 1 in. across, opening one per day from a tight one-sided cluster of buds at the tip of the stem, blooming spring into summer. Fruit is a small pod, maturing in early summer.
PROPAGATION seeds, sown as soon as ripe; division of clump **LANDSCAPE USES** specimens in borders and beds, naturalized in open woods and along streams **EASE OF CULTIVATION** easy **AVAILABILITY** frequent.
NOTES Smooth spiderwort is an amazing plant. It grows elegantly erect, flowering day after day for weeks. Cut it back, and it will sprout out and keep on blooming. I sit and watch—mesmerized—a big clump by my front door. I believe the hairy stamens help the bumblebees hold on for their rapid-fire visits.

There is one selection, 'Mrs. Loewer', with very narrow threadlike leaves and purple flowers; it needs full sun to grow well. You might like it.

Virginia spiderwort (*T. virginiana*) is more common as the major parent of many garden hybrids known collectively as *T.* ×*andersonii*. These have larger, gorgeous flowers in a variety of colors. 'Concord Grape' and 'Purple Dome' are purple; 'Red Cloud' is rosy red; 'Innocence' is white; 'Sweet Kate' is a knockout, with deep blue flowers standing out against shocking chartreuse leaves on a vigorous 1-ft. tall plant.

One of my favorite species is hairy spiderwort (*T. hirsuticaulis*), from rock outcrops. It is 1 ft. tall, very compact, and disappears in summer. The flowers are blue-lavender and the whole plant is very hairy and likes much sun and very well drained soil. I use it as a winter-green border edging.

Tiarella wherryi

Tradescantia ohiensis

Tradescantia hirsuiticaulis

Trillium cuneatum

Trillium cuneatum ★★★★

Toadshade trillium, sweet Betsy trillium

Trilliaceae (trillium family), formerly in Liliaceae (lily family)

HABITAT & RANGE rich woods and bottomlands throughout most of the Southeast, except absent in Virginia and Florida **ZONES** 4–8 **SOIL** moist, well drained **LIGHT** shade to part shade.

DESCRIPTION Deciduous, single stem, from a short, stout rhizome, 3–12 in. tall, eventually clumping. Leaves three, symmetrically arranged at the top of the stem, usually mottled different shades of green, each blade 6–8 in. long, diamond-shaped. Flowers showy, solitary, fragrant, three maroon petals are tubular-upright tapering towards the tip with a purple ovary, 2 in. long, not stalked (sessile), blooming for many days in late winter to mid spring. Fruit is a berrylike pod, maturing in early summer.

PROPAGATION seeds, collected just before ripening (or the ants will get them), cleaned and cold stratified for three months; division of clump immediately after bloom **LANDSCAPE USES** groups in wildflower beds, naturalized in woodland gardens **EASE OF CULTIVATION** easy **AVAILABILITY** infrequent.

NOTES Trilliums as a group are the jewels of the spring garden, and like gems, they come in a range of colors, shapes, and sizes. Flowers are either sessile (stalkless) or pedicillate (stalked). Trilliums are definitely iconic eastern American plants, and there are some thirty different ones that are easy to grow, but they take a long time to propagate (typically five to seven years from seed to flower) and so are rare and expensive. They are becoming more readily available, so ask for them. I have never killed a trillium (well, hardly ever), even transplanting in bloom; that means they are adaptable. Cheap ones ($2–$3 each) in little packages are always collected from the wild (and should be a no-no) and are less likely to survive the improper handling they receive.

Trillium decipiens

Trillium decumbens

Trillium underwoodii

Trillium luteum

Trillium recurvatum

Toadshade trillium is a favorite because the leaves are so gorgeous, ranging from two-tone mottled to a more complex pattern, or, very rarely, solid silver with no mottling at all. The plant is very easy to grow and, like all the sessile trilliums, self-sows much more readily than the pedicillate trilliums if you start with two or three plants that can cross-pollinate. It is one of the most satisfying trilliums in that respect. It is also more prone to making clumps.

Other desirable sessile trilliums vary in flower color but otherwise are all basically like toadshade trillium and all equally attractive.

Chattahoochee trillium (*T. decipiens*) has maroon flowers and leaves that are spectacularly mottled in shades of green and purple (like *T. underwoodii*). This tall and stately plant is my favorite.

Decumbent trillium (*T. decumbens*) has red-maroon flowers with long petals. The mottled leaves sit flat on the ground and form colonies. This trillium is the first one to bloom, first to die down. Choice.

Lanceleaf trillium (*T. lancifolium*) has maroon flowers, often with twisted petals. The leaves are narrow, and the branching rhizome forms a small colony.

Yellow trillium (*T. luteum*) is the bright yellow, sweet-smelling, classic giant trillium of the Smoky Mountains and can have leaves 1 ft. long.

Mottled trillium (*T. maculatum*) has red-maroon flowers with petals that are wider towards the tip. The leaves are marked by exquisite mottling (like *T. cuneatum*). This trillium is one of my favorites.

Prairie trillium (*T. recurvatum*) has bright red-maroon flowers with petals widest in the middle, and with three sepals bending downwards.

Twisted trillium (*T. stamineum*) has maroon flowers with twisted petals that are splayed-out flat to reveal the stamens as the showiest part of the flower. This whimsical-looking species forms clumps readily.

Underwood's trillium (*T. underwoodii*) has dark maroon flowers, wonderful leaf mottling, and a short stem.

Trillium grandiflorum ★★★★

White trillium

Trilliaceae (trillium family), formerly in Liliaceae (lily family)

HABITAT & RANGE rich woods, mostly in the southern Appalachian Mountains and Cumberland Plateau of Tennessee (and northward) **ZONES** 3–7 **SOIL** moist, well drained **LIGHT** shade to part shade.

DESCRIPTION Deciduous, single stem, from a short-stout rhizome, 6–18 in. tall. Leaves three, symmetrically arranged at the top of the stem, solid green, each blade 6–8 in. long, diamond-shaped. Flowers showy, white, solitary, fragrant, three petals tubular-flaring with white ovary, 2 in. long on an erect 2-in. stalk (pedicel), blooming in mid to late spring. Fruit is a reddish berrylike pod, maturing in early summer. **PROPAGATION** seeds, collected just before ripening (or the ants will get them), cleaned and cold stratified for three months **LANDSCAPE USES** groups in wildflower beds,

Trillium grandiflorum

Trillium catesbaei

naturalized in woodland gardens **EASE OF CULTIVATION** easy **AVAILABILITY** infrequent.

NOTES White trillium is probably the single most widely recognized wildflower, and rightly so, for it's simply elegant. There is nothing as breathtaking as a spring woodland with a solid carpet of white trilliums as far as you can see. The flower fades to a nice pinkish purple, extending the show. The double-flowered white forms known collectively as 'Multiplex' ('Flore Pleno', 'Plenum') are spectacular and are asexually propagated to make an expensive garden treasure. They are usually of northern stock and may not do well south of Zone 7.

Variety *rosea* opens with a distinctive pink color.

Other desirable pedicillate trilliums come in a variety of flower colors but otherwise are all like white trillium and all equally attractive.

Catesby's trillium (*T. catesbaei*) has pink flowers with a white ovary. The flowers are usually slightly to strongly nodding and they are fragrant.

Trillium flexipes

Trillium sulcatum

Red trillium (*T. erectum*), also known as wake robin or stinking Benjamin, has an erect, white or red flower with a purple ovary, and a fetid odor (similar to *T. sulcatum*).

Bent white trillium (*T. flexipes*) has white flowers with a large keeled white ovary, erect to strongly nodding, sometimes fragrant.

Sweet white trillium (*T. simile*) is pure white with a purple ovary. It is strongly erect and usually fragrant.

Southern red trillium (*T. sulcatum*) has maroon-red flowers with a purple ovary. It is strongly erect and has a fetid odor.

Vasey's trillium (*T. vaseyi*), also known as sweet Beth, has maroon flowers with a purple ovary and wide petals. The stem is strongly nodding below the leaves. The very large flower (similar to *T. erectum* in structure) is up to 3 in. across, has a fetid odor, and is the largest trillium. Unbelievable!

Least trillium (*T. pusillum*) is white to pinkish with a white ovary. The stem is erect, the flower fragrant. This is the smallest Southeastern trillium at 3–5 in. tall and often forms small colonies.

Painted trillium (*T. undulatum*) is the most charming (white petals with a pink marking at the base) and tempting, but it can't be grown, so don't waste time and money.

Uvularia grandiflora

Large bellwort

Colchicaceae (meadow saffron family), formerly in Liliaceae (lily family)

HABITAT & RANGE rich woods, mostly throughout the Upper South, not getting into the coastal plain **ZONES** 3–8 **SOIL** moist, well drained **LIGHT** shade to part sun.

DESCRIPTION Deciduous, clump forming, 1–2 ft. tall. Leaves alternate, simple, elongate, seemingly pierced through by the stem at one end (perfoliate), 2–3 in. long. Flowers showy, bright yellow, six tepals, 1–2 in. long, pendulous, blooming in spring. Fruits are three-angled pods, maturing in early summer.

PROPAGATION seeds, cold stratified for three months; division of clump **LANDSCAPE USES** specimens in wildflower beds, naturalized in woodland gardens **EASE OF CULTIVATION** easy **AVAILABILITY** infrequent.

NOTES Large bellwort is a delightful plant, pushing its way up quickly and showing early color in earliest spring. The stems and leaves appear to have just come out of the washing machine and are twisted and wrinkled, but that is part of the charm. As the plant matures, it becomes a neater-looking clump. It is long-lived; I have had the same clump for thirty years. This bellwort should be more widely grown, but asexual propagation is slow.

Perforated bellwort (*U. perfoliata*) has pale yellow flowers and spreads by underground rhizomes.

Carolina bellwort (*U. puberula*, syn. *U. pudica*) is a smaller cousin that does not have perforated leaves. It forms a simpler clump with one cute pale yellow flower; subtle pleasure on a small scale.

The species that some people call sessile bellwort (*U. sessifolia*), I love to call wild-oats because the young shoots look like little oat seedlings as they first come up. A cheerful little yellow flower emerges. This bellwort spreads thickly, but does not choke out stronger plants.

Uvularia grandiflora

Uvularia sessifolia

Vernonia noveboracensis ★★★

Ironweed

Asteraceae (aster family)

HABITAT & RANGE meadows, bottomlands, and streamsides throughout the Southeast, but less common westward and not reaching Mississippi **ZONES** 3–9 **SOIL** moist, well drained to seasonally wet **LIGHT** sun.

DESCRIPTION Deciduous, clump forming, slowly enlarging from a tough crown, 4–7 ft. tall. Leaves alternate, simple, elongate-tapering, toothed, 4–10 in. long. Flowers showy, purple, ½ in. long, floriferous heads in large, loosely flat-topped inflorescences atop leafy stems, blooming in late summer. Fruits are nutlets with brown-purple fluff, maturing in autumn.

PROPAGATION seeds, cold stratified for three months; division of smaller clumps **LANDSCAPE USES** specimens in back of border, clumps in open, moist meadows **EASE OF CULTIVATION** easy **AVAILABILITY** frequent.

NOTES Ironweed is a splendid garden plant. The huge clumps are glorious when in bloom and attract much wildlife, so plant them where you can see the flowers. You can also cut the stems back halfway by June 1 (like you would do for Joe-pye weed), so they can branch and grow shorter; although you will lose architectural appeal, you will gain access to the luscious flowers.

Vernonia noveboracensis

Viola pedata

Gray-leaved ironweed (*V. glauca*) is a little smaller, the leaves more blue-green.

The sandhills ironweed (*V. angustifolia*) might just be the ticket, growing only 4 ft. tall in drier conditions, with very narrow leaves.

For something completely different, try leafless ironweed (*V. acaulis*). It grows 2–4 ft. tall in dry, sandy soil and has naked stems with large basal leaves flat on the ground, same purple flowers.

Viola pedata ★★★

Bird's-foot violet

Violaceae (violet family)

HABITAT & RANGE dry, rocky or clay roadbanks and open, sandy woods throughout the Southeast, except absent in Florida **ZONES** 3–9 **SOIL** dryish, well drained, sandy to clay-loam **LIGHT** sun.

DESCRIPTION Deciduous, stemless, clump forming, from a thick, erect rootstock, 3–5 in. tall. Leaves alternate, basal, deeply cut into three to five segments that are further subdivided, 2–3 in. long. Flowers are very showy, light to dark blue-lavender, rarely white, five petals not identical in shape, 1 in. across, blooming in spring. Fruits are small pods, ripening in early summer by bursting open.

PROPAGATION seeds, cold stratified for one month or allowed to self-sow **LANDSCAPE USES** groups in sunny beds, gritty rock gardens, rocky or clay banks **EASE OF CULTIVATION** moderate **AVAILABILITY** frequent.

NOTES Bird's-foot violet is the most charming of the many violets, and one of the most beautiful of all wildflowers. If it just were not so small and tricky to grow, it would be an all-American winner. Unlike other violets, bird's-foot *must* have practically full sun. They are not fussy about soil as long as it's well drained, except it's sure death to plant them in rich woodland soil. I have seen it best on raw, eroded, clay roadbanks where there is little competition. Start with several plants to get cross-pollination, then seedlings will establish on your own site.

Viola hastata

The choicest example of this species is a two-tone form (*V. pedata* var. *bicolor*) with the two upper petals purple and the lower three blue. It is harder to come by, but no more difficult to grow. The cultivar 'Eco Artist's Palette' is easier to acquire and grow, with narrow-petaled two-tone purple flowers.

Of course, there are plenty of other nice woodland violets. Canada violet (*V. canadensis*) makes a large clump approaching 1 ft. tall and has numerous white flowers with purple veins on the lower yellow lip.

Two nice yellow violets are yellow forest violet (*V. pubescens*) and halberd-leaf violet (*V. hastata*). The former makes a clump in moist soil, the latter likes drier sites and has a simple stem with two triangular leaves that are often marked with silver blotches.

Creamy white violet (*V. striata*) is a vigorous clump-former that self-sows to make a groundcover. Wild white violet (*V. pallens*) spreads by delicate rhizomes to form a colony in damp soil with small sweet white flowers.

Zephyranthes atamasco

Zephyranthes atamasco ★★★

Atamasco-lily

Amaryllidaceae (amaryllis family)

HABITAT & RANGE wet meadows and bottomland forests throughout the Piedmont and coastal plain of the Southeast **ZONES** 7–9 **SOIL** moist, well drained to seasonally wet or continuously damp **LIGHT** shade to sun.

DESCRIPTION Deciduous, clump forming, from a bulb, 12–18 in. tall. Leaves basal, three to five, grasslike, 8–16 in. long. Flower showy, white, solitary, on a leafless stalk, one per bulb, 10–12 in. tall, blooming in early to late spring depending on location. Fruits are a three-chambered pod with flat, black seeds, maturing in early summer. Plants disappear in summer.

PROPAGATION seeds, collected and sown immediately; division of clump **LANDSCAPE USES** groups in moist soil in beds or borders with rain-lilies **EASE OF CULTIVATION** easy **AVAILABILITY** infrequent.

NOTES Atamasco-lily, also called Easter-lily by some locals, is a delightful native amaryllis relative. The large white flowers are hauntingly beautiful when you see them blooming *en masse* in a shady seep or wet woods. They do need extra moisture to thrive. An open shady, permanently moist site would be ideal. It is magical to see them come up so suddenly in early spring.

Zizia aurea

Zizia aurea ★★

Golden-Alexanders

Apiaceae (carrot family)

HABITAT & RANGE damp woods, mostly throughout the Southeast **ZONES** 3–8 **SOIL** moist, well drained to seasonally wet **LIGHT** part shade to part sun.

DESCRIPTION Deciduous, one to several erect stems from a fleshy taproot, 1–2 ft. tall. Leaves alternate, three-times compound, leaflets ovate, toothed, 1–2 in. long. Flowers showy, tiny, yellow, arranged in a flat-topped umbel (radiating inflorescence), blooming in spring. Fruits are nutlets, maturing in summer.

PROPAGATION seeds, cold stratified for three months **LANDSCAPE USES** colonies naturalized in open woodlands, especially in damp spots **EASE OF CULTIVATION** easy **AVAILABILITY** infrequent.

NOTES Because its flowers are individually small, golden-Alexanders needs to be planted in groups to draw your attention. Once established, it will spread readily by seeds and form a colony. The tiny flowers are delightful, and the plants are interesting with their fine-cut leaves. I would grow them for the novelty of having a native member of the carrot family that is not Queen-Anne's-lace.

Heart-leaved golden-Alexanders (*Z. aptera*) is similar but thicker textured and the basal leaves are simple. Grow in open woods for best results.

Meadow-parsnip (*Thaspium barbinode*) is very similar to golden-Alexanders and difficult to differentiate. It makes a more robust clump. I love its fine-textured leaf and flower combination in a damp open woodland situation.

Vines

Have you ever been walking through the woods, spied a large grape vine hanging down from a tree and had an urge to swing from it? Well, while you're considering that, think about what's holding that vine up in the tree (and how much you weigh) not to mention how it got way up there in the first place and grew so large as to inspire your inner Tarzan. The vine is attached and growing there in the canopy, sharing the sunlight with the tree leaves. It had to grab on to that tree crown when it was just a sapling, and grow up with the tree, foot-by-foot, all those years to reach the sky. As the tree grows, new stems and leaves form only at the tips of last year's branches, and so it is with the vine. As the tree trunk grows in diameter each year, adding growth rings and forming bark, so does the vine. Therefore, you see, the high-climbing woody grape vine really is . . . a tree.

In growing vines for ornament, they need two things: plenty of sunlight and a sturdy support. Give them an arbor, a fence, or a post, and they will create a crown of growth you can see and be proud of.

Is the clinging vine bad for the tree? Well, it does add weight to the tree canopy, so it could cause excess wind resistance. Moreover, a vigorous vine, which so many are, can grow faster than the tree branches and cover the canopy with leaves, blocking out some sun. Robust vines can certainly smother smaller trees. You may have seen wild grape vines covering roadsides in open sun. Kudzu and Japanese wisteria can kill small trees by strangling them as they twine, like a boa constrictor. Nevertheless, a big tree can support a big vine. When I see a large fallen tree in the woods with a big vine on it, I rarely say, "The vine did this."

Vines are like trees—they seek the sun. Since vines can't hold themselves up, they have to reach the canopy in one of three ways: twining around a tree trunk, climbing up the side of the tree with clinging roots, or using tendrils to wrap around the tree branches when both are young, growing up with the tree as it ascends. As the lower branches of the tree become older and weaker they break off, and the vine branches get older too, falling away with the tree limbs they have held on to for so many years. Again you see, a woody vine really is a tree in so many ways.

Twining vines may wind clockwise, or counterclockwise; you can tell by looking down on them from above, or

Masses of beautiful cross vine (*Bignonia capreolata*) flowers are captivating—to people and to hummingbirds.

VINES BY CLIMBING MECHANISM

TWINERS
Apios americana (groundnut)
Aristolochia macrophylla (Dutchman's-pipe vine)
Berchemia scandens (American rattan vine)
Cocculus carolinus (Carolina moonseed)
Gelsemium sempervirens (yellow jessamine)
Gelsemium rankinii (swamp jessamine)
Ipomoea pandurata (wild sweet potato)
Lonicera spp. (honeysuckles)
Schisandra glabra (magnolia-vine)
Wisteria frutescens (American wisteria)

CLIMBERS BY ROOTS
Bignonia capreolata (cross-vine)
Campsis radicans (trumpet-creeper)
Decumaria barbara (climbing hydrangea)
Parthenocissus quinquefolia (Virginia creeper)
Toxicodendron radicans (poison ivy)

WRAPPERS BY TENDRILS
Passiflora spp. (passionflowers)
Vitis spp. (grapes)

OTHERS
Ampelaster carolinianus (climbing aster)
Clematis viorna (northern leatherflower)
Clematis virginiana (virgin's-bower)
Smilax smallii (Jackson-vine)

Yellow passionflower (*Passiflora lutea*) climbs by tendrils into low-growing supportive vegetation in part shade.

following the direction of the growing tip. While it's not vital to know what direction a vine twines, it's interesting: American wisteria and Dutchman's-pipes twine right, while the nonnative Chinese wisteria and native honeysuckles twine left. The tightness of the twine helps the vine to cling as it climbs. As both the tree and the woody vine grow from year to year and increase in diameter, they begin to squeeze against each other. Sometimes the vine wins, especially if it starts out on a small sapling less than 2 in. in diameter. A vine can kill or mar the trunk of the tree as it digs into and constricts the expanding wood. Rustic walking sticks are fashioned out of saplings that had twists in their wood from tightly twinned vines. In cultivation, it's best to place tight twiners on a post or fence, not a valuable tree trunk.

A second group of vines climbs up the side of a trunk using clinging roots (trumpet vine is an example) or tendrils that have claws that grab the rough bark or become sucker disks and attach like gecko feet to the bark. These vines grow straight up the side of a mature tree—no twining necessary. Only one vine climbs by claws on the tips of its tendrils and that is cross vine; it grabs the bark and climbs, later using roots for a tighter grasp. Other vines, like Virginia creeper, get their initial start up a tree by producing branched tendrils that have sucker disks on the tips. These fuse to the bark (or even a brick or concrete wall) and help the vine to climb. Later, true roots grow out along the length of the vine and penetrate the bark for permanent support. All root climbers will penetrate to some degree the flat surface of the support they are on, and that may make them tenacious and difficult to remove. However, these vines generally don't do as well trying to climb on a spindly trellis or wire fence.

A third group of climbers uses tendrils without roots, claws, or sucker disks. Grapes and maypops are two examples. Their typically multibranched tendrils form abundantly at most nodes and act like watch-springs to wrap around any object they encounter as they grow. Thus, they wrap themselves tightly and tenaciously around small branches of trees and keep themselves in the canopy as the tree grows. These tendrils can become woody and be very difficult to remove. Theoretically, grape vines have a tendril opposite each bunch of heavy grapes to help hold them to a support, allowing, for example, easy training in vineyards with wires to cling to.

Finally, there are the inevitable special cases that are fascinating. Greenbriers (*Smilax*) climb by tendrils modified

Passionflower (*Passiflora*) tendril.

from the pair of leaflike stipules at the base of each leaf. A second special case is the genus *Clematis*, which includes the woody virgin's-bower (*C. virginiana*) and several herbaceous species that mostly climb over low-growing vegetation using only the petioles of their compound leaves. It would be like a person using their elbows to reach out and grab hold of a pole or rope to add stability where the footing is tenuous. This method only works for clematis when the petioles are young and just forming. As the petioles mature, they harden and can't twine at that point.

The last special case is climbing aster (*Ampelaster carolinianus*). This strange vine grows in marshes and shrubby wetlands and does not really climb. It has no specialized devices, and it does not even really twine. Yet, it's a vine for all practical purposes. It merely grows up through the lower vegetation and scrambles over and out the top to create a spreading mass that blooms heavily in late fall, even into November. It makes a wonderful landscape plant, but you have to provide the framework to support it.

High-climbing woody vines can have beautiful flowers, nice leaves, and interesting fruits, but if you grow them on a tree, they will quickly ascend to the top and you will never see them again. This is what they are adapted to do. To get the best landscape effect with these vines, it's imperative that they be grown on a relatively short post, trellis, arbor, fence, bannister, or other support where they can't climb up very far and can be enjoyed for many years. In such positions, they can be cut back hard periodically, or even every year as with grapes, to the older woody branches. That forces them to grow out profusely in the spring with a new set of young stems bearing flowers, foliage, and eventually fruits. You will need to learn when best to prune, whether in the late winter or immediately after they bloom in the spring, as you don't want to cut the buds off spring-flowering vines like yellow jessamine and cross-vine, or the fruits off fall-ripening vines like grapes.

The descriptions that follow include several good perennial herbaceous vines to use in the landscape. I have chosen ones that are more ornamental and can climb to at least 3 ft. or so, often more, then totally die down to the ground each year and come back from roots or rhizomes. These avoid the problems of annual pruning or climbing out of control, though some may be rambunctious as vines are want to do. They are fast growing usually, and many of them have showy flowers, or simply interesting leaves that serve to hide an unsightly fence or arbor—exactly what vines are useful for in the home landscape.

Ampelaster carolinianus ★★★

Climbing aster

Syn. *Aster carolinianus, Symphyotrichum carolinianum*
Asteraceae (aster family)

HABITAT & RANGE swamps, thickets, marshes, and streambanks in the coastal plain of southeastern North Carolina to Peninsular Florida **ZONES** 7–10 **SOIL** moist, well drained, tolerates seasonal flooding **LIGHT** part sun to sun.
DESCRIPTION Slender semiwoody deciduous vinelike shrub with no specializations for climbing but growing up through and scrambling over itself and surrounding vegetation, stems 8–12 ft. long. Leaves alternate, simple, 2½ in. long. Flowering heads 1–2 in. across, numerous, arranged in branched inflorescences, ray (outer) flowers pinkish to purplish, disk (central) flowers yellow to red, blooming very late fall. Fruits are small, one-seeded nutlets, maturing with fluff for dispersal.
PROPAGATION seeds, cold stratified for one month; cuttings in summer **LANDSCAPE USES** perennial borders, fences, posts, or corrals, with support, and mixing with open shrubs through which it can grow **EASE OF CULTIVATION** moderate **AVAILABILITY** common.
NOTES Climbing aster blooms heavily late in the season, sometimes surviving light frosts. Its habit of sprawling and climbing without means of attachment requires that you support it in some way, but it's worth the effort. Although it prefers moist soil, it's adaptable. It has become a popular garden subject and is the best thing in bloom into November. You will come to anticipate its spectacular garden show.

As with many other traditional *Aster* species, this one is a double victim of the recent splitting and renaming within the group that has led to confusion for gardeners. The prefix *ampel* simply means "climbing" in Latin. Prune in late winter.

Apios americana ★★

Groundnut

Fabaceae (legume family)

HABITAT & RANGE streambanks, pond margins, and damp thickets throughout the Southeast **ZONES** 5–10 **SOIL** moist, well drained to wet **LIGHT** part shade to sun.
DESCRIPTION Herbaceous perennial vine, climbing and sprawling by twining, stems 8–16 ft. long, spreading underground by extensive rhizomes with oval swollen tubers formed periodically like oversized knots in a rope. Leaves alternate, compound like a feather, to 6 in. long with five to seven leaflets, fall color yellow. Flowers attractive, reddish maroon to purple, fragrant, in full clusters, blooming sporadically all summer. Fruits are slender green bean-pods, 4 in. long, maturing continuously, edible.
PROPAGATION seeds, cold stratified for three months; division of tubers **LANDSCAPE USES** specimens on trellises or fences, naturalized in thickets **EASE OF CULTIVATION** easy **AVAILABILITY** common.
NOTES This interesting vine is not robust, but pervasive. It will easily twine on itself to fill a sunny trellis, or mix in with other plants by climbing among them. It is probably best to keep it out of the formal shrub border, but site it where you can enjoy it all summer. The nutty-tasting tubers were an important early food source for humans and are making a comeback as a nutritious supplement, used somewhat like Jerusalem artichoke (*Helianthus tuberosus*). Once established, the tuberous rhizomes spread vigorously (and are not easily eradicated), especially in damp places, so you may want to confine it to a large container. It is the host plant for the larvae of the silver-spotted skipper butterfly, and the fragrant flowers attract butterflies. Prune it in late winter.

Ampelaster carolinianus

Apios americana

Aristolochia macrophylla, flower

Aristolochia macrophylla ★★★

Dutchman's-pipe vine

Syn. *Isotrema macrophylla, Aristolochia durior*
Aristolochiaceae (birthwort family)

HABITAT & RANGE rich woods and streambanks, mostly in the mountains from western Pennsylvania to northern Georgia **ZONES** 5–8 **SOIL** moist, rich, well drained **LIGHT** shade to part shade.

DESCRIPTION High-climbing, robust deciduous woody vine, climbing by twining, stems 20–80 ft. long. Leaves alternate, simple, heart-shaped, to 12 in. long and wide, fall color weak yellow. Flowers borne under the very young leaves, single, consisting of three sepals fused into a tubular S shape, about 1 in. long with mouth flaring, yellow-brown marked with red-brown, petals lacking, with mildly fetid odor which attracts pollinators, blooming in spring. Fruits are seedpods that hang down, drying brownish black, 3–4 in. long, splitting open to shake out flat seeds (stacked like Pringles potato chips).

PROPAGATION seeds, cold stratified for one month; cuttings in summer; layering **LANDSCAPE USES** bold texture cover on supports, as screening **EASE OF CULTIVATION** easy **AVAILABILITY** common.

NOTES The peculiar fly-pollinated flowers are often hidden under the developing leaves, but are worth a look. The foliage makes a wonderful mass. This is a vital host plant for the larvae of the pipe-vine swallowtail butterfly. Prune it in spring after flowering or keep stems with fruits.

Aristolochia macrophylla, leaf

Berchemia scandens ★★

American rattan vine, Alabama supplejack

Rhamnaceae (buckthorn family)

HABITAT & RANGE swamp forests, bottomlands, and streambanks, often in less-acidic soils, throughout the Southeast **ZONES** 7–10 **SOIL** moist, well drained to wet, tolerates seasonal flooding **LIGHT** shade to part sun.

DESCRIPTION Robust deciduous woody vine, scandent or climbing by twining, stems to 20 ft. or more long and up to 1 in. or more in diameter, with distinctive smooth, dull-green bark. Leaves alternate, simple, about 3 in. long, with distinctive pinnate venation, fall color rich yellow to red. Flowers in terminal clusters, small, not showy, blooming in spring. Fruits are fleshy drupes, ¼ in. elongate, blue-black, maturing in fall, inedible (poisonous to people and pets).

Berchemia scandens

PROPAGATION seeds, cold stratified for one month; cuttings in summer **LANDSCAPE USES** screening, ornamental specimen **EASE OF CULTIVATION** easy **AVAILABILITY** infrequent.
NOTES This fast- and rampant-growing vine does not necessarily always climb trees, but it does scramble over all vegetation. Trained on a fence or old tree trunk, the vine makes an outstanding display of distinctive bark and leaf texture. It is perhaps my favorite vine for foliage. Prune it in late winter.

Bignonia capreolata ★★★★

Cross-vine

Syn. *Anisostichus capreolata*
Bignoniaceae (bignonia family)

HABITAT & RANGE swamp forests, bottomlands, and woodlands throughout the Southeast **ZONES** 7–9 **SOIL** moist, dryish to wet, tolerates seasonal flooding **LIGHT** part sun to sun.
DESCRIPTION Robust high-climbing semievergreen woody vine, climbing by hooked leaf tendrils. Leaves in pairs, once-divided with 3-in. leaflets creating a distinctive unit of four leaflets with two branching tendrils per stem node, turning purple in winter. Flowers in clusters of two to five, tubular, about 2 in. long, red-orange outside and yellow inside, blooming in midspring. Fruits are dry pods with papery seeds.
PROPAGATION seeds, without treatment; cuttings in summer; layering **LANDSCAPE USES** specimens on supports, as screening **EASE OF CULTIVATION** easy **AVAILABILITY** common.
NOTES Cross-vine is named after the continuous pith that appears as a cross when the stem is sliced open exactly where the two leaves attach, and for how the two pairs of divided leaves can form a cross. Masses of the beautiful flowers are captivating, even if relatively brief. Fall and winter leaf color is a bonus. There are several cultivars, including the popular 'Dragon Lady' with dark ruby red flaring flowers and 'Tangerine Beauty' with reddish orange flowers. Hummingbirds love the flowers. Prune the vine immediately after flowering.

Bignonia capreolata

Campsis radicans ★★

Trumpet-creeper, cow-itch vine

Bignoniaceae (bignonia family)

HABITAT & RANGE bottomlands, swamp forest, forests, old fields, fencerows, thickets, and disturbed places throughout the Southeast **ZONES** 5–9 **SOIL** moist, well drained, tolerates seasonal flooding **LIGHT** part sun to sun.
DESCRIPTION Robust high-climbing deciduous woody vine with rough bark, climbing by roots. Leaves in pairs, divided like a feather, to 1 ft. long, the 2-in. leaflets with large teeth. Flowers tubular, 3 in. long, bright red-orange, in clusters terminating the extending leafy branches, blooming in midsummer. Fruits are dry pods to 8 in. long, shedding winged seeds.
PROPAGATION seeds, cold stratified for three months; cuttings in summer; layering **LANDSCAPE USES** specimens on sturdy supports, as screening **EASE OF CULTIVATION** easy **AVAILABILITY** common.
NOTES Trumpet-creeper is a long-flowering beauty that attracts hummingbirds and occupies three-dimensional space as its flexuous branches can extend several feet out from the main "trunk" if left unpruned. Several color forms and hybrids are available. This coarse vine is characteristic of pastureland fence posts throughout the Southeast. Grows from a deep, tough root that is difficult to eradicate once established. Prune it just after flowering in summer.

Clematis viorna ★★

Northern leatherflower

Ranunculaceae (buttercup family)

HABITAT & RANGE rich woods and open thickets in less-acidic soil, mostly throughout the Southeast, rarer in the coastal plain **ZONES** 6–8 **SOIL** moist, well drained, tolerates seasonal flooding **LIGHT** part sun to sun.
DESCRIPTION Sprawling herbaceous vine, clinging by leaf stalk (petiole) tendrils. Leaves in pairs, divided twice into lobed leaflets. Flowers 1 in. long, numerous but borne singly on long stalks, the four very thick sepals barely opening (lacks petals), rose to reddish purple. Fruits are a cluster of dry "seeds," each bearing a 2-in. feathery plume for wind dispersal.

Campsis radicans

Clematis viorna

PROPAGATION seeds, cold stratified for three months; cuttings in summer **LANDSCAPE USES** specimens in informal borders or trained on low supports **EASE OF CULTIVATION** moderate **AVAILABILITY** infrequent.
NOTES These flowers are unbelievable thick textured and showstopping attractive. The fluffy head of seeds may be more exciting than the flowers. The plants are definitely informal in growth form. Prune in late winter.

Marsh clematis (*C. crispa*), also known as Southern leather-flower, is a widespread southeastern herbaceous species (one of five common ones). It has much more open pinkish-blue-lavender flowers and similar sprawling habit.

Clematis virginiana ★★

Virgin's-bower, devil's darning needles

Ranunculaceae (buttercup family)

HABITAT & RANGE low woods, openings, thickets, and streambanks throughout the Southeast **ZONES** 4–9 **SOIL** moist, well drained, tolerates seasonal flooding **LIGHT** part sun to sun.
DESCRIPTION Sprawling deciduous woody vine, climbing over other vegetation by leaf stalk (petiole) tendrils. Leaves in pairs, each divided into three leaflets, the leaflets 1–2 in. long, stiff, and toothed. Flowers in showy clusters scattered along the stems, each flower about 1 in. wide with four white sepals and many showy stamens, male and female flowers generally on separate plants (so some plants will not have fruits), blooming in late summer. Fruits are a showy cluster of dry "seeds," each bearing a 2-in. feathery plume for wind dispersal.

Clematis crispa

PROPAGATION seeds, cold stratified for three months; cuttings in summer; layering **LANDSCAPE USES** specimens on supports, naturalized to spread on other low-growing shrubs **EASE OF CULTIVATION** easy **AVAILABILITY** common.
NOTES This prolific vine is very effective for late-summer flowering, but the best sight is the maturing fluffy seeds head backlighted for weeks during fall. Keep the vine in check by pruning so as not to overwhelm supporting vegetation. Prune in late winter.

Almost identical is the widely grown nonnative sweet autumn clematis (*C. terniflora*, syn. *C. paniculata*), which is more vigorous and easily naturalizes. It usually has five or more less-coarse, often mottled green leaflets per leaf and each flower has both male and female parts.

Clematis virginiana

Cocculus carolinus

Cocculus carolinus

Carolina moonseed, snailseed

Menispermaceae (moonseed family)

HABITAT & RANGE woods, fields, roadside thickets, maritime forests, rocky hillsides, and limestone cliffs throughout the Southeast **ZONES** 6–9 **SOIL** moist to dryish, well drained **LIGHT** part sun to sun.
DESCRIPTION Semievergreen woody vine from a long-running root system, climbing by twining, to 15 ft. long. Leaves alternate, thick, more or less three-lobed, without teeth, turning yellowish in autumn. Flowers in spikelike clusters along the stems, not showy, male and female flowers on separate plants (so some plants will not have fruits), blooming all summer. Fruits are showy, red, fleshy, long-lasting, 1/4 in. "beads" in grapelike clusters, the seeds curiously coiled like a snail shell.
PROPAGATION seeds, cold stratified for three months; division **LANDSCAPE USES** specimens on supports, especially fences, as screenings **EASE OF CULTIVATION** easy **AVAILABILITY** frequent.
NOTES Carolina moonseed is worth growing for its striking persistent red fruits, but it self-sows and sprouts from its extensive root system unless kept in check by mowing or pruning. Grow it on a fence or trellis where the fruits can be readily seed for a long period. Prune in late winter after effective fruiting.

Its more northern relative, Canada moonseed (*Menispermum canadense*), can be a vigorous vine in floodplains and thickets across the Upper South, with large interesting three- to five-lobed leaves and purple-black grapelike dangling fruits. I have found it more difficult to establish south of Zone 7. The fruits of both species are poisonous.

Decumaria barbara

Climbing hydrangea

Hydrangeaceae (hydrangea family), formerly in Saxifragaceae (saxifrage family)

HABITAT & RANGE swamp forests, bottomlands, and moist forests in the low mountains, from southeastern Virginia to Mississippi and central Peninsular Florida to Louisiana **ZONES** 7–9 **SOIL** moist to wet, well drained, tolerates seasonal flooding **LIGHT** shade to part sun.
DESCRIPTION High-climbing deciduous woody vines with rough brittle stems, climbing by roots. Leaves simple, shiny dark green, in pairs, ovate, somewhat succulent, 2–4 in. long, turning yellow in autumn. Flowers small, 1/2 in. wide,

but showy by virtue of the numerous white stamens, in very prominent 4-in. clusters on the tips of branches held out from their main support, blooming briefly in late spring. Fruits are clusters of interesting tiny urnlike pods with dustlike seeds.

PROPAGATION seed, cold stratified for three months; cuttings in summer; layering **LANDSCAPE USES** specimens on supports, tree trunks, or wooden fences **EASE OF CULTIVATION** easy **AVAILABILITY** common.

NOTES I love this vine; it has subtle details in all of its structure which are worth examination. It can be distinguished from other vines with opposite leaves by its broad, unlobed somewhat fleshy leaves and stems climbing by roots. It is quite effective on a small limbed-up tree (say a dogwood to 20 ft.) where it clothes the trunk in twiggy vines, handsome shiny leaves, and elegant white flowers suddenly all in bloom for a week. Unlike with many other high-climbers, even the lowest branches continue to produce abundant leaves and flowers on side branches. This vine is very well behaved, except for the constant onslaught of vigorous basal shoots travelling rapidly across the ground in search of another tree to climb. Just prune them off anytime, or watch out, they will sneak up somewhere distant. Prune immediately after flowering, if you have to prune at all.

Decumaria barbara

Dioscorea villosa

Dioscorea villosa ★★

Wild yam

Dioscoreaceae (yam family)

HABITAT & RANGE moist forests and woodlands throughout the Southeast **ZONES** 4–10 **SOIL** moist to dryish, well drained, tolerates seasonal flooding **LIGHT** shade to part shade.

DESCRIPTION Flexuous deciduous herbaceous vine from a single knobby tuber, climbing to 12 ft. long. Leaves alternate, simple, broadly ovate to heart-shaped, tapering to a prolonged apex, 2–4 in. long, with conspicuous parallel veins arching towards the tip, fall color yellow. Flowers minute, in surprisingly delicate branching clusters at the leaves, male and female flowers on separate plants (so some plants will not produce seedpods), blooming in spring to early summer. Fruits are in simple clusters on the upper part of the stem, several three-winged papery pods splitting to release wafer-thin seeds, long lasting when dried.

PROPAGATION seeds, cold stratified for three months **LANDSCAPE USES** specimens on supports or shrubs **EASE OF CULTIVATION** easy **AVAILABILITY** infrequent.

NOTES Architecturally, these plants are fascinating, the way they grow so delicately on other plants, form their flowers, and then ripen their nifty seed capsules. Grow them just to get the dried pods! Remember to plant several, as you need individuals of both sexes to get fruits.

Gelsemium sempervirens 

Yellow jessamine, Carolina jessamine

Gelsemiaceae (jessamine family), formerly Loganiaceae (strychnine family)

HABITAT & RANGE swamp forests, thickets, and upland woodlands throughout the Piedmont and coastal plain of the Southeast **ZONES** 7–9 **SOIL** moist to dry, well drained, tolerates seasonal flooding **LIGHT** part sun to sun.

DESCRIPTION High-climbing or trailing evergreen woody

Gelsemium sempervirens

vine, scrambling and mounding by twining. Leaves in pairs, simple, narrow, lustrous dark green, 1–2 in. long. Flowers in small clusters with the leaves on branch tips, lemon yellow, 1 in. long, with flaring petals, fragrant, blooming in early spring. Fruits are nonshowy elongate-blunt dry pods.

PROPAGATION seeds, cold stratified for three months; cuttings in summer; layering **LANDSCAPE USES** specimens on supports, as screening **EASE OF CULTIVATION** easy **AVAILABILITY** common.

NOTES Yellow jessamine is one of the great southern vines: easy to grow, with evergreen foliage that is not too dense or coarse, and with very showy early spring flowers. The vine can climb, creep, crawl, mound, hang, strangle, wrap, cover, and otherwise engulf almost anything—and it still looks great. You don't really want to get rid of it. The most captivating specimen I have seen was on a simple fan-shaped trellis in semishade where a see-through mass of leafy growth was evenly covered with cheerful yellow flowers. It *will* find its way into your attic if you let it climb over the house, and it will self-sow in a garden situation.

A close relative is swamp jessamine (*G. rankinii*), growing in blackwater swamps from the coastal Carolinas to the Florida Panhandle and Louisiana. Swamp jessamine differs in having nonfragrant golden-yellow flowers and pointed seedpods. Prune the vine immediately after flowering. All parts of both plants, including their nectars, are poisonous to people and livestock.

Ipomoea pandurata

Ipomoea pandurata ★★

Wild sweet potato, man-of-the-earth

Convolvulaceae (morning glory family)

HABITAT & RANGE roadsides, fencerows, and disturbed areas throughout the Southeast **ZONES** 6–9 **SOIL** moist to dryish, well drained **LIGHT** sun.

DESCRIPTION Gentle herbaceous vine from an enlarged root, trailing or climbing by twining. Leaves alternate, ovate or often having rounded ends and a contracted center (pandurate), 2–3 in. long. Flowers large, 3 in. wide, in clusters near the leaves, broadly bell shaped, white with red-purple inside the tube. Fruits are globose seedpods, cracking open to release five angular hairy seeds.

PROPAGATION seeds; root sprouts **LANDSCAPE USES** specimens on supports, especially fences, naturalized groundcovers **EASE OF CULTIVATION** easy **AVAILABILITY** infrequent.

NOTES While wild sweet potato is abundant in weedy places, I find it charming and cheerful on ugly wire fences, and along roadsides where it looks like the abundant flowers are sitting upright on the ground. It would certainly be an alternative to nonnative morning glories. The starchy roots can be quite large and were eaten by native peoples.

Lonicera sempervirens ★★★★

Coral honeysuckle

Caprifoliaceae (honeysuckle family)

HABITAT & RANGE dry forests, maritime forests, and roadside thickets throughout the Southeast **ZONES** 6–9 **SOIL** moist, well drained, tolerates seasonal flooding **LIGHT** part sun to sun.

DESCRIPTION Semievergreen nonhairy brittle woody vine, climbing and scrambling by twining. Leaves in pairs, simple, narrow, 1–2 in. long, shiny green above and whitish beneath, turning yellow in fall. Flowers narrowly tubular, 2 in. long, bright red outside and yellow inside, in rings of tight clusters on stem tips, the last pair of leaves below them usually fused together around the stem, blooming in spring or sporadically all summer. Fruits are 1/4 in. red berries, in clusters, long persistent if not eaten by wildlife.

PROPAGATION seeds, cold stratified for three months; cuttings in summer; layering **LANDSCAPE USES** specimens on supports, for screening **EASE OF CULTIVATION** easy **AVAILABILITY** common.

NOTES Coral honeysuckle is certainly one of the most entertaining and showy native plants of the Southeast. The long sinuous vines grow rapidly and searchingly as if anxious to find their support and clamber on up or over. If you try to move them to grow in one direction, they will spring back as if to defy your suggestion. They are quintessential hummingbird flowers, because of their bright red-and-yellow pattern, and perhaps because they place themselves right out there in your face as the birds do.

When not in bloom, coral honeysuckle can be distinguished from the ubiquitous exotic invasive Japanese honeysuckle (*L. japonica*) by the latter's hairy stems and often-lobed leaves that are not whitish underneath and flowers in twos with unfused leaves.

Coral honeysuckle has several excellent cultivar selections: 'Cedar Lane' is a profuse bloomer, 'John Clayton' has orange-yellow flowers, and 'Major Wheeler' blooms all summer.

Prune immediately after flowering for rejuvenation; flowers only on new soft green stems that grow from older woody stems.

Another interesting species with fused leaf pairs and flowers in terminal clusters is yellow honeysuckle (*L. flava*), but its flowers are two-lipped and less tubular.

Lonicera sempervirens, flower

Lonicera sempervirens, fruit

Lonicera flava

Parthenocissus quinquefolia

Passiflora incarnata

Parthenocissus quinquefolia ★★★

Virginia creeper

Vitaceae (grape family)

HABITAT & RANGE bottomlands, wet to dry forests, thickets, rock outcrops, and fencerows throughout the Southeast **ZONES** 4–10 **SOIL** moist to dry, well drained **LIGHT** part shade to sun.

DESCRIPTION Rapidly growing and high-climbing deciduous woody vine, creeping and rooting, climbing by tendrils with sucker disks, and clinging by roots. Leaves alternate, uniquely divided into five- to seven-toothed leaflets arising from a single point (palmately), fall color strikingly yellow-red-purple. Flowers tiny, not showy, in flat-topped branching clusters, blooming early to mid summer. Fruits are ¼ in. grapelike berries, blue-black, on bright red stems, readily eaten by birds.

PROPAGATION seeds, cold stratified for three months; cuttings in summer; layering **LANDSCAPE USES** specimens on supports, as screening, as groundcover **EASE OF CULTIVATION** easy **AVAILABILITY** common.

NOTES Virginia creeper has probably the best fall color of any vine and is as good as the best shrubs as well. Furthermore, it's a uniquely double-duty vine, acting as a great-looking groundcover in the woodland garden, and then a sun-loving climbing vine with great foliage and fruits. It can be used on more massive arbors, pillars, and posts (think utility pole), or wood or brick-and-mortar walls as it can attach to broad, flat surfaces with its sucker disk tendrils. Prune it in late winter after effective fruiting.

People confuse Virginia creeper, which has five all-green leaflets, with poison ivy, which has three shiny leaflets with red little leaf stalks (petiolules) where the three leaflets join. Remember: "Leaves of three, let it be; leaves of five, let it thrive." Poison ivy (*Toxicodendron radicans*) is a creeping shrub up to waist high or a high-climbing vine by roots only, no tendrils. It can easily be in your garden or natural area, often occupying living or, especially, dead trees held tightly to the trunk by a massive cloak of wiry roots emanating all along the robust stems, sending out limber branches many feet long such that it appears to be the tree branches themselves. It has beautiful red and yellow fall foliage and abundant white berries. All parts are toxic, especially the roots, and even burning smoke can carry the irritant. Be careful!

Passiflora incarnata ★★

Maypops, passionflower

Passifloraceae (passionflower family)

HABITAT & RANGE fencerows, fields, roadsides, and thickets throughout the Southeast **ZONES** 5–10 **SOIL** moist to dry, well drained **LIGHT** sun.

DESCRIPTION Vigorous herbaceous perennial vine with stems to 12 ft. long, clambering by stem tendrils. Leaves alternate, simple, three-lobed with fine teeth, to 6 in. long. Flowers single at each leaf axil, 2–3 in. wide, with many green sepals, bluish lavender petals, and numerous whitish filaments (corona), flowering and fruiting all summer, six stout stamens bend downwards to rub pollen on pollinators. Fruits are egg-shaped fleshy leathery berries.

PROPAGATION seeds, cold stratified for three months; stem cuttings in summer; root cuttings anytime **LANDSCAPE USES** mounding groundcover, fence cover **EASE OF CULTIVATION** moderate **AVAILABILITY** common.

NOTES Everyone would agree that passionflower has one of our most interesting and attractive flowers. The foliage

Schisandra glabra

Smilax smallii

is clean and handsome. Elements in the flower structure reminded the early Catholic priests of the passion (crucifixion) of Jesus. The flowers last only one day, and then develop into the familiar green egg-sized fruits that POP! when you stomp on them. You can nibble the ripe fruity insides, too. While spreading aggressively from its extensive underground root system, it can be useful in a naturalized or managed situation. An all-white variant exists. They are larval host plants for several butterflies. Prune any time to keep in check. Resprouts from roots in spring.

The equally widespread yellow passionflower (*P. lutea*) is much smaller and more delicate with three-lobed toothless silver-mottled leaves and pale flowers to 1½ in. wide, blooms all summer, and makes small black berries. It climbs by tendrils into low-growing supportive vegetation in part shade (see photo on page 200). I like to see it bloom and allow it to persist here and there.

Schisandra glabra ★★★

Magnolia-vine, star-vine

Syn. *Schisandra coccinea*
Schisandraceae (star-vine family)

HABITAT & RANGE rich forests and floodplain slopes, very widely scattered across the Southeast, especially in the Deep South **ZONES** 7–9 **SOIL** rich, moist, well drained, tolerates seasonal flooding **LIGHT** shade to part shade.
DESCRIPTION High-climbing deciduous woody vine with rough stems, climbing by twining. Leaves alternate, simple, widely spaced on long shoots, closely spaced on short shoots, somewhat ovate, 3–5 in. long, fall color yellow. Flowers borne on short branchlets with crowded leaves, each ½ in. wide, striking shiny bright red, hanging on its own thin stalk, either male or female but both sexes on the same plant, blooming in early summer. Fruits are ¼-in. round berries, translucent candy-apple red, crowded on elongating "stems" as they mature, forming attractive bunches.
PROPAGATION seeds, cold stratified for three months; cuttings in fall, layering not effective **LANDSCAPE USES** specimens on supports such as small trees or arbors **EASE OF CULTIVATION** moderate **AVAILABILITY** rare.
NOTES Among the rarest and most enigmatic plants in the Southeast, star-vine grows in rich woodlands at widely scattered localities across its range, but is most frequent in the Mississippi coastal plain, in rich beech-magnolia forests. It grows easily from seeds and could certainly be more widely distributed in cultivation. The plant's odd characteristics and year-around appeal will grow on you. Birds devour the berries immediately. You might want to get to them first yourself. Prune in late winter.

Smilax smallii ★★

Jackson-vine

Smilacaceae (greenbrier family), formerly Liliaceae (lily family)

HABITAT & RANGE bottomland forests in the coastal plain throughout the Southeast **ZONES** 7–9 **SOIL** average to wet **LIGHT** shade to part sun.
DESCRIPTION High-climbing evergreen woody monocot vine with hard, sharp prickles, climbing by leaf tendrils, stem arising from a massive crown of tuberous rhizomes. Leaves alternate, simple, shiny green, 2–4 in. long in attractive flattened sprays along the upper branches. Flowers in clusters with the leaves, greenish, not showy. Fruits are ¼-in. black dry berries, maturing the second year after formation.

PROPAGATION seeds, cold stratified for three months; division of clump **LANDSCAPE USES** screening **EASE OF CULTIVATION** easy **AVAILABILITY** frequent.

NOTES Normally I would not recommend a woody *Smilax* because the species are difficult to manage, impossible to eradicate, and normally viciously prickly. However, this famous species is used widely throughout the Southeast as an attractive evergreen screening, fence adornment, and eave cover on a house. It has minimal prickles on the upper branches that bear the leaves. If you start with a young plant and keep it short, it can be trained to form a magnificent foliage cover and can be cut judiciously for indoor decoration and florist greens. It flowers only when it reaches high in the trees, and then it will make so many seeds they will rain down everywhere in your garden. Prune for foliage removal anytime and remove old stems as they die. Don't cut main living stems.

Coral greenbrier (*S. walteri*) has the most beautiful red winter-persistent berries and is found in very wet swamps. I covet this vine but have not been able to grow it, perhaps luckily because all smilax form thickets.

Smilax walteri

Vitis rotundifolia ★★

Muscadine, scuppernong

Syn. *Muscadinia rotundifolia*
Vitaceae (grape family)

HABITAT & RANGE forests, swamps, thickets, and dunes throughout the Southeast **ZONES** 6–10 **SOIL** moist to dryish, well drained, tolerates seasonal flooding **LIGHT** part sun to sun.

DESCRIPTION Tough deciduous woody vine from a massive root, climbing by unbranched stem tendrils that become hardened. Leaves alternate, simple, almost circular heart-shaped, 2–3 in. wide, with large teeth along the edges and not lobed, fall color yellow. Flowers not showy, in clusters, blooming in early summer. Fruits are ¾-in. round, purple, juicy "berries," in few-fruited clusters.

PROPAGATION seeds, cold stratified for three months; cuttings **LANDSCAPE USES** specimens on arbors or supports, naturalized in thickets **EASE OF CULTIVATION** easy **AVAILABILITY** common.

NOTES This species is one of the most ubiquitous grapes, both wild and cultivated, in the Southeast. There are many cultivar selections of muscadine (purple) and scuppernong (white) grapes. Whether or not you grow this vine for its delectable fruits, the foliage turns a handsome yellow fall color and can be quite attractive. The vines can certainly smother small trees and shrubs that they grow on in full sun along roadsides. Muscadine is our only grape with unbranched tendrils and few-fruited clusters. Prune it in late winter on a cold day well before new growth begins or it will harmlessly "bleed" sap.

Vitis rotundifolia

Two other widespread Southeastern grapes are frost grape (*V. vulpina*), which has leaves that are hardly lobed and that are green on the undersides, and summer grape (*V. aestivalis*), which has deeply lobed leaves that have rounded sinuses and are often waxy-whitish (glaucous) or with whitish cobwebby hairs underneath. The ripe fruits in large clusters may be attractive and tolerably tasty, or not.

These long-lived woody vines are frequently seen hanging from large trees and even partly lying on the ground, their bark shredding. In some cases, they make exciting landscape statements with their architectural twists and curves, and could be left to grow in a natural setting.

Wisteria frutescens

Wisteria frutescens ★★★

American wisteria

Fabaceae (legume family)

HABITAT & RANGE lowland woods, streambanks, and pocosins throughout the Southeast **ZONES** 5–9 **SOIL** moist, well drained, tolerates seasonal flooding **LIGHT** part shade to sun. **DESCRIPTION** Deciduous woody vine climbing by twining. Leaves alternate, divided like a feather, 4–12 in. long, with an odd number (nine to fifteen) of 2-in. leaflets, fall color yellow. Flowers pealike, blue-purple, fragrant, densely arranged on horizontal short stalks on the tips of newly grown leafy branches, blooming in late spring and sporadically later. Fruits are nonhairy 4-in. pods with hard seeds. **PROPAGATION** seeds, cold stratified for three months; cuttings; layering **LANDSCAPE USES** specimens on supports, for screening **EASE OF CULTIVATION** easy **AVAILABILITY** common.

NOTES This wisteria is a delightful flowering vine and certainly a much tamer vine than the Chinese and Japanese species that are known to strangle trees and whole forests to death. American wisteria differs from its Asian counterparts in having smaller, more hairy leaves; shorter, denser, nonhanging flower clusters; less fragrant flowers; and nonhairy seedpods. Three good cultivars of the several available ones are 'Amethyst Falls', with bluer flowers; 'Longwood Purple', with more-purple flowers; and 'Alba', with white flowers. Prune American wisteria in late winter.

Kentucky wisteria (*W. macrostachya*) seems to have longer flower clusters consistently. 'Clara Mack' is a beautiful white form.

Shrubs

My family goes on a traditional August trek to the mountains of North Carolina to pick blueberries along the Blue Ridge Parkway at a famous site called Graveyard Fields. This area of several square miles burned in the 1940s and has never grown back into forest but instead remains an open shrub-dominated habitat. It is a picturesque, gentle-sloping broad valley cut by a fast-flowing mountain stream. Two or three species of blueberries *(Vaccinium)* form extensive stands, through which paths have been worn in the soft, organic soil. All produce abundant delicious berries that more than fulfill our annual desire to eat fresh from the wild and take a few back home.

In this meadowlike setting are many other shrubs and small trees that produce colorful, if not edible, fall berries such as bearberry, fire cherry, mountain ash, black chokeberry, serviceberry, and withe-rod viburnum. The black bears must love this smorgasbord of berried delights as they fatten for winter. It is a paradise of shrubs where you can't get lost because most of the vegetation is below head-high.

What is even more spectacular about this site is that in mid-October the whole vista turns into a blaze of fiery leaves in all shades of red and yellow. The scattered pockets of evergreen trees and rhododendrons provide a bit of green to break the swathe of crimson, and it can be an unbelievable site in a good year.

Throughout natural habitats in the Southeast, the diversity of shrubs is generally greater than that of trees in any given district. Shrubs certainly are plentiful in open habitats such as old fields and roadsides in the Piedmont and coastal plain. They appear early in the sequence of plant succession when an old field slowly changes into a mature forest. Among the earliest species needing full sun and average to dry habitats are blackberries, Carolina rose, Chickasaw plum, and smooth sumac. Such plants are particularly important for wildlife where they form vegetation edges between two habitats, such as when a grassy meadow abuts an older shrub-dominated site. Birds and small mammals can move between habitats to find food and cover. If

Fields of blueberries and other colorful shrubs line the boardwalk at Graveyard Fields, a high-elevation valley on the Blue Ridge Parkway.

Shrubs in a pocosin-savanna habitat.

the vegetation were all grasses, for example, the collection of animals and insects would be more uniform. Then, as the forest canopy develops, the shrub-dominated habitat provides another opportunity for the edge effect, where diversity of plants and animals is greater due to variations in light, moisture, nutrition, and temperature.

Shady woodlands are home to various azaleas, blueberries, hydrangeas, hazelnuts, and more. In many cases, these shrubs can tolerate shade once the forest closes in around them in later stages of succession, but they prefer some sun and so are at their best in forest openings and along the edges. They are deciduous themselves, dropping their leaves in winter to conserve moisture during the coldest weather when photosynthesis is less efficient.

In the mountains, certain shrubs thrive in full shade of deciduous trees. Rhododendrons and mountain laurel are two examples. Although these evergreen shrubs are able to photosynthesize in the shade, they receive much more sunlight for half the year when the forest has lost its leaves. This ability is very significant, and the shrubs depend on this open-canopy period for photosynthetic activity as long as the soil is not frozen and the air temperature is above 40°F. Many of these species of mountain plants would generally not do well in full afternoon sun in the lowlands of the South. Rarely do you find a diversity of shrubs, or wildflowers for that matter, in continuously shady evergreen forests, such as under hemlocks.

Other shrub-dominated habitats have plants that do require lots of sun, and this is in marsh wetlands with damp but well-drained soil (except during periods of flooding). Here you find an abundance of shrubs such as choke-berries, St.-John's-worts, summer sweet, Virginia-willow, button-bush, fothergilla, hollies, and many more. The famous pine flatwoods and shrubby pocosins of the coastal plain fall into this category.

Just as shrubs fill many niches in nature, they are invaluable in the garden. Visible at eye level, they can be used as foundation plantings, as beautiful evergreens that help define space, for privacy screening, or as accent plants of striking beauty in flower and fruit in the mixed border. Some of the smaller or dwarf shrubs can be used for groundcover in some circumstances, and others remain compact enough for a rock garden. More so than with trees, shrubs are chosen for their year-round ornamental features.

Shrubs can form clumps; others spread slowly by creeping underground stems (rhizomes) or roots, sending up suckers that eventually create a large, dense colony. A few shrubs are rapid growers, with rhizomes spreading 1 ft. or more in one year. In nature, where they have to compete with other plants in their habitats, they grow more slowly, but when placed in a garden with less competition, more water, and more care, they can take over. In such a situation, you can thin out the growth by removing the outer sections of the colony. This will rejuvenate the growth. Occasionally you may have to remove the plant entirely and relegate it to the "back forty" or marginal areas just outside the garden where it can spread and be enjoyed without harm. It is best, of course, to know what your shrub is going to do so you can site it properly from the beginning.

Many shrubs grow well with each other, forming a mixed border with informal intermingling. Consider the texture (coarse or fine leaves and twigs) and growth rate when deciding what to mix, so one shrub does not overtake another.

Most shrubs do well as discrete specimens. Often they have multiple stems, as opposed to trees that usually have one main trunk growing taller than 25 ft., but some treelike shrubs, such as fever tree and fringe tree, produce a larger main trunk with smaller root sprouts. It can be somewhat arbitrary to designate a plant as a tree or shrub, but I have made decisions in this book based on whether I think the species in question is shorter and more likely to spread (shrubs with treelike traits) or regularly grows into larger single-trunked individuals (trees than can have shrublike characteristics).

Like trees, deciduous shrubs can also turn beautiful fall colors if they have the innate ability to produce red anthocyanin pigment to mix with the yellow and orange pigments. In wetlands, shrubs can turn beautiful mixed colors, whereas trees rarely do so. The shrub must be in strong sunlight to make the best colors, even though shrubs in the shade will develop some fall color. Try to site your shrub in the best light, avoiding hot afternoon sun unless the shrub can take it. The fall weather must then cooperate to give you cool nights and bright sunny days to manufacture the red pigment. Some understory shrubs may actually delay color change until deciduous forest trees have lost their leaves, thereby affording more sun to the shrub. In an effort to allow for the best fall color development, it's best not to fertilize shrubs after July 1 so that they are overstimulated to make lush growth as late summer and fall come along. Lush growth could be a detriment in the hottest and driest time of year (hurricane season notwithstanding), and such growth is not conducive to producing the strongest fall color.

Pruning shrubs is an important practice to help direct their growth as they mature. Older shrubs may need to be shaped or rejuvenated. Knowing when to prune and how to prune are the two key questions. Shrubs that bloom in spring on last year's twigs have already formed their flower buds the previous summer, just as with trees. These must

WHEN TO PRUNE SHRUBS

PRUNE SPRING-BLOOMERS IN VERY EARLY SUMMER	PRUNE SUMMER-BLOOMERS IN SPRING
Azaleas (*Rhododendron*)	Beauty berry (*Callicarpa americana*)
Buckeyes (*Aesculus*)	Hollies (*Ilex*)
Sweet shrub (*Calycanthus floridus*)	Hydrangeas (*Hydrangea*)
Viburnums (*Viburnum*)	New Jersey tea (*Ceanothus americana*)
Witch-alder (*Fothergilla gardenii*)	Summer sweet (*Clethra alnifolia*)

not be pruned until after they bloom or you will cut off the spring flower buds. Instead, prune these as soon as they have finished flowering so they have time to grow new twigs and buds for next year. Shrubs that bloom on new growth in early summer can be pruned in spring and then not again or you may risk removing the colorful fall berries. Know your shrub's flowering and fruiting behavior before you prune. Even so, you may forfeit a season of flowering for the hope of a better future show.

As for pruning to shape or rejuvenate, there are two approaches. In the first, a technique called heading back, you can make a shrub thicker with more blooms by pruning off a third or more of each twig just above a bud. This stimulates one or more new twigs to branch out from lateral buds and produce more flowers. A second technique, thinning out, is useful when you have a dense overgrown shrub that is not blooming well. Here you want to open up the whole shrub by removing a third of the stems by cutting them entirely to the ground. Do this each year and you will stimulate a new set of stems to grow each time you remove the oldest. These new stems will branch and bloom vigorously for two or three years, then be ready to be removed to make way for new ones.

Not all shrubs need regular pruning. Many may only require the removal of broken or wayward branches from time to time. In general, don't plan to prune your evergreen shrubs regularly as you have seen done with foundation plants against a house. These "green meatballs," as they are called, can be unsightly, time consuming to maintain, and detrimental to the health of the shrub in the end. It is better to choose a shrub that grows the size and shape you want without having to perform routine maintenance pruning. The slow-growing dwarf selections of our native yaupon holly (*Ilex vomitoria* 'Nana' or 'Schillings') are perfect examples of cultivars that never need to be pruned and will make a 4 ′4 foot shrub over ten to fifteen years. Furthermore, rhododendron, azaleas, and other shrubs that flower on new growth every year without becoming overgrown are examples of shrubs that can't easily be pruned without adversely altering their desirable informal shapes.

As with small trees, if you are planting shrubs for their colorful fruits and berries, you may need several genetically different individuals to cross-pollinate. Planting in groups of three or five is a good way to utilize the shrubs in an effective design configuration. If your shrub is an asexually propagated clone, it may not produce abundant fruit even if several specimens are present because they will all be the exact same genetic type. Of course, don't forget that several shrubs, most especially hollies, have separate sexes on separate plants and require a mixture of males and females to make berries. You will need only one male for several females in close proximity, but it must be of a compatible mating type and one that blooms at the same time. Finding a compatible pollinizer can be frustrating (for you and the shrub), but it must be done properly to create the berried treasures that ornamental shrubs can provide. Good nurseries can advise you on which cultivars to pair together for effective pollination.

In the write-ups that follow, I have indicated whether the shrub is deciduous or evergreen and whether clumping or spreading as these factors are of primary importance in selecting for a particular role in a landscape. Equally important is to determine the light levels for the shrub. Too much shade and you get fewer flowers and fruits, or poorer fall color. Too much sun and the shrub may develop scorched leaves or die during times of heat stress.

Aesculus parviflora ★★★★

Bottlebrush buckeye

Sapindaceae (soapberry family), formerly in the Hippocastanaceae (horse chestnut family)

HABITAT & RANGE moist to dry woods, typically in less-acidic soil, throughout most of Alabama, just into west-central Georgia, and in Aiken County, South Carolina **ZONES** 5–8 **SOIL** moist, well drained **LIGHT** part shade to part sun.
DESCRIPTION Deciduous multistemmed shrub forming a suckering, symmetrical mass with branches to the ground, to 12 ft. tall and 15 ft. wide or wider. Leaves opposite, compound, with five (to seven) leaflets attached at one central point (palmate), each leaflet 4–8 in. long, turning a beautiful bright butter yellow in autumn. Flowers showy, numerous, white, arranged like a bottle brush on erect 1- to 2-ft.-long stalks, long-lasting, attracting butterflies, blooming in late spring to early summer. Fruits are attractive, shiny, brown, 1 in. spherical nutlike seeds splitting out of leathery husks in late summer.
PROPAGATION seeds, stratified for one month and not allowed to dry out; suckers; stem cuttings **LANDSCAPE USES** specimens, massing, borders, naturalized in woodland garden, kept as an understory shrub **EASE OF CULTIVATION** easy **AVAILABILITY** common.

NOTES Bottlebrush buckeye is a unique and first-rate garden plant with four-seasons of interest. The plumes of white flowers are exciting, but the bright yellow fall color is exceptional. Although this shrub can be used in light shade, it's best with as much light as possible. In the warmer zones of the Southeast, it should not have full, direct afternoon sun, but is best on the sunny edge of a woodland border. *Aesculus parviflora* var. *serotina* leafs out later and blooms two to four weeks later than the species.

Aesculus parviflora, flower

Aesculus parviflora, habit

Aesculus pavia, flower

Aesculus pavia, fruit

Aesculus pavia ★★★

Red buckeye

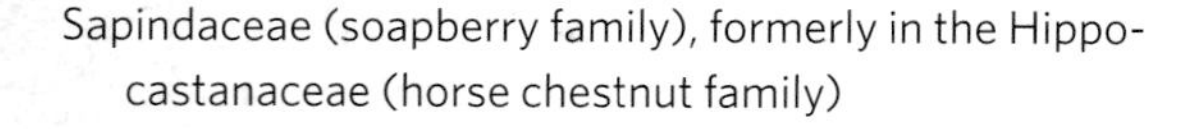

Sapindaceae (soapberry family), formerly in the Hippocastanaceae (horse chestnut family)

HABITAT & RANGE floodplains and swamp forests, often in less-acidic soils, throughout the Southeast, absent from the Piedmont and mountains of North Carolina **ZONES** 6–9 **SOIL** moist, well drained, tolerates seasonally wet soil **LIGHT** part shade to sun.

DESCRIPTION Deciduous small tree or shrub to 20 ft. tall or taller, irregularly branched, with large, conspicuous tip buds. Leaves opposite, compound with five (to seven) shallowly toothed leaflets attached at one central point (palmate), emerging reddish in spring but lacking fall color and often dropping by late summer. Flowers showy, bright scarlet-red, usually with some yellow, tubular, 1 in. long, in branched clusters, blooming in spring. Fruits are attractive, shiny, tan-brownish, 1-in. spherical nutlike seeds, splitting out of leathery husks in late summer.

PROPAGATION seeds, cold stratified for one month and not allowed to dry out **LANDSCAPE USES** specimens, borders, accents, groupings, parks and estates, naturalized in woodland garden, kept as an understory shrub **EASE OF CULTIVATION** easy **AVAILABILITY** common.

NOTES Red buckeye is one of our best native shrubs or small trees. The spring flowers are striking, and the shiny brown seeds look like polished woodwork. Carry a seed in your pocket for good luck. This shrub grows slowly but is long-lived. I would find a place to have one somewhere in the garden; its spring growth is fascinating when the husky stem, big leaves, and large flowering cluster emerge from the solitary tip bud in just a few days.

Agarista populifolia ★★★

Florida hobblebush

Syn. *Leucothoe populifolia*
Ericaceae (heath family)

HABITAT & RANGE hummocks in swamps and wet woodlands, along the extreme outer coastal plain from southeastern North Carolina to Peninsular Florida, and then around Mobile, Alabama **ZONES** 7–9 **SOIL** moist, well drained, tolerates seasonally wet soil **LIGHT** shade to part sun.

DESCRIPTION Large evergreen multistemmed shrub to 18 ft. tall, with gracefully upward-arching stems. Leaves alternate, simple, elongate-pointed, finely toothed, 2–4 in. long; flowers white, tubular, ½ in. long, fragrant, inconspicuous in clusters under the foliage, blooming in early summer. Fruits are inconspicuous small round pods with many tiny seeds, maturing in autumn.

PROPAGATION cuttings in summer **LANDSCAPE USES** groupings, accents, borders, privacy screenings, to help define space and views in woodland gardens **EASE OF CULTIVATION** easy **AVAILABILITY** common.

NOTES While its white flowers are of no ornamental consequence, Florida hobblebush is one of the best evergreen shrubs for screening or growing along a stream. It makes a large loose clump that may benefit from, but does not require, periodic thinning of the older stems. The glossy dark green foliage looks good year-round, and is clean and disease-free (compared to common leucothoes). Don't grow this shrub in full sun or in dry soil. It can tolerate very wet soils. If cut to the ground, it sprouts back quickly and rejuvenates.

A dwarf form sold under the trade name Leprechaun, grows about 4 ft. tall and wide and looks like it will become an excellent garden plant.

Agarista populifolia

Amelanchier canadensis

Eastern serviceberry, shadbush

Rosaceae (rose family)

HABITAT & RANGE shrubby pocosins, bogs, and acidic wetlands in the coastal plain and some Piedmont regions throughout the Southeast, especially common in the Carolinas, absent in Florida **ZONES** 3–8 **SOIL** moist, well drained **LIGHT** part sun to sun.

DESCRIPTION Deciduous multistemmed shrub 6–20 ft. tall, with smooth gray bark. Leaves alternate, simple, toothed, 1–2 in. long, turning wonderful shades of gold, yellow, orange, and red in autumn. Flowers showy, white, 1 in. wide, in loose erect clusters, blooming in early spring. Fruits are purplish black berries in loose clusters, ripening in summer, loved by birds.

PROPAGATION seeds, cold stratified for one month **LANDSCAPE USES** flowering specimens, accents, mixed borders, in informal woodland settings **EASE OF CULTIVATION** easy **AVAILABILITY** common.

NOTES This large multistemmed shrub is a superior landscape choice, producing abundant fruits. It is an excellent bloomer and has some of the best fall color you will see, especially if grown in almost full sun.

Amelanchier obovalis

You may encounter similar but smaller and spreading-rhizomatous species, which can be charming in dry, sunny parts of the garden. They are coastal plain serviceberry (*A. obovalis*) and dwarf serviceberry (*A. spicata*, including *A. stolonifera*).

Amorpha herbacea ★★

Dwarf indigo-bush

Fabaceae (legume family)

HABITAT & RANGE sandy fields and woodlands, mostly in the coastal plain from the Carolinas to Peninsular Florida **ZONES** 5–9 **SOIL** moist, well drained **LIGHT** sun.

DESCRIPTION Deciduous shrublet 1–4 ft. tall. Leaves alternate, compound, with many untoothed leaflets about 1 in. long arranged like a feather, whitish gray soft-hairy. Flowers small, bluish purple, with conspicuous white stamens and yellow pollen, densely arranged on erect spikes to 8 in. long, blooming throughout spring. Fruits are one-seeded little grayish pods, ripening throughout autumn.

PROPAGATION seeds; cuttings **LANDSCAPE USES** specimens, groupings, borders, kept as a native woodland shrub **EASE OF CULTIVATION** easy **AVAILABILITY** common.

NOTES Dwarf indigo-bush is an interesting shrub, likely to become a showpiece in the garden, because it blooms over a long period and the distinctive purple flowers contrast against the overall gray appearance. This species is widespread in dryish habitats throughout North America.

Amorpha herbacea

Aronia arbutifolia ★★★★

Red chokeberry

Syn. *Sorbus arbutifolia, Photinia pyrifolia*
Rosaceae (rose family)

HABITAT & RANGE bogs, savannahs, and low woods throughout the Southeast **ZONES** 5–9 **SOIL** moist, well drained to wet **LIGHT** part sun to sun.

DESCRIPTION Semi-evergreen multistemmed shrub 5–10 ft. tall, often forming colonies by suckering. Leaves alternate, simple, hairy, with black-tipped teeth, 2–4 in. long, a few staying evergreen, most turning shades of brilliant scarlet in autumn. Flowers showy, white, five-petaled, ½ in. wide in flat-topped clusters on stem tips, short-lived, blooming in spring. Fruits very showy, rich red, ¼ in. berries in clusters, maturing in autumn and persisting through winter unless eaten by birds.

PROPAGATION seeds, cold stratified for three months; stem cuttings; suckers **LANDSCAPE USES** specimens (colony), borders, accents, kept as a native understory plant

NOTES Red chokeberry is a fabulous shrub, especially if you like red. Even the winter buds are red. This shrub grows fast and lush, adapts well to average or wet soil, light shade or sun—though you get many more berries in the sun—and has the richest red fall colors I have seen in a shrub. The berries are striking and last better than most through the cold of winter. Site this shrub in the garden so that you can enjoy it every day of the year.

Aronia arbutifolia, flower

An excellent cultivar is 'Brilliant' (syn. 'Brilliantissima'), which produces slightly larger berries and even glaringly scarlet-red fall leaf color.

Black chokeberry (*A. melanocarpa*) is common in the mountains. It is practically identical to red chokeberry except it has no hairs on the twigs or leaves and the abundant big ⅜ in. berries are purple-black. I find that the berries can look good up close, but don't show up as well in the garden. The shrub has good fall color and is widely available. Don't accidently grab it thinking the berries will turn red.

Asimina angustifolia ★★

Slimleaf pawpaw, dwarf pawpaw

Syn. *Asimina longifolia*
Annonaceae (custard-apple family)

HABITAT & RANGE dry pinelands and roadside sand scrub in northern Florida and adjacent Georgia and Alabama **ZONES** 8–9 **SOIL** sandy, very well drained **LIGHT** sun.
DESCRIPTION Deciduous clump-forming shrub 3–4 ft. tall, with leaves in flattened rows, one on each side of the twig. Leaves alternate, simple, leathery, narrow to 1 in. wide, 4–6 in. long, turning yellow in late autumn. Flowers very showy, three outer petals white and 1–3 in. long, three inner petals shorter and white to pink or maroon, fragrant, blooming in early spring. Fruits are fleshy potato-like berries, 2–4 in. long, with several flattened seeds in a fragrant pulp, maturing in late summer.
PROPAGATION seeds, scarified **LANDSCAPE USES** specimens for beauty and curiosity **EASE OF CULTIVATION** moderate **AVAILABILITY** infrequent.
NOTES Slimleaf pawpaw is one of several species known collectively as the Florida pawpaws. This group differs from woodland pawpaw, which has maroon-brown flowers and is found widely throughout the Southeast. The Florida pawpaws grow mainly in northern Florida in sand scrub and pine forests, and are so beautiful that they should be wider known and grown. They are found in deep well-drained sandy soil, and should be cultivated on a raised sand pile to accommodate an initial 14 in. taproot in full sun. All pawpaws can serve as the larval food plant for the zebra swallowtail butterfly.

A new nursery propagates the Florida pawpaws (Pietro's PawPaws, contact *1bushwoman@embarqmail.com*) and advises to shade them the first year and keep them watered. They are most unusual and stunningly beautiful in flower.

Other Florida pawpaws to try are woolly pawpaw (*A. incana*), scrub pawpaw (*A. obovata*), and netleaf pawpaw (*A. reticulata*).

Bejaria racemosa ★★

Tar-flower

Syn. *Befaria racemosa*
Ericaceae (heath family)

HABITAT & RANGE pine flatwoods and sand scrub along the outer coastal plain of eastern Georgia and Peninsular Florida **ZONES** 8–9 **SOIL** moist, sandy, well drained **LIGHT** sun.
DESCRIPTION Evergreen, clumping shrub to 8 ft. tall, twigs hairy. Leaves alternate, simple, elongate-ovate, 2–4 in. long. Flowers very showy, fragrant, white, the petals gelatinous-sticky, about 2 in. wide, on long stalks. Fruits are small round pods, maturing in autumn.
PROPAGATION seeds **LANDSCAPE USES** specimens, also in mixed shrub borders **EASE OF CULTIVATION** moderate **AVAILABILITY** infrequent.
NOTES This strikingly showy plant, rarely seen in the wild or in cultivation, has beautiful flowers; then you are surprised that they are sticky to the touch (hence the common name),

Aronia arbutifolia, fruit

Asimina angustifolia

Bejaria racemosa

Callicarpa americana and gray catbird

Calycanthus floridus

presumably to trap insects before they can damage the floral parts. The shrub is somewhat azalea-like overall, but has seven petals and twelve to fourteen stamens, not five as is typical of azaleas. Tar-flower would mix in well with other coastal shrubs.

Callicarpa americana ★★

Beautyberry

Lamiaceae (mint family), formerly in Verbenaceae (verbena family)

HABITAT & RANGE rocky and sandy woods, thickets, and waste areas throughout the Southeast, just reaching into southeastern Virginia **ZONES** 7–9 **SOIL** moist to dryish, well drained **LIGHT** part sun to sun.

DESCRIPTION Deciduous, clump forming, fast-growing, informal shrub to 10 ft. tall. Leaves opposite, simple, toothed, 2–4 in. long, tapering at each end, turning yellowish in autumn. Flowers not very showy, small, pink, clustered around the leaves, blooming in summer. Fruits showy, bright purple, 1/4-in. berries in dense clusters, maturing in late summer to early autumn, highly regarded by birds.

PROPAGATION seeds; cuttings **LANDSCAPE USES** specimens, groupings, massing, mixed borders, naturalized in woodlands or thickets **EASE OF CULTIVATION** easy **AVAILABILITY** common.

NOTES The main reason to grow this ungainly shrub is for the extraordinary show of gaudy purple berries in late summer and early autumn, persisting until the birds eat them, which is usually very quickly. The plant will soon grow into an unmanageable thicket, but may be cut back hard every year—even to the ground—and it will flower and make fruits on new growth. Because the berries drop off readily when branches are cut, beautyberry is not desirable for flower arrangements. It self-sows readily, and seedlings should be removed from the garden. Despite these issues, it's always a treat to see beautyberry's colorful fruits (especially in someone else's yard). Gray catbird is one of many birds that eat the fruits. *Callicarpa americana* var. *lactea*, an elegant, heavily fruiting white-berried form, is also available.

Calycanthus floridus ★★★

Sweet-shrub, Carolina allspice

Calycanthaceae (sweet-shrub family)

HABITAT & RANGE moist to dry woodlands and streambanks throughout the Southeast, but not in the outer coastal plain from Virginia to Peninsular Florida **ZONES** 6–8 **SOIL** moist, well drained **LIGHT** shade to part sun.

DESCRIPTION Deciduous colony-forming shrub to 8–10 ft. tall, the brittle brown twigs spicy when scratched. Leaves opposite, simple, ovate, not toothed, with tapering tip, 2–6 in. long, turning butter yellow in autumn. Flowers showy, maroon, uniquely spicy-fragrant, about 1 in. long, with numerous narrow petals, one flower per leaf pair, lasting two days, blooming in spring. Fruits are nonsplitting fibrous pods with several 1/2 in. brown seeds, maturing in late summer, eaten by rodents.

PROPAGATION seeds, cold stratified; root suckers; stem cuttings are difficult **LANDSCAPE USES** specimens, foundations, hedges, mixed borders, naturalized in woodland **EASE OF CULTIVATION** easy **AVAILABILITY** common.

NOTES This species is one of the most widely recognized shrubs in the Southeast, often affectionately known as "bubbie bush," perhaps because rural ladies would place a few flowers in their bosoms as perfume. It is usually seen as an informal, slowly spreading shrub, but tight clumps can be found. What is fun is to show children the seedpods on woodland walks and point out pods must be chewed open by mice in winter.

The cultivar 'Athens' has attractive greenish yellow flowers with intense fragrance.

There is some confusion about the presence of the wonderful scent: it's produced only on the first day the flower opens, and then only in the evening, to attract small beetles. The flower then closes, to open again the next day, but odorless. Some specimens are more fragrant than others, so check before you buy. I look forward to this enjoyable aromatic experience each spring. Without the pleasant odor, you have just another broad-leaved shrub that can get lost in the woods.

Ceanothus americana ★★

New Jersey tea

Rhamnaceae (madder family)

HABITAT & RANGE dry woods and woodland borders, roadsides, and rocky areas around granitic outcrops throughout the Southeast **ZONES** 4–9 **SOIL** moist to dryish, well drained **LIGHT** part sun to sun.

DESCRIPTION Deciduous, delicate shrub to 3–4 ft. tall, often massed on sunny roadbanks. Leaves alternate, simple, narrowly ovate, with three main veins from the base, toothed, 1–3 in. long. Flowers small, white, in showy dense clusters, blooming in summer. Fruits are small, round, 1/4 in. brownish dry pods shed in autumn, borne on interesting little pedestals that remain into winter.

PROPAGATION seeds, cold stratified for one month; cuttings in summer **LANDSCAPE USES** massing **EASE OF CULTIVATION** moderate **AVAILABILITY** frequent.

NOTES New Jersey tea was used as a tea substitute during the Revolutionary War and may have other medicinal qualities. It is best grown in full sun in well-drained soil. Although it's tricky to establish, it may be worth the effort in situations where a long-flowering, low-growing plant is wanted. The leaf texture and fruit stalks are distinctive.

Ceanothus americana

Cephalanthus occidentalis ★★

Buttonbush

Rubiaceae (madder family)

HABITAT & RANGE swamps, marshes, and lake margins throughout the Southeast **ZONES** 4–9 **SOIL** moist, well drained to wet, or standing water **LIGHT** part sun to sun.

DESCRIPTION Deciduous, informal rounded clump-forming shrub to 6–8 ft. tall and wide. Leaves normally in whorls of three or four (but may be just opposite), simple, ovate, 3–6 in. long, turning yellowish in autumn. Flowers small, white, in dense round 2-in. clusters with the styles sticking out like a pincushion, blooming profusely in summer. Fruits are 1/4 in. long, dry nutlets disassociating from the round mass in late summer or persisting through the winter, eaten by waterfowl.

Cephalanthus occidentalis

PROPAGATION seeds, without pretreatment; cuttings in summer **LANDSCAPE USES** specimens, borders, edges of large ponds, wetlands **EASE OF CULTIVATION** easy **AVAILABILITY** frequent.

NOTES Buttonbush is a big green bush, but the Sputnik-like flower spheres are amazing butterfly magnets. In the larger landscape, or near a pond, actually growing in standing water, the bush is worth the space. It can be cut to the ground to rejuvenate when it becomes overgrown.

Clethra alnifolia 

Coastal sweet-pepperbush, summersweet

Clethraceae (clethra family)

Clethra alnifolia

HABITAT & RANGE shrubby pocosins, swamp forests, and pine barrens in the coastal plain throughout the Southeast and in some Piedmont locations in Alabama and Georgia **ZONES** 5–9 **SOIL** moist, well drained to wet **LIGHT** part sun to sun.

DESCRIPTION Deciduous, colony-forming shrub 3–10 ft. tall. Leaves alternate, simple, elongate-ovate, toothed, 2–3 in. long, turning rich yellow in fall. Flowers showy, white, very fragrant, ½ in. wide, in dense spikes, blooming in summer. Fruits nonshowy, small, ⅛ in. round pods, maturing in autumn.

PROPAGATION seeds, without pretreatment; cuttings in summer; division of clump **LANDSCAPE USES** specimens, accents, borders, massing **EASE OF CULTIVATION** easy **AVAILABILITY** common.

NOTES No garden should be without coastal sweet-pepperbush. In summer, the fragrance is glorious, attracting butterflies readily. The upright growth, bright white flowers, and handsome foliage make for multiple uses. The shrub does not spread rapidly and is quite manageable.

Pink forms are especially nice, the brightest being 'Ruby Spice'. 'Hummingbird' grows wider than tall, is very twiggy and floriferous, but tends to fall over. A less floppy selection is 'Sixteen Candles'. For something a little unusual, try 'Creel's Calico' with white-spotted leaves.

Downy sweet-pepperbush (*C. tomentosa*, syn. *C. alnifolia* var. *tomentosa*) is very similar to coastal sweet-pepperbush and quite heat tolerant, with hairier leaves and twigs. It has named cultivars 'Cottondale', with exceptionally long flower stalks, and 'Woodlander's Sarah', with strongly white-spotted leaves.

Mountain sweet-pepperbush (*C. acuminata*) grows in rich moist woods and rock outcrops throughout the southern Appalachian Mountains. It gets taller than coastal sweet-pepperbush, reaching 10–15 ft. tall, and the 6-in. leaves are long pointed with equally good yellow fall color.

Clethra acuminata

The real treat here is the brown peeling bark on stout trunks of this strongly clump-forming robust shrub. It likes a cooler climate (Zones 6–7) and may be shy to grow well in the warmer parts of Zone 8, but is worth a try.

Clinopodium coccineum

Clinopodium coccineum ★★★

Scarlet calamint, scarlet wild basil

Syn. *Calamintha coccinea, Satureja coccinea*
Lamiaceae (mint family)

HABITAT & RANGE sandhills and pine flatwoods in three distinct regions of the coastal plain—eastern Georgia, central Peninsular Florida, and the Florida Panhandle to Mississippi **ZONES** 7–9 **SOIL** sandy, well drained **LIGHT** sun.
DESCRIPTION Evergreen, open, airy subshrub to 3 ft. tall, with thin, much-branched stems. Leaves opposite, simple, narrow-elongate, ½ in. long. Flowers showy, scarlet, 1–2 in. long, with two lips (sagelike), blooming in spring and peaking in summer, sporadically later. Fruits inconspicuous tiny dark seeds, ripening continuously.
PROPAGATION seeds; cuttings in summer; division of clump **LANDSCAPE USES** specimens or masses in open sandy soils, naturalized in sandy sites **EASE OF CULTIVATION** easy **AVAILABILITY** common.
NOTES This beautiful open, delicate shrub blooms almost continuously and requires full sun and perfect drainage, preferably a sand pile or open, gravelly rock garden setting. It mixes well with other delicate shrubs and should be more widely grown as the flowers are striking. The scientific name has gone through many changes—just call it scarlet calamint. 'Amber Blush' and 'Ohoopee Yellow' are lovely yellow-flowered cultivars.

Related to this species, but looking quite different, is Georgia calamint (*C. georgianum*), a smaller shrub 1–2 ft. tall, with smaller rounded minty-smelling leaves and abundant bright pink rosemary-like flowers in late summer. Found on the coastal plain and Piedmont from southeastern North Carolina to Mississippi, it would be wonderful in a rock garden setting or in a raised bed with other small slow-growing plants.

Clinopodium georgianum

Conradina canescens ★★

False rosemary

Syn. *Calamintha canescens, Conradina puberula*
Lamiaceae (mint family)

HABITAT & RANGE dunes and sandy scrub in the coastal plain from western Panhandle Florida to Mississippi **ZONES** 7–9 **SOIL** dryish, well drained **LIGHT** sun.
DESCRIPTION Dwarf evergreen subshrub 1–3 ft. tall, stems, leaves, and flower buds more or less densely hairy. Leaves opposite, simple, very narrow and needlelike, gray-green, less than 1 in. long. Flowers showy, lavender-pink and white with purple spots, two-lipped like rosemary, less than 1 in. long, clustered with the leaves on the upper stems, blooming in summer. Fruits small, nutlike, hidden in the sepal cups, maturing continuously.
PROPAGATION seeds; cuttings **LANDSCAPE USES** specimens or massing in rock gardens, raised beds, containers, or naturalized in sandy soils **EASE OF CULTIVATION** easy **AVAILABILITY** frequent.
NOTES This minty-fragrant gray-leaved plant needs very sandy, well-drained soil in full sun. Water regularly when newly planted. Once established, it's drought tolerant. Although the plant is somewhat short-lived, it's easy to propagate by cuttings. Perhaps the fragrant foliage could be used *en masse* to repel rabbits or even deer.

Conradina canescens

Cumberland rosemary (*C. verticillata*) is more of a low, trailing shrub that roots at the nodes. It comes from the Cumberland Plateau of northeastern Tennessee and adjacent Kentucky.

Cornus amomum ★★

Silky dogwood

Cornaceae (dogwood family)

HABITAT & RANGE margins of streams, shores, and floodplains throughout most of the Southeast, rare in southern Georgia and Mississippi **ZONES** 4–8 **SOIL** moist, well drained, to wet **LIGHT** part sun to sun.
DESCRIPTION Deciduous colony-forming shrub to 20 ft. tall, new branches rusty or silvery hairy. Leaves opposite, simple, broadly ovate, nontoothed, veins arching towards the tip (arcuate), 3–5 in. long, lower surface with reddish hair, turning yellowish or reddish in fall. Flowers small, white, four-petaled, in somewhat dense flat-topped clusters at the tips of the leafy branches, blooming in early summer. Fruits showy, ¼ in. round fleshy drupes, porcelain-blue, in clusters, maturing in late summer, readily eaten by birds.
PROPAGATION seeds, cold stratified for three months; softwood cuttings in summer; hardwood cuttings in winter **LANDSCAPE USES** masses, naturalized in wetlands **EASE OF CULTIVATION** easy **AVAILABILITY** common.

Cornus amomum

NOTES Silky dogwood is a common riverbank and floodplain species throughout the region, and has been successfully used in restorations of streams and wetlands for erosion control, in heavy wet soils, and in rain gardens. It is a handsome foliage plant (the flowers are short-lived) for these informal settings where it can spread and hold the soil. In a formal landscape setting, it will become too large and spread too vigorously. The fruits are attractive and long-effective. This very adaptable shrub is good for naturalizing and for wildlife use.

Corylus americana ★★

American hazelnut

Betulaceae (birch family)

HABITAT & RANGE rich to rocky woodlands, moist thickets, and floodplains throughout the Upper Southeast, absent (or rare) in the coastal plain south of South Carolina **ZONES** 4–8 **SOIL** moist, well drained **LIGHT** part shade to sun.
DESCRIPTION Deciduous clump-forming shrub to 18 ft. tall, with many stout stems emerging from the crown and forming a large rounded thicket. Leaves alternate, simple, broadly ovate, soft-hairy, toothed, 3–6 in. long, turning weakly yellow in autumn. Flowers tiny, male and female flowers on the same plant, males in elongate dangling catkins, females near the leaves, blooming in late winter or very early spring well before the leaves. Fruits are ¾ in. rounded brown nuts enclosed in leafy bracts, maturing in late summer.
PROPAGATION seeds, cold stratified for three months **LANDSCAPE USES** informal specimens, masses, naturalized in woodlands, kept as a native understory shrub **EASE OF CULTIVATION** easy **AVAILABILITY** frequent.
NOTES I love to see the catkins bloom on hazelnuts and other members of the birch family (alders, birches) in late winter. They dangle and remind me of fuzzy worms. For a few

Corylus americana

Croton alabamensis, male flower

Croton alabamensis, leaves

weeks, they are showy, then the leaves emerge and everything goes green and leafy for the summer. Use American hazelnut informally in the woodland garden or edge for its texture and wildlife food. Don't expect any nuts for yourself.

Beaked hazelnut (*C. cornuta*) forms low-growing colonies in rocky woods. It is amusing to find the nuts that have elongated bracts forming a tube.

Tag alder (*Alnus serrulata*), a related species in the birch family, grows along virtually every stream and pond margin in the Southeast. It forms large multitrunked clumps up to 15 ft. tall with handsome dark green wavy-edged leaves. It is great as an erosion control plant and wildlife food. Its early spring catkins are wonderful in bloom for a week, and the twigs with their dark red buds are worth a look in winter. Keep this shrub along natural stream edges at least to hold the banks.

Croton alabamensis ★★★★

Alabama croton

Euphorbiaceae (spurge family)

HABITAT & RANGE rocky, dry to moist limestone glades, bluffs, and woodlands, known only from central Alabama **ZONES** 6–8 **SOIL** moist, well drained, less acidic **LIGHT** part shade to part sun.

DESCRIPTION Semievergreen informal shrub 8–10 ft. tall. Leaves alternate, simple, elongate, not toothed, 2–5 in. long, apple-green, uniquely covered with silvery scales beneath, many turning a distinctive bright orange in autumn and persisting into winter. Flowers small, showy, yellow, sexes separate but grouped together in clusters, blooming in late winter or early spring. Fruits small, silvery, three-seeded pods 3/8 in. wide, maturing in autumn.

PROPAGATION seed, cold stratified for three months; cuttings root well in autumn **LANDSCAPE USES** specimens, accents, groupings, borders **EASE OF CULTIVATION** easy **AVAILABILITY** infrequent.

NOTES This distinctive and as yet under-promoted species needs more friends in the gardening world. It is a gem for the garden and should be used more. Its unique orange fall foliage is exciting, and the silvery winter leaves, twigs, and buds are subtle but beautiful. Alabama croton is easy to propagate and grow. It can be left alone or pruned informally. Everyone should have one. You have to try new things.

Diervilla sessilifolia ★★

Southern bush-honeysuckle

Caprifoliaceae (honeysuckle family)

HABITAT & RANGE rocky open places and streambanks in the mountains of the southern Appalachians into the Cumberland Plateau of northern Alabama **ZONES** 4–7 **SOIL** moist, well drained **LIGHT** part sun to sun.

DESCRIPTION Deciduous clump-forming or suckering shrub to 5 ft. tall. Leaves opposite, simple, ovate, wavy-toothed, dark green with reddish veins, 3–6 in. long, turning somewhat reddish purple in autumn. Flowers showy, sulfur yellow, tubular, ¾ in. long, in small clusters at branch tips, blooming in summer. Fruits not showy, elongated pods, maturing in autumn.

PROPAGATION cuttings in summer **LANDSCAPE USES** groupings, borders **EASE OF CULTIVATION** easy **AVAILABILITY** frequent.

NOTES While at its best in the mountains above 3000 ft., southern bush-honeysuckle can be tried in Zone 7. It makes a surprisingly neat and tidy shrub with little pruning in the sun, and the flowers are delightful.

Dirca palustris ★★★

Leatherwood

Thymeliaceae (mezereum family)

HABITAT & RANGE rich woods and bottomlands, usually less acidic soils, mostly in the Piedmont and mountains throughout the Southeast **ZONES** 4–8 **SOIL** moist, well drained **LIGHT** shade to part sun.

DESCRIPTION Deciduous twiggy shrub 8–10 ft. tall, the brown twigs very flexible and appearing jointed with enlarged nodes. Leaves alternate, simple, not toothed, broadly ovate to rounded, 2–3 in. long, turning yellow in autumn. Flowers small, yellow, in clusters of two or three, appearing before the leaves, blooming in early spring. Fruits not showy, ¼ in. round greenish yellow berrylike drupes, maturing in early summer.

PROPAGATION seeds, cold stratified for three months **LANDSCAPE USES** specimens, groupings, naturalized in woodlands, kept as an understory shrub **EASE OF CULTIVATION** easy **AVAILABILITY** infrequent.

NOTES Leatherwood is my favorite shrub. It is not showy, ever, but its charm will captivate you in subtle ways. It is always rare throughout the East. I like the unusual early spring flowers, tiny though they are, and the delicate frill around the new leaves. The mature leaves seem indistinctive, but their shape is unique, subtly angled while being rounded. The little greenish fruits come and go without much notice—a colony of plants can form around a fertile parent plant. The form of the shrub can be upright and formal, handsome if you will, or one-sided and scraggly. The fall color is a nice yellow. The twigs are unusually jointed and very flexible, being used as ties or thongs by American Indians. I love to show people that the twigs can be tied into knots right on the plant where they will grow on that way for years. That is weird.

Diervilla sessilifolia

Dirca palustris

Elliottia racemosa ★★★

Georgia plume

Ericaceae (heath family)

HABITAT & RANGE sandy woods in the coastal plain of east-central Georgia **ZONES** 6–8 **SOIL** moist, well drained **LIGHT** part sun to sun.

DESCRIPTION Deciduous shrub to 15 ft. tall, forming large colonies in sandy soil. Leaves alternate, simple, elongate 2–5 in. long, turning reddish purple in autumn. Flowers showy, white, 1 in. long, slightly fragrant, arranged in elongate plumes 1 ft. long or longer, conspicuously displayed from the top of the plant, blooming in summer. Fruits are small nonshowy pods with tiny seeds, rarely formed, maturing in autumn.

PROPAGATION seeds, cold stratified for one month; sprouts from root pieces **LANDSCAPE USES** specimens, borders **EASE OF CULTIVATION** difficult **AVAILABILITY** rare.

NOTES Georgia plume is one of our most striking native plants, a delight to see in the wild where colonies can cover many square yards. Butterflies love the distinctive flowers. This shrub is a challenge to grow, but it can perform well in cultivation. A notable example grows at Brookgreen Gardens, Murrell's Inlet, South Carolina, and others have grown and flowered well as far north as the Arnold Arboretum near Boston.

Elliottia racemosa

Epigea repens ★★

Trailing arbutus

Ericaceae (heath family)

HABITAT & RANGE eroded rocky banks and dryish woods in all provinces throughout most of Virginia, the Carolinas, the mountains of Georgia, east Tennessee and Alabama, rare elsewhere **ZONES** 3–9 **SOIL** moist to dryish, very well drained, very acidic **LIGHT** part shade to part sun.

DESCRIPTION Evergreen shrublet with ground-hugging creeping stems, forming dense mats. Leaves alternate, simple, elongate-oval, gray-green, leathery, 1–2 in. long. Flowers delicate, pink, ½ in. long, in clusters, very fragrant, blooming in very early spring. Fruits nonshowy dry pods maturing in summer.

PROPAGATION seeds; cuttings **LANDSCAPE USES** specimens, groundcover **EASE OF CULTIVATION** very difficult **AVAILABILITY** rare.

NOTES Trailing arbutus is one of the most beloved American spring wildflowers, blooming as early as a warm day in February in coastal areas. It is often seen in dryish

Epigea repens (right) *and Pyxidanthera brevifolia* (left)

heath-dominated mountains forests and eastward in dry rocky woods along eroded road banks and trail edges, or in thin organic soil over rocks and roots under blueberries (*Vaccinium* spp.) and mountain laurel (*Kalmia latifolia*). Notoriously difficult to propagate and grow, and nearly impossible to transplant, it has delicate roots that are easily disturbed and overwatered. It probably has a close mycorrhizal (fungal) relationship that requires exacting conditions. Try container-grown plants in very acidic, very well drained organic soil. Don't forget to go looking for them early; you will never forget your first encounter.

Erythrina herbacea ★★★

Coral-bean

Fabaceae (legume family)

HABITAT & RANGE sandy woods, dune edges, and sandy open areas in the coastal plain from extreme southeastern North Carolina to Florida and Texas **ZONES** 7–10 **SOIL** sandy, well drained **LIGHT** part sun to sun.

DESCRIPTION Deciduous prickly herb or shrub 3–12 ft. tall, taller and woodier farther south, growing from a large woody rootstock. Leaves alternate, compound, 6–8 in. long, bright green, each with three leaflets broadly lobed with tapering tips (pandurate). Flowers very showy, bright scarlet, tubular, 2–3 in. long, borne on erect stems, blooming from late spring through summer. Fruits are dry bean-pods 3–6 in. long, opening to reveal showy red seeds, maturing in late summer, drying well.

PROPAGATION seeds, scarified; mature stem cuttings **LANDSCAPE USES** specimens, clumps, sunny borders, naturalized in open sandy woods **EASE OF CULTIVATION** easy **AVAILABILITY** frequent.

NOTES Coral-bean is a gorgeous plant, unique in its growth form, reminiscent of its more tropical *Erythrina* relatives that can grow to large trees. It is an herbaceous perennial in the Carolinas in Zones 7–8, becoming a woody shrub further south and west into Texas. It is late to sprout in spring, so don't give up too soon. The flowers are most unusual for a member of the bean family, being tubular and attractive to hummingbirds. You may open the flowers, count the stamens, and convince yourself it's a legume, but the bean pods give it away. The uniquely lobed leaves, similar to those of the wild potato vine (*Ipomoea pandurata*), make it a cinch to identify. Coral-bean does not love shade, but will grow and flower in open woods. In open sandy soil, it makes a stunning clump 3–6 ft. across. In bloom, it's a showstopper.

This species hybridizes with the subtropical *E. crista-galli* to form *E.* ×*bidwillii*, a robust, clump-forming shrub that produces the most beautiful and abundant bright red flowers all summer in a sunny location. This natural hybrid is easily propagated from 3-in. stem cuttings and is hardy in Zone 7 with heavy mulch. Not as hardy when container grown.

Erythrina herbacea

Eubotrys racemosa

Euonymus americanus

Eubotrys racemosa ★★

Coastal fetterbush, sweetbells leucothoe

Syn. *Leucothoe racemosa*
Ericaceae (heath family)

HABITAT & RANGE moist to wet swamps, ponds, and thickets, mostly in the Piedmont and coastal plain throughout the Southeast, absent from Tennessee and northwestern Mississippi **ZONES** 5–9 **SOIL** moist, well drained to wet **LIGHT** part sun to sun.
DESCRIPTION Semievergreen thicket-forming shrub to 10 ft. tall. Leaves alternate, simple, elongate, 1–3 in. long, tardily deciduous, turning reddish purple in autumn. Flowers numerous, white-pinkish, waxy, ½ in. long, tubular urn-shaped, hanging downward on striking one-sided spikes, in masses on naked stems, blooming in late spring. Fruits are nonshowy dry pods maturing in autumn.
PROPAGATION seeds; division of clump **LANDSCAPE USES** specimens, informal borders, naturalized in damp areas in woods or thickets **EASE OF CULTIVATION** easy **AVAILABILITY** frequent.
NOTES It is a glorious sight to see the delicate one-sided spikes of flowers suddenly appear amid a tangle of naked stems. This species is the epitome of an informal shrub, so place it where it can grow and roam a little. It may have limited garden appeal, but it's one of my favorite flowers, and it should be grown more.

Mountain fetterbush (*E. recurva*, syn. *Leucothoe recurva*) is its attractive highland cousin, blooming in early spring at higher elevations in rocky acidic woods, roadbanks, and outcrops. The leaves are longer and the flowers are on shorter, distinctly curved spikes. Mountain fetterbush is a much more compact shrub than coastal fetterbush is and has a more colorful fall showing of bright purple-red leaves. The problem is, it's difficult to grow warmer than Zone 7.

Euonymus americanus ★★½

Strawberry-bush, hearts-a-burstin'

Celastraceae (bittersweet family)

HABITAT & RANGE moist to dry woods throughout the Southeast **ZONES** 6–9 **SOIL** moist, well drained **LIGHT** part shade to part sun.
DESCRIPTION Deciduous clump-forming to straggling shrub 3–8 ft. tall, younger stems bright green. Leaves opposite, simple, elongate, finely toothed, rich green, 1–2 in. long, tardily deciduous, turning yellowish in autumn. Flowers flat, five-parted, subtle yellowish green, ½ in. wide, in clusters of one to three on stalks at each leaf, blooming in spring. Fruits are striking, 1-in. round fleshy pods, bright red and knobby, splitting open to reveal attractive orange seeds, maturing in early autumn, relished by birds.
PROPAGATION seeds, warm stratified, then cold stratified; cuttings; division **LANDSCAPE USES** specimens, groupings, borders, naturalized in informal woodlands, kept as an understory shrub **EASE OF CULTIVATION** easy **AVAILABILITY** common.
NOTES This well-known and adaptable species has a place in every woodland garden. There is no more wonderful sight in the fall to show people, especially children, than the ripening fruits of hearts-a-burstin'. They are mouthwatering, but inedible. The plant is a popular one among those establishing a native woodland garden because it's so easy to grow and it provides attractive green stems for winter interest. Just plant and forget—until fall. A lush shrub in part sun will usually put on an amazing show. You can see the relationship with the showy bittersweet vine (*Celastrus scandens*). The showy red pods of heart-a-burstin' pop open to reveal bright orange seeds—what fun. And then that's about it.

Fothergilla gardenii ★★★★

Dwarf fothergilla, coastal witch-alder

Hamamelidaceae (witchhazel family)

HABITAT & RANGE wet savannahs and shrubby pocosins, uncommon in the coastal plain from North Carolina to Alabama, rarer in Georgia **ZONES** 5–8 **SOIL** moist, well drained to seasonally wet **LIGHT** part sun to sun.
DESCRIPTION Deciduous, slowly spreading shrub 2–5 ft. tall, with hairy twigs and fat buds. Leaves alternate, solitary, ovate, 1–3 in. long, coarsely toothed or lobed, hairy, turning magnificent shades of red, orange, and yellow in autumn. Flowers showy, brushlike with conspicuous white stamens, clustered at stem tips, 1 in. long, blooming in spring. Fruits are interesting, hard, ½ in. fat pods, in clusters, maturing in late autumn, slowly drying and opening, then suddenly spitting their four shiny black seeds several feet.
PROPAGATION seeds are tricky, warm stratification followed by cold stratification; cuttings are easy; division of suckering clumps **LANDSCAPE USES** specimens, groupings, accents, mixed borders, foundation plantings **EASE OF CULTIVATION** easy **AVAILABILITY** common.
NOTES Dwarf fothergilla is one of our showiest shrubs with year-round appeal. The clean white early spring flowers are delicate and delightful, the summer foliage is handsome, and the fall color is one of the best, if not the best, especially when grown in sun. The plants are neat, upright, and adaptable, mix well with other deciduous shrubs, and rarely

Fothergilla major

Fothergilla 'Mount Airy'

Hamamelis virginiana

become too large. Plant several in your garden for viewing from different vantages.

'Blue Mist' and Woodlanders Selections (as a group) have handsome blue-green foliage in summer.

The large fothergilla or mountain witch-alder (*F. major*) is never common, but is found scattered in the Piedmont and mountains from North Carolina to Alabama. It is larger than dwarf fothergilla in all ways, growing to 10 ft. tall. It forms slowly spreading clumps in rocky, dryish deciduous woods with mountain laurel (*Kalmia latifolia*) and other acid-loving plants. It should be displayed in a prime partly sunny spot.

Fothergilla 'Mount Airy', which is widely available and vigorous, grows to 5–8 ft. tall (it can be cut back hard), has larger flower clusters and consistently good fall color, even in Zone 8.

Hamamelis virginiana ★★★

Witchhazel

Hamamelidaceae (witchhazel family)

HABITAT & RANGE dry to moist woods throughout the Southeast **ZONES** 3–8 **SOIL** moist, well drained **LIGHT** shade to part sun.

DESCRIPTION Deciduous large shrub or small tree 8–30 ft. tall, normally multitrunked with wide-spreading canopy, twigs show a zigzag pattern due to enlarged leaf ledges, and the little round flower buds are in threes. Leaves alternate, simple, widely ovate with rounded or wavy lobelike teeth, 3–5 in. long, turning yellow in autumn. Flowers showy, yellow, each about 1 in. across, with four narrow spidery petals, blooming late autumn into early winter. Fruits are interesting, noticeable, hard, ½-in. fat pods, in clusters of one to three, maturing in late autumn, slowly drying and opening, then suddenly spitting their four shiny black seeds several feet.

PROPAGATION seeds difficult, warm stratified then cold stratified; cuttings root with difficulty **LANDSCAPE USES** specimens, informal groupings in woodland gardens, kept as an understory shrub **EASE OF CULTIVATION** easy **AVAILABILITY** infrequent, check identity

NOTES Witchhazel is a well-known large shrub. I hesitate to call it a tree, even though it can become taller than most shrubs, because it always seems to be multitrunked. This plant is the source of the astringent ointment called "witchhazel." I look forward every year to the shrub's weird behavior of blooming so late, into November and December. It is the last native species to bloom, however, and not the first. That would be its Ozarkian cousin, vernal witchhazel (*H. vernalis*), which is not very ornamental with its tiny pale flowers blooming in mid to late winter. Annoyingly, I have found that the majority of the specimens I purchase as *H.*

virginiana turn out to be *H. vernalis*. I think this is because *H. vernalis* is much easier to grow from seed and has gotten into the nursery trade by mistake. My advice is to purchase witchhazel only when in bloom in the late fall. The good fall color, cute flowers, strange fruits, interesting winter twigs, and overall folklore still make witchhazel an all-around worthwhile plant.

Hydrangea radiata

Hydrangea quercifolia ★★★★

Oakleaf hydrangea

Hydrangeaceae (hydrangea family), formerly in Saxifragaceae (saxifrage family)

HABITAT & RANGE dryish wooded ravines, hammocks, moist forests, and thickets, mostly throughout Mississippi and Alabama, just getting into adjacent states **ZONES** 6–9 **SOIL** moist, well drained **LIGHT** part shade to part sun.

DESCRIPTION Coarse, deciduous, clump-forming to slowly spreading shrub 4–8 ft. tall, older stems have nice exfoliating bark, buds very large. Leaves very large, opposite, simple, deeply three- to seven-lobed and toothed, 3–8 in. long, turning exquisite rich yellow-purple-red in autumn. Flowers very showy, white, 1 in. wide, in elongate clusters 1 ft. long, blooming in early summer, old flowers persistent. Fruits are not showy, small dry pods, maturing in autumn, fruiting clusters persisting with some dried flowers through winter.

PROPAGATION seeds, without treatment; cuttings in summer not kept too wet; suckering sprouts **LANDSCAPE USES** specimens, borders, foundations, groupings, massing, screening, kept as an understory shrub **EASE OF CULTIVATION** easy **AVAILABILITY** common.

NOTES Although oakleaf hydrangea is one of the great Southeastern plants, it's very large and coarse and thus difficult to use in the small landscape. It is heat and cold tolerant, drought tolerant, and can take full sun (but does not like it), but is a little tricky to be established. This shrub has one of the finest floral and fall color displays of any plant—on a grand scale. The winter stems are interesting with flaking bark and large tip buds. Oakleaf hydrangea is certainly one of the best all-around deciduous shrubs. No garden should be without it. It is especially suited to the informal more open woodland garden.

Many wonderful selections improve on the species. 'Alice' is very floriferous and robust, becoming 12 ft. tall and wide; Snow Queen™ is large and robust with many doubled flowers held strongly upright on larger inflorescences. For smaller plants in the 3–4 ft. range that keep all the other good qualities of the species, use 'Sikes Dwarf' or 'Pee Wee'.

Hydrangea quercifolia

Hydrangea radiata ★★★

Silverleaf hydrangea

Hydrangeaceae (hydrangea family), formerly in Saxifragaceae (saxifrage family)

HABITAT & RANGE forests, streambanks, and roadside seeps, confined to the mountains of the Carolinas and Tennessee **ZONES** 4–8 **SOIL** moist, well drained **LIGHT** part shade to part sun.

DESCRIPTION Deciduous, twiggy clump-forming shrub 3–8 ft. tall, with older bark peeling off. Leaves opposite, simple, toothed, ovate, white beneath, 3–6 in. long. Flowers tiny, white, aggregated into showy, large flat-topped clusters 3–6

in. wide, with five to fifteen sterile flowers around the margin, blooming in early summer. Fruits are tiny dry pods in a flat cluster, maturing in autumn, persisting into winter.
PROPAGATION seeds; cuttings **LANDSCAPE USES** specimens, borders, groupings in informal woodlands, kept as an understory shrub **EASE OF CULTIVATION** easy **AVAILABILITY** frequent.
NOTES Silverleaf hydrangea is a wonder shrub, blooming well in shade in summer when little else is in flower in the woods. It would even be attractive as a native "lace-cap" in the semiformal shrub border with other hydrangeas in part sun. The white undersides of the leaves are striking when the wind flutters them over. This shrub can be thinned out annually for a neater look or cut back heavily if it becomes too overgrown. Be aware that it blooms only on new growth that comes from the previous year's woody stems, so don't cut every stem completely to the ground. This hydrangea grows beautifully well beyond the mountains. A new cultivar, 'Samantha', looks promising and has dense heads of sterile florets like 'Annabelle'.

More widespread throughout the Southeast is wild hydrangea (*H. arborescens*), which is widely grown in woodland gardens, and is heat and drought tolerant. It is identical to silverleaf hydrangea except the leaves are green underneath and there are usually no sterile flowers or up to two.

If you like the big "mophead" types of hydrangeas, *H. arborescens* 'Annabelle' is one that has performed well. It has a huge ball of all-sterile white flowers, and is best in part sun with afternoon shade.

Hypericum prolificum ★★★

Shrubby St.-John's-wort

Hypericaceae (St.-John's-wort family)

HABITAT & RANGE meadows, seeps, and rocky woods throughout the Piedmont of the Southeast, including central Tennessee, rarer in the mountains and coastal plain, very rare in extreme northern Georgia, not found in Florida **ZONES** 4–8 **SOIL** moist, well drained **LIGHT** part sun to sun.
DESCRIPTION Deciduous clump-forming or spreading shrub 1–4 ft. tall. Leaves opposite, simple, narrow, 1–2 in. long. Flowers showy, yellow 3⁄4 in. wide, five petals with a "powder-puff" of yellow stamens, in few-flowered clusters covering the shrub, blooming all summer. Fruits are interesting but not showy, 1⁄2-in. cone-shaped dry pods, maturing all summer.
PROPAGATION seeds, without pretreatment; cuttings **LANDSCAPE USES** specimens, accents, borders, foundations, masses **EASE OF CULTIVATION** easy **AVAILABILITY** frequent.
NOTES The shrubby St.-John's-worts as a group are being used more widely because they are sun loving, compact, long-flowering, and adaptable to a wide range of conditions. This particular species is especially floriferous and compact, but will be more open and spreading in shade. The bright yellow flowers and light green leaves offer a diversion from the all-green foundation shrubs usually seen, especially in municipal plantings. The several species are difficult to identify, and many look alike, so I have indicated here some representative species.

Hypericum frondosum 'Sunburst', a widely grown cultivar of a species that occurs wild throughout Alabama and middle Tennessee, grows to 4 ft. tall and has narrow bluish green leaves (a few may remain through mild winters) and very large and showy yellow powder-puff flowers. It has been excellent in full-sun rain gardens.

A low shrubby hypericum with four petals instead of five is called St. Andrew's-cross (*H. stragulum*). This small

Hypericum prolificum

Hypericum stragulum

clumping shrublet to 1 ft. tall is becoming more widely available. It seems adaptable as an informal groundcover growing in rock gardens or dryish, shady to sunny sites.

Other species of interest are Atlantic St.-John's-wort (*H. tenuifolium*, syn. *H. reductum*), and sandhill St.-John's-wort (*H. lloydii*), which have small, needlelike leaves to 1 in. long and make dense clumping shrublets that are useful in dryish rock gardens and well-drained sunny sites.

Dense St.-John's-wort (*Hypericum densiflorum*) is a taller shrub (2–4 ft.), with handsome yellow flowers, flaking bark, and narrow leaves. It thrives in moist, well-drained soils in part shade or sun. The dwarf Buckley's St.-John's-wort (*H. buckleyi*) is from the high mountains and may not be heat tolerant warmer than Zone 7.

Ilex glabra

Ilex decidua

Ilex glabra 

Inkberry, little gallberry

Aquifoliaceae (holly family)

HABITAT & RANGE savannas, shrubby pocosins, pine flatwoods, and wetland margins in sandy soils, found in the coastal plain throughout the Southeast **ZONES** 6–10 **SOIL** moist, well drained, acidic **LIGHT** part sun to sun.

DESCRIPTION Evergreen, slowly spreading shrub 6–8 ft. tall with greenish stems. Leaves alternate, solitary, elongate, with a few nonsharp teeth near the tip, 2–3 in. long. Flowers not showy, white, 1/4 in. wide, male and female on separate plants, blooming in summer. Fruits conspicuous black berries (drupes) 1/4 in. in diameter, maturing in autumn, persisting through the winter.

PROPAGATION seeds are difficult, requiring a long time; cuttings root readily with a rooting hormone **LANDSCAPE USES** foundation plantings, masses, borders, hedges **EASE OF CULTIVATION** easy **AVAILABILITY** common.

NOTES Inkberry holly is never flashy, but it's a good evergreen for foundations and other situations where you want something green year-round that you don't have to prune (much). Even the black berries are not very showy, and like the fruit of most hollies, are not eaten by birds until they have to, in late winter. Many cultivars are available, 'Shamrock' being one most favored for its shiny dark green leaves. Some cultivars (for example, 'Compacta') have been reportedly somewhat deer resistant.

Big gallberry (*H. coriacea*) is similar with very shiny dark green leaves and black berries, but it spreads much more readily and would be good only for naturalizing.

Ilex verticillata 

Winterberry

Aquifoliaceae (holly family)

HABITAT & RANGE bogs, swamps, pocosins, and shrubby wetlands throughout the Southeast, absent in southern Georgia and Peninsular Florida **ZONES** 3–9 **SOIL** moist, well drained to seasonally wet **LIGHT** part sun to sun.

DESCRIPTION Deciduous clump-forming shrub, becoming suckering and slowly spreading, to 10 ft. tall. Leaves alternate, simple, broadly elongate, toothed but not sharp, 2–3 in. long, turning a rich yellow in autumn. Flowers not particularly showy, though when in full bloom can be noticeable and even fragrant, white, 1/4 in. wide, male and female on separate plants, in tight clusters near the leaves on new growth, blooming in late spring. Fruits are very showy, bright red berries (drupes) 1/4 in. in diameter, tightly clustered, maturing in late summer but persisting all winter if not eaten by birds.

PROPAGATION seeds require two seasons of dormancy; cuttings root readily **LANDSCAPE USES** specimens, mixed borders, masses, naturalized in damp woods **EASE OF CULTIVATION** easy **AVAILABILITY** common.

NOTES This deciduous holly is one of the best winter-color fruits because the bright berries are quite visible and persistent. A delight in the garden, they are not highly nutritious to birds, so are eaten later in winter as emergency food. A flock of cedar-waxwings, robins, or cardinals can wipe them out quickly, though. You need a territorial male mockingbird to "adopt" your holly bush and keep other birds at bay so you can enjoy the berries for a long time.

There have been many, many cultivar selections of winterberry holly throughout eastern North America (see Dirr 2009 for a listing). Because the male and female flowers of hollies are on separate plants—one of nature's little tricks, it's not a conspiracy on the part of nurseries to sell more plants—two compatible selections must be planted near each other. Herein lies a quandary. Different male and female selections (from different parts of the country) bloom at slightly different times. You must match the two plants to have a compatible pair. So, the female selection 'Winter Red' (one of the most widely planted) requires 'Southern Gentleman' as a pollinator. The dwarf, large-berried 'Red Sprite' (my favorite) can use 'Jim Dandy' or 'Apollo', but not 'Southern Gentleman'. And so on. If your holly does not make fruits, it's either a male, or a female without a male, and unless you know what you bought, you can't easily determine its "mate." Keep those labels, folks! You only need one male for six (or so) females, and it can be in the background. It can be cut back regularly and will regrow to bloom on new growth each year.

Possumhaw (*I. decidua*) is very similar to winterberry. It has lustrous red berries that can look great with the tardily deciduous shiny green leaves. 'Warren's Red' is one of the best selections, growing to 10 ft. or so in part shade to sun. Use 'Red Escort' as the male pollinator.

Ilex verticillata in a rare heavy snowfall in Charlotte

Ilex vomitoria 

Yaupon

Aquifoliaceae (holly family)

HABITAT & RANGE maritime and other dry sandy forests, generally throughout the coastal plain of the Southeast **ZONES** 6–9 **SOIL** moist, well drained to wet **LIGHT** part sun to sun.
DESCRIPTION Evergreen clump-forming shrub (often suckering) to 20 ft. tall, though rarely attaining that height in cultivation. Leaves alternate, simple, shiny dark green with regular rounded teeth (crenations), 1–1½ in. long. Flowers not particularly showy, white, ¼ in. wide, male and female on separate plants, in tight clusters near the leaves, blooming in late spring. Fruits are very showy, bright candy-apple red translucent berries (drupes) ¼ in. in diameter, tightly clustered, maturing in late summer but persisting all winter if not eaten by birds.
PROPAGATION seeds require two seasons of dormancy; cuttings **LANDSCAPE USES** specimens, foundation plantings, groupings, hedges, screening, naturalized in moist woodlands **EASE OF CULTIVATION** easy **AVAILABILITY** common.
NOTES In the wild, yaupon holly is typical of the maritime forest of the Carolinas outer coastal plain, moving farther inland further South. In cultivation, yaupon is a dense upright shrub with dark evergreen foliage. It makes an excellent landscape specimen for many uses and has an informal appearance. It is very attractive in winter with snow and red berries on green leaves. It can be pruned and shaped as a hedge. It is a superior evergreen shrub.

Yaupon holly has been used for decades as the widely known (and perhaps overused) dwarf yaupons 'Nana' and 'Schillings'. The latter has reddish purple new growth. Both cultivars are landscape staples as low-growing, tough, never-needs-pruning foundation and border shrubs. These great plants are heat and drought tolerant, very dense and slow-growing (4–5 ft. tall in twenty years), and regrow well when damaged or cut back. There are no finer native evergreen dwarf plants. They look like dwarf boxwoods, but never do anything. No (or few) berries, no perceptible change. That is fine. Keep on using them.

A weeping form, 'Pendula', grows to 20 ft. tall. It needs a male pollinator nearby, like all hollies. There are none named so get one when you get your berried specimen—plants within the nursery row without berries will likely be males.

Other cultivars include 'Yawkey' with yellow berries and 'Virginia Dare' with orange. A new cultivar, 'Carolina Ruby', can produce a very heavy crop of luscious red berries. Use 'Schillings' as a male pollinator for all.

There are two other evergreen hollies in the Southeast you may want to grow. Although they are not closely related to yaupon, they are mentioned here for convenience. Dahoon holly (*Ilex cassine*) is actually a small tree to 30 ft. It has broad leaves 1–4 in. long and red berries. Its close relative is myrtle holly (*I. myrtifolia*) with very narrow needle-like leaves. I love these trees and their red berries, and wish they were more commonly seen in cultivation.

Ilex vomitoria

Ilex myrtifolia

Illicium floridanum ★★★★

Florida anisetree

Illiciaceae (star-anise family)

HABITAT & RANGE acidic woods, swamps, and sandy floodplains, mostly in southern Alabama and Mississippi **ZONES** 6–9 **SOIL** moist, well drained **LIGHT** part shade to part sun.
DESCRIPTION Evergreen colony-forming shrub 8–12 ft. tall. Leaves alternate, simple, not toothed, elongate-pointed, leathery, dark green, 2–6 in. long, very aromatic, on red-purple leaf stalks (petioles). Flowers showy, dark red to maroon, 1–2 in. wide, with twenty to thirty long strap-shaped petals, in clusters near the leaves, blooming in spring and sporadically after. Fruits are noticeable dry pods, 1 in. in diameter, with eight to ten pointed one-seeded chambers arranged in a circle, maturing in autumn.
PROPAGATION seeds, without pretreatment; cuttings **LANDSCAPE USES** specimens, accents, foundations, borders, screening, informal clumps in woodland gardens **EASE OF CULTIVATION** easy **AVAILABILITY** common.
NOTES Since coming into cultivation in the 1980s, this species has become a very popular plant with several good cultivar selections. In its best form, the attractive but informal shrub resembles a broad-leaved rhododendron from a distance and may serve that textural niche as the Southeastern summers become warmer and drier (and more inhospitable to rhododendrons). The strong anise odor from the brushed or crushed foliage is unpleasant to some, acceptable to others. The flowers are unusual, with many radiating narrow petals resembling a flattened octopus. As an adaptable, heat- and drought-tolerant, shade-loving, broad-leaved native evergreen, it is, in my opinion, one of the very best. In sun, the leaves yellow and scorch. This shrub can be sheared, pruned, thinned, cut back, and otherwise manipulated, and it grows back rapidly anytime. Give it some room (8 ft. or more in width) and let it flourish. It is closely related to the Chinese star-anise spice.

Some of the cultivars are more compact. Among the more refined cultivars are 'Head's Compact', 'Pebblebrook', and 'Woodland Ruby'. 'Pink Frost' has light-green variegated leaves. My favorite is 'Halley's Comet', which reblooms almost continuously. I find them all outstanding.

The small anisetree (*I. parviflorum*) has been used for years as a heat-tolerant, tough evergreen foliage plant for light shade. The flowers are small, yellowish, and inconspicuous; the leaves are a lighter green and lack the red leaf stalks of Florida anisetree, and are held at a 45-degree angle to the stem. The plant spreads slowly by suckering and gets to about 8 ft. tall and wide. Some suggest the anise odor is more pleasant than that of Florida anisetree. Small anisetree is somewhat rare in wet areas of central Peninsular Florida. The cultivar 'Forest Green' has darker leaves; 'Florida Sunshine' has striking yellow foliage.

Illicium floridanum

Itea virginica

Itea virginica 

Virginia-willow, sweetspire

Iteaceae (sweetspire family), formerly in Saxifragaceae (saxifrage family)

HABITAT & RANGE swamps, wet woodlands, and streams banks throughout the Southeast, rarer in the highest mountains **ZONES** 6–9 **SOIL** moist, well drained to wet **LIGHT** shade to sun.
DESCRIPTION Semievergreen rhizomatous-spreading shrub 3–5 ft. tall, with gracefully arching stems. Leaves alternate, simple, toothed, elongate, 1–4 in. long, turning beautiful

colors of red-purple-orange-yellow in autumn. Flowers showy, white, ½ in. wide, in elongate-curved spikes, abundantly blooming in late spring. Fruits are dry pods, not showy, maturing in autumn.
PROPAGATION seeds, without pretreatment; cuttings; suckers **LANDSCAPE USES** specimens, borders, masses, naturalized in informal woodlands and wet areas **EASE OF CULTIVATION** easy **AVAILABILITY** common.
NOTES This species produces a wonderful informal, spreading colony that needs to be managed to keep within bounds. I have seen it growing well in a South Carolina swamp in 3 ft. of floodwater and then in average-to-dry landscape settings. The flowers are fragrant and graceful, attracting butterflies. The exciting fall colors can be a uniform red-purple or a mixture of colors depending on sun exposure (more sun means more purple). It could be grown well in difficult situations, such as wet areas, moderate shade, and uneven terrain in a ravine, ditch, or rain garden. Its only fault is that it spreads relentlessly and is difficult to eradicate completely if you ever want to. While the species is good, the selected cultivars are improvements.

One of the best selections is 'Henry's Garnet', with consistent reddish purple fall color. Newer forms are sold under the trade name Little Henry and as 'Merlot', both of which are dwarf and compact (to 3 ft.) in Zone 8. More upright and with good yellow-orange-red fall color in Zone 7 is 'Saturnalia'.

Kalmia latifolia ★★

Mountain laurel

Ericaceae (heath family)

HABITAT & RANGE acidic forests, sandy bluffs, and rocky streambanks throughout the Southeast, rarer in the outer Atlantic Coastal Plain and Florida Panhandle **ZONES** 4–8 **SOIL** moist, acidic, well drained **LIGHT** part shade to part sun.
DESCRIPTION Evergreen twiggy shrub 3–15 ft. tall, often with gnarled and crooked branches in age. Leaves alternate, simple, leathery, elongate-ovate, 2–3 in. long, older leaves turning bright yellow in autumn. Flowers very showy, pink to white, cup-shaped about 1 in. wide, in dense, rounded clusters, blooming in late spring. Fruits are small round dry pods in clusters, maturing in autumn.
PROPAGATION seeds, without pretreatment; cuttings are very difficult **LANDSCAPE USES** specimens, accents, borders, foundation plantings, informal woodland specimens, kept as an understory shrub **EASE OF CULTIVATION** moderate **AVAILABILITY** common.
NOTES This beautiful and well-known shrub is ubiquitous in dry rocky forest and exposed roadbanks in the mountains, but surprisingly it grows on rocky outcrops in the Piedmont and sandy bluffs in the outer coastal plain as well. As a young specimen, it makes a symmetrical green bush covered in flower clusters. The flowers are among the most beautiful and offer observers the unique experience of watching each stamen (there are ten of them) spring up and throw its pollen when triggered from its notch in the petal bowl. Mountain laurel does not grow in clay soils and

Kalmia latifolia

Kalmia angustifolia

must have perfect drainage and even moisture until well established. Planting in a loamy soil with leaf compost, perhaps in a raised bed, would be an ideal way to grow it. Older plants can be rejuvenated by cutting them to near the ground just after spring bloom. The foliage is supposedly poisonous to livestock and humans.

Many cultivars exist with various flower colors and patterns, but few of them thrive in the heat warmer than Zone 7. The best bet would be 'Ostbo Red'. The dwarf forms 'Minuet' and 'Elf' have done reasonably well in Charlotte in well-drained beds.

Northern sheep-laurel (*K. angustifolia*) is a spreading compact evergreen shrub of moist sites in the mountains. Its narrow leaves and smaller rich pink flowers are lovely. The coastal form Southern sheepkill (*K. carolina*) is a better choice south of Zone 7.

Leiophyllum buxifolium ★★

Sand-myrtle

Syn. *Kalmia buxifolia*
Ericaceae (heath family)

HABITAT & RANGE locally abundant in rocky outcrops in the mountains and again in sandy soil in the coastal plain, both in the Carolinas, but very rare in between, also found in New Jersey pine barrens **ZONES** 6–8 **SOIL** moist, acidic, well drained **LIGHT** part sun to sun.
DESCRIPTION Evergreen shrub, erect or prostrate and spreading, 6–24 in. tall. Leaves alternate, simple, small, 1⁄8 to 1⁄2 in. long, oblong-oval, shiny dark green, crowded on the plant. Flowers small but showy, pink in bud then white, 1⁄4 in. wide, in dense clusters, blooming in late spring. Fruits tiny, dry pods, maturing in autumn.
PROPAGATION seeds, without pretreatment; cuttings **LANDSCAPE USES** groundcover in rock gardens or terrace beds **EASE OF CULTIVATION** difficult **AVAILABILITY** infrequent.
NOTES Sand-myrtle is a remarkably variable plant, growing as a creeping groundcover or upright shrub, in what seem to be the harshest bare-rock habitats in the high mountains or in the coastal plain in sand. It is very beautiful and dainty, engendering thoughts of elfin landscapes. Although it's tricky to establish, it's worth trying.

Leucothoe axillaris ★★

Coastal doghobble

Ericaceae (heath family)

HABITAT & RANGE pocosins, swamp forests, and moist slopes, mostly in the coastal plain throughout the Southeast, also in the Piedmont of South Carolina **ZONES** 5–9 **SOIL** moist, acidic, well drained **LIGHT** part shade to part sun.
DESCRIPTION Evergreen spreading-suckering shrub 2–3 ft. tall, with arching stems and zigzag spreading branches. Leaves alternate, simple, sharply toothed, shiny green, narrowly elongate, tapering, 2–4 in. long, with leaf stalks (petioles) to 2⁄5 in. long. Flowers showy, white, tubular urn-shaped (urceolate), densely arranged in elongate 1- to 2-in.-long spikes hanging somewhat out-of-view under each leaf, blooming in spring. Fruits are round dry pods, not showy, maturing in autumn.
PROPAGATION seeds, without pretreatment; cuttings; suckers
LANDSCAPE USES groupings, massing, foundation plantings, groundcover, filler under taller plants, kept as an understory

Leiophyllum buxifolium

Leucothoe fontanesiana

shrub **EASE OF CULTIVATION** moderate **AVAILABILITY** common. **NOTES** Coastal doghobble is a common understory shrub in damp habitats where it spreads and forms a tangle of stems (hence makes the hunting dogs stumble). The glossy green foliage is beautiful, and the arching stems are graceful. This shrub is tricky to establish, as keeping it too wet or too dry stresses it. The nifty flowers are hidden under the leaves, so use it on a bank or hill where you walk along and can look somewhat up under the foliage. Some forms have knockout fall colors (even though they hold their leaves) of deep burgundy when grown in some sun. I have used it best in natural settings along creekbanks in loamy soil.

Mountain doghobble (*L. fontanesiana*) is identical except for technical differences such as longer flower spikes, longer petioles, and longer leaves. It is very common in damp and boggy habitats throughout the southern mountains, so does not do well south of Zone 7. This very attractive plant begs to be used to create evergreen masses along streambanks and slopes in the woodland garden. It can get up to 5 ft. tall, creating thickets that thwart even larger dogs.

Lindera benzoin ★★★

Spicebush

Lauraceae (laurel family)

HABITAT & RANGE rich alluvial forests, floodplains, and swamps, in less-acidic soils, throughout the Southeast, absent from southeastern Georgia and Peninsular Florida **ZONES** 5–8 **SOIL** moist, well drained **LIGHT** shade to part sun. **DESCRIPTION** Deciduous shrub to 12 ft. tall, with upwardly spreading branches, aromatic twigs, and round flower buds evident in winter. Leaves alternate, simple, oblong-ovate, 2–4 in. long, turning rich butter yellow in autumn. Flowers small, yellow, in tight clusters, appearing before the leaves, male and female on separate plants, blooming in spring. Fruits are showy bright red elongate berries to ½ in. long, maturing in late summer to autumn, quickly eaten by many birds.

PROPAGATION seeds, warm stratified, then cold stratified for three months **LANDSCAPE USES** specimens, borders, groupings, naturalized in informal woodlands, kept as an understory shrub. **EASE OF CULTIVATION** easy **AVAILABILITY** common. **NOTES** While it's a harbinger-of-spring bloomer with year-round appeal, its showiest time is when the red fruits are ripening against the yellow foliage in early autumn. I like to scratch and sniff the twigs in winter and enjoy the conspicuous yellow-green buds. This open shrub can harbor wildflowers beneath it. A native tea can be made from the buds, leaves, and flowers, and the dried seeds were used like nutmegs. The plant is a larval food source for spicebush swallowtail and eastern swallowtail butterflies.

Lyonia lucida ★★★

Shining fetterbush

Ericaceae (heath family)

HABITAT & RANGE pocosins, wet woodlands, swamps, and peaty wetlands in the coastal plain throughout the Southeast **ZONES** 7–9 **SOIL** moist, acidic, well drained **LIGHT** part shade to sun.

DESCRIPTION Evergreen, loosely clumped or slowly spreading-suckering upright shrub 5–6 ft. tall, with arching, green, distinctly angled stems. Leaves alternate, simple, leathery, shiny dark green, elongate-ovate, 1–3 in. long. Flowers

Lindera benzoin

Lyonia lucida

showy, white to pink, 3⁄8 in. long, tubular (urceolate), in small clusters with each leaf such that the entire branch appears to have flowers hanging down from it, blooming in early summer. Fruits are round pods, not showy, maturing in autumn.
PROPAGATION seeds, without pretreatment; cuttings; suckers **LANDSCAPE USES** specimens, accents, borders, informal groupings, naturalized in sunny wooded gardens, especially wetlands **EASE OF CULTIVATION** easy **AVAILABILITY** infrequent.
NOTES This handsome evergreen shrub has an excellent growth form, showy flowers, and an overall outstanding performance at UNC Charlotte Botanical Gardens. I consider it one of the best evergreen native shrubs we have tried. It is relatively unknown in cultivation and needs wider testing.

Lyonia mariana, one of several species called stagger-

Lyonia mariana

Myrica cerifera with myrtle warbler

Myrica pumila

bush, is a spreading deciduous shrub to 4 ft. found in sandy dry to moist pine woodlands along the Atlantic Coastal Plain. It has large attractive tubular white flowers to 1 in. long produced in spring on leafless stems.

Myrica cerifera ★★★

Common wax-myrtle, southern bayberry

Syn. *Morella cerifera*
Myricaceae (bayberry family)

HABITAT & RANGE pocosins, brackish marshes, swales between sand dunes, and other wetland habitats, primarily in coastal plain throughout the Southeast and in the lower Piedmont of Alabama and Mississippi **ZONES** 8–10 **SOIL** moist, well drained to sandy and dryish **LIGHT** part sun to sun.

DESCRIPTION Evergreen clump-forming to suckering fragrant shrubs to 10 ft. or more tall, open and spreading in age. Leaves alternate, simple, elongate, broadest and toothed nearer the tip, dark green, fragrant when crushed, 2–3 in. long. Flowers tiny, not showy, clustered along the older stems, male and female on separate plants, blooming in spring. Fruits are berrylike, 1/4 in. wide, clustered, bluish wax-coated, fragrant, maturing in late summer.

PROPAGATION seeds; cuttings **LANDSCAPE USES** specimens, borders, foundations, screening **EASE OF CULTIVATION** easy **AVAILABILITY** common.

NOTES Wax-myrtle and its northern counterpart, bayberry, have long been known as sources of fragrant candle-wax produced by melting the fruits. The plants grow variably and can be dense and leafy or thin and semievergreen. They can thus make airy, see-through screenings. They also can be pruned anytime; limbing up exposes the interesting bark on the trunks. In coastal windswept sites, they can grow into old, gnarled specimens. Only the females produce the interesting blue waxy fruits, so males must be part of the mixed planting. I love everything about this plant: the forms, the fragrant foliage, and the colorful fruits. When old, the shrubs can be cut down to regrow. They are best in full sun and well-drained soil. Keep them watered during drought to reduce stress. They can self-sow into moist cultivated habitats and woodlands.

Of great value in landscaping is dwarf wax-myrtle (*M. pumila*), sometimes treated as a variety of *M. cerifera*. It is smaller, to 4 ft., with smaller leaves, and tends to spread more, but the evergreen foliage, delicate texture, and smaller size make it very useful in the landscape. I am very fond of the selections 'Don's Dwarf' and 'Don's Other Dwarf' as truly low-growing specimens.

Nevieusia alabamensis ★★½

Alabama bridalwreath

Rosaceae (rose family)

HABITAT & RANGE thin limestone soils in woods and on bluffs, mostly in the northern half of Alabama and adjacent Tennessee and Georgia **ZONES** 4–8 **SOIL** moist, well drained **LIGHT** shade to sun.

DESCRIPTION Deciduous, spreading shrub to 5–8 ft. tall, with gracefully arching stems and running rhizomes. Leaves alternate, simple, doubly toothed, ovate, 1–2 in. long, turning yellow in autumn. Flowers with numerous showy white stamens, lacking petals, 1/2 in. wide, in wispy branching clusters along the stems, appearing just before the leaves, blooming in early spring. Fruits are dry achenes (1-seeded), not showy, maturing in later summer.

PROPAGATION softwood cuttings; separate rhizomes from the clump **LANDSCAPE USES** specimens, borders, naturalized in woodland gardens or edges **EASE OF CULTIVATION** easy **AVAILABILITY** infrequent.

NOTES This rare shrub is hardly seen in gardens. It is interesting for its early spring flowers, but then it fades into the background much like an old-fashioned spiraea. I think it's worth having somewhere in the larger garden, even as a curiosity to challenge your gardening friends to identify. A substantial border of it is well established at the Coker Arboretum in Chapel Hill, North Carolina. Don't plant near other desirable shrubs as it spreads relentlessly. You will not have to baby it once it's established.

Nevieusia alabamensis

Opuntia humifusa ★★

Eastern prickly-pear cactus

Syn. *Opuntia compressa*
Cactaceae (cactus family)

HABITAT & RANGE sandy or rocky open habitats and dunes, scattered literally throughout the Southeast **ZONES** 7–10 **SOIL** well drained **LIGHT** part sun to sun.
DESCRIPTION Leafless creeping succulent with jointed green pads to 1 ft. tall, stems with tufts of irritating hair-like spines (glochids) and sometimes longer spines up to 1-in. long. Leaves absent. Flowers very showy, large, yellow, with many petals, 2–3 in. broad, opening only in bright sunlight, blooming in early summer and sporadically later. Fruits are juicy red berries with many black seeds, maturing periodically.
PROPAGATION division of the clump into single (or more) pads, which root readily **LANDSCAPE USES** clumping specimens in rock gardens, on ledges, in raised beds, in rocky places **EASE OF CULTIVATION** easy **AVAILABILITY** common.
NOTES Did you realize that cacti were woody shrubs? They do have a tough internal woody skeleton, and their "branches" are these weird flattened pads. Eastern prickly-pear cactus would work well in a dry, open setting. The flowers are attractive, and the pads stand out in winter. The spines and hairs are barbed and can be very irritating, so you don't want to grow this cactus where you might accidentally touch it when weeding, or where a child might fall on it or try to play with it. One source has suggested this species is the best cactus for green roofs in the Southeast.

The sand-bur prickly-pear (*O. pusilla*, syn. *O. drummondii*) is a smaller clumping plant with longer spines and short, cylindrical pads (forming a nestlike colony) that disarticulate and attach to your shoes and skin—and to your fingers when you try to remove them. These two species hybridize, and the result is no less interesting or painful. Watch out.

Osmanthus americanus ★★

Wild olive, devilwood

Oleaceae (olive family)

HABITAT & RANGE swamp margins and borders of streams, bluffs, and sandy maritime forests in the coastal plain mostly throughout the Southeast **ZONES** 6–9 **SOIL** moist, well drained **LIGHT** part sun to sun.
DESCRIPTION Evergreen nonsuckering shrub or small tree 6–12 ft. tall, developing a loose and open form, producing an extremely hard wood. Leaves opposite, simple, shiny olive green, narrowly oblong widest near the end, 2–4 in. long. Flowers small, not particularly showy nor fragrant, creamy white, in clusters, male and female on separate plants, blooming in summer. Fruits are dark blue-purple one-seeded drupes, ½ in. long, maturing in late summer.
PROPAGATION cuttings **LANDSCAPE USES** borders, foundations, hedges, screening, plantings in informal woodlands **EASE OF CULTIVATION** easy **AVAILABILITY** frequent.
NOTES Wild olive is not commonly seen in landscapes, and it has no single striking trait. Because it is heat tolerant and maintains its green color year-round, in the right setting it would be a very useful background shrub. It would be good to have a dwarf selection. Certainly worth considering.

Philadelphus inodorus ★★

Appalachian mock-orange

Hydrangeaceae (hydrangea family), formerly in Saxifragaceae (saxifrage family)

NATURAL RANGE widely distributed across the Southeast, but not in the Atlantic Coastal Plain or Peninsular Florida **ZONES** 6–8 **SOIL** moist to dry, well drained **LIGHT** part shade to sun.
DESCRIPTION Deciduous informal shrub 4–8 ft. tall, forming a tangled thicket with long-arching stems. Leaves opposite, simple, three-veined, with widely spaced small teeth, ovate 1–3 in. long. Flowers showy, white, 1½–2 in. wide, nonfragrant, in clusters of three on short branches, blooming briefly in late spring. Fruits are dry pyramidal pods, nonshowy, maturing in late summer.
PROPAGATION seeds; cuttings **LANDSCAPE USES** mixed borders, informal shrub in woodlands, on banks **EASE OF CULTIVATION** easy **AVAILABILITY** rare.
NOTES This sprawling, very informal shrub with arching stems produces a good but short-lived crop of very attractive flowers, and then fades away, leaving the interesting peeling bark for winter interest. This mock-orange can flower in full shade, but is much better in some sun.

Physocarpus opulifolius 

Ninebark

Rosaceae (rose family)

HABITAT & RANGE streambanks, thickets, and rock outcrops, usually in less-acidic soil, primarily in the Piedmont and mountains southwestward to Alabama coastal plain, then up into Tennessee and northward **ZONES** 2–8 **SOIL** moist, well drained to occasionally wet **LIGHT** part sun to sun.

Opuntia humifusa

Philadelphus inodorus

Osmanthus americanus

Physocarpus opulifolius

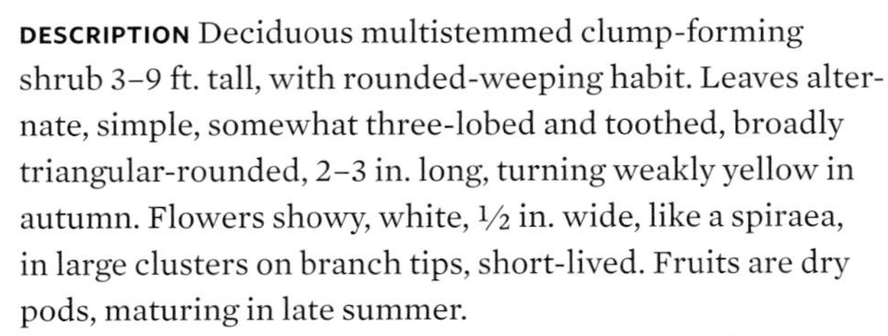

DESCRIPTION Deciduous multistemmed clump-forming shrub 3–9 ft. tall, with rounded-weeping habit. Leaves alternate, simple, somewhat three-lobed and toothed, broadly triangular-rounded, 2–3 in. long, turning weakly yellow in autumn. Flowers showy, white, ½ in. wide, like a spiraea, in large clusters on branch tips, short-lived. Fruits are dry pods, maturing in late summer.

PROPAGATION seeds; cuttings **LANDSCAPE USES** specimens, accents, borders, groupings, informal woodlands **EASE OF CULTIVATION** easy **AVAILABILITY** common.

NOTES Ninebark gets its name from the peeling and shredding bark, in nine layers. The unique growth form is that of a very broad, rounded shrub with branches that gracefully weep nearly to the ground when open-grown. The clusters of flowers are at the tips of the branches. Flower production is brief; the fruits are persistent but not particularly ornamental. Still, it's an interesting and worthwhile shrub through the cooler parts of Zone 8.

Many selections of ninebark are available. A striking black-purple dark-leaved cultivar is often sold under the trade name Diablo. It is not likely to hold its dark color through the hot summer south of Zone 7, but is still a nice shrub with striking white-flower/black-leaf contrast when in spring bloom. The leaves on 'Seward' (trade name

Summer Wine), emerge red and then darken to purple. 'Dart's Gold' has bright yellow leaves.

Ninebark, which looks like a giant mounding-arching spiraea, is related to the familiar garden spiraea; the flowers being practically identical in basic structure. Meadowsweet (*Spiraea alba*) is a northern species with flat-topped clusters of white flowers that barely gets into the North Carolina mountains, so it's probably best in Zones 6 and 7. Hardhack or steeplebush (*S. tomentosa*) is a shrub 4–8 ft. tall with elegant, elongate plumes of pink flowers and grows in various wetland habitats all across the Carolinas and Tennessee. It would be a beautiful and worthwhile plant in a sunny border.

Pieris phillyreifolia ★★

Vine-wickey, climbing fetterbush

Ericaceae (heath family)

HABITAT & RANGE swamp forests in the outer coastal plain from southeastern South Carolina and adjacent Georgia, then skipping to southeastern Georgia and northern Peninsular Florida, then westward to Mississippi **ZONES** 7–8 **SOIL** moist to dryish, well drained **LIGHT** part sun to sun.
DESCRIPTION Evergreen colony-forming shrub to 3 ft. tall, often vinelike. Leaves alternate, simple, narrow with a few teeth near the sharply pointed tip, 1–2 in. long, leathery. Flowers showy, white, tubular, to 1⁄3 in. long, in few-flowered hanging clusters, blooming in late winter. Fruits are dry pods, nonshowy, maturing in late summer.
PROPAGATION seeds; cuttings **LANDSCAPE USES** groundcover, groupings, massing **EASE OF CULTIVATION** moderate **AVAILABILITY** infrequent.

NOTES I like vine-wickey because of its curious behavior of growing underneath the outer bark of pond cypress (*Taxodium ascendens*) trees to emerge up the trunk like a flowering branch of the cypress. In this sense, it's a strange type of vine. I did not treat it with the vines because in the garden it does not seem to want to climb as it does on hammocks in very wet swamps, such as the Okefenokee in southeastern Georgia. In cultivation, I have seen it form exceptional low informal masses of dwarf shrubs. In the UNC Charlotte Botanical Gardens, it has survived for decades in average loamy soil in a very shady woodland site, producing a few flowers, but never spreading much or attempting to climb. You would never know what a secretive life it has.

In stark contrast to vine-wickey is one of our loveliest native shrubs, mountain fetterbush (*Pieris floribunda*), found at high elevations (above 3500 ft.) in just a few mountain counties from southwestern Virginia through North

Pieris floribunda

Prunus angustifolia, flowers

Carolina. It produces stiffly erect clusters of white pendant tubular (urceolate) flowers in early spring on 3- to 6-ft. compact shrubs with handsome dark evergreen leaves. It is best in Zone 7 and colder. Until heat-tolerant strains are developed, try the beautiful little hybrid cultivar *P.* 'Brower's Beauty', a cross between this species and *P. japonica* that is much more easily grown in warmer climates.

Prunus angustifolia ★★

Chickasaw plum

Rosaceae (rose family)

HABITAT & RANGE roadsides, fencerows, and abandoned fields throughout the Southeast **ZONES** 5–9 **SOIL** moist, well drained **LIGHT** sun.

DESCRIPTION Deciduous shrub or small tree to 18 ft. tall, normally shorter, suckering to form dense thickets, stem with 1 in. thorns. Leaves alternate, simple, ovate, toothed, each tooth with a tiny swollen gland on the tip, somewhat folded longitudinally, 1–2½ in. long, turning red-yellow-orange in autumn. Flowers showy, white, small, about ½ in. wide, in clusters of four or five, blooming briefly in early spring. Fruits are small plums, ½–1 in. long, turning yellow to red, maturing in summer, remaining somewhat firm.

PROPAGATION seeds, cold stratified for three months; suckers **LANDSCAPE USES** borders, informal plantings, kept in open natural areas **EASE OF CULTIVATION** easy **AVAILABILITY** frequent.

NOTES This species is one of several native wild plums that are often seen in large dense thickets along roadsides and overgrown fields. Native animals eat the fruits; humans simply make jam out of them. Flowering is brief; the fruits are attractive; the thickets have good fall color. The shrubs are part of the rural American landscape. I may not plant one in the landscaped garden, but I would love to see one nearby.

Hog plum (*P. umbellatus*) is a similar shrub or small tree to 20 ft. or more tall. It grows in dry, sandy woods and is less prone to making dense thickets. It produces a sudden mass of tiny white flowers in earliest spring. The fruits are usually black.

Wild plum (*P. americana*) is a larger shrub or small tree with red plums.

Prunus angustifolia, fruit

Ptelea trifoliata

Ptelea trifoliata ★★

Wafer-ash, hoptree

Rutaceae (citrus family)

HABITAT & RANGE dry, rocky, less acidic woods and bluffs, widely scattered throughout the Southeast **ZONES** 3–9 **SOIL** moist to dry, well drained **LIGHT** part sun to sun.

DESCRIPTION Deciduous shrub or small tree to 18 ft. tall, usually suckering and forming an informal rambling colony, stems pungent when scratched. Leaves alternate, compound with three leaflets (trifoliate), each leaflet 2–5 in. long, turning yellow in autumn. Flowers not particularly showy, yellowish, ½ in. wide, in clusters, very fragrant, flowering in late spring. Fruits are interesting, papery round wafers (winged samaras) with a bulging seed in the middle, about 1 in. wide, hanging in conspicuous clusters, maturing in late summer.

PROPAGATION seeds, cold stratified for three months; cuttings **LANDSCAPE USES** specimens, borders, informal groupings, informal woodlands, as a curiosity **EASE OF CULTIVATION** moderate **AVAILABILITY** infrequent.

NOTES I like wafer-ash. When it's in bloom, you can find it using your nose. It is not a flashy plant, but fun to scratch and smell, and to see the nifty flat fruits, so named because they were used as a substitute for hops. Being in the citrus family explains where this shrub gets its floral and pungent odors. The leaves somewhat resemble those of bladdernut (*Staphylea trifolia*), but the latter are opposite with toothed leaflets. Wafer-ash will make a stronger, more formal shrub in bright light, but does well in shade.

Rhododendron austrinum

Rhododendron alabamense

Rhododendron calendulaceum

Rhododendron austrinum ★★★★

Florida flame azalea

Ericaceae (heath family)

HABITAT & RANGE hammocks, bluffs, and floodplain forests in the Gulf Coastal Plain from southwestern Georgia to Mississippi **ZONES** 6–9 **SOIL** moist, acidic, well drained **LIGHT** part shade to sun.

DESCRIPTION Deciduous clump-forming shrub to 16 ft. tall, crown spreads with age, large flower buds form at stem tips above leaf clusters. Leaves alternate, simple, narrowly oblong, 2–3 in. long, clustering near the tips of older stems, turning red-purple-orange-yellow in autumn. Flowers very showy, orange to yellow, radiating in clusters of seven to fifteen or more at the tips of main stems and branches, opening before or with the leaves, with five stamens longer than the flaring petal tubes, normally not fragrant, blooming in early spring. Fruits are dry, elongate pods (capsules) to 1 in. long, erect in clusters, with many small seeds, maturing in autumn.

PROPAGATION seeds, without pretreatment; stem cuttings are difficult; root cuttings are easier **LANDSCAPE USES** specimens, accents, borders, massing, scattered in woodland gardens, kept as an understory shrub **EASE OF CULTIVATION** easy **AVAILABILITY** common.

NOTES There are no finer flowering shrubs than our native azaleas, often referred to as deciduous azaleas. At least sixteen species originate in the Southeast, more than everywhere else on Earth combined. They are world famous, virtually iconic in the Great Smoky Mountains and along the Blue Ridge Parkway, where as many as four species may grow together.

I selected Florida flame azalea as the main entry for native azaleas, since it's often the first to flower. All the species will be briefly described in alphabetical order as they all have the same basic characteristics. They are all easy to grow, except that those from high elevations may not do well warmer than Zone 7 and certainly should not be planted in full sun in warmer zones. Deciduous azaleas provide year-round interest: flowers, growth form, fall color, and large winter buds. All the shrubs have outstanding fall color, especially when grown in half a day of sun. You also get more flowers when in sun. The flowers attract butterflies, moths, and hummingbirds. Native azaleas are excellent for naturalizing in an open woodland garden or massing in semisunny borders. Be sure to water them the first summer until they are established; thereafter they can take severe droughts by dropping their leaves and going dormant. They must have acidic soil or the leaves will turn yellow-green with dark green veins, a condition termed "chlorotic." Countless hybrids and named cultivars are known (Dirr 2009, Galle 1987, Towe 2004), as most of the species are interfertile and have been favorite subjects of plant breeders worldwide for centuries. In some cases, it may be difficult to find a pure species in cultivation; nevertheless, there are no ugly ones.

Our deciduous native azaleas are not to be confused with the evergreen Japanese azaleas that are mainstays in the Southeastern home landscape as small to large foundation plants and very large border and screening shrubs. These typically have less tubular, much wider flaring flowers, and ten stamens that are shorter than the petals.

For each species that follows, the description gives the height of the species, from tall (8–16 ft.) to medium (4–8 ft.) to short (3–4 ft.); the blooming time, from early spring (March), mid spring (April), or late spring (May) to summer (July–August); flower color, which includes every color except green and blue; whether or not the flower is fragrant; if growth habit is spreading, and origin of the species.

Alabama azalea (*Rhododendron alabamense*), medium, mid spring, white with yellow blotch, fragrant, throughout Alabama and adjacent Tennessee.

Sweet azalea (*R. arborescens*), tall, late spring, white with red stamens, fragrant, throughout the mountains and Piedmont.

Dwarf azalea (*R. atlanticum*), short, mid spring, white to pink, very fragrant, spreading, primarily coastal plain

Florida flame azalea (*R. austrinum*), tall, early to mid spring, orange to yellow, not fragrant, from the coastal plain of southeastern Georgia to Mississippi.

Flame azalea (*R. calendulaceum*), tall, late spring, orange to yellow, not fragrant, Piedmont and mountains.

Piedmont azalea (*R. canescens*), tall, mid spring, white to pink, very fragrant, throughout the Southeast, very rare in North Carolina, absent from Virginia.

Red Hills azalea (*R. colemanii*), tall, late spring, white with yellow blotvch, fragrant, rare, two counties in extreme southern Alabama.

Cumberland azalea (*R. cumberlandense*, syn. *R. bakeri*), short, very late spring, orange to red-orange, not fragrant, mostly high mountains.

May white azalea (*R. eastmanii*), medium, late spring, white with yellow blotch and red stamens, very fragrant, rare, Piedmont and coastal plain of South Carolina.

Oconee azalea (*R. flammeum*, syn. *R. speciosum*), medium, mid spring, red-orange to yellow shades, not fragrant, mostly central Georgia into southwestern South Carolina.

Pinxter-bloom azalea (*R. periclymenoides*, syn. *R. nudiflorum*), tall, early to mid spring, pink shades, very fragrant, throughout much of the Southeast.

Rhododendron canescens

Rhododendron cumberlandense

Rhododendron eastmanii

Rhododendron periclymenoides

Rhododendron vaseyi

Roseshell azalea (*R. prinophyllum*, syn. *R. roseum*), medium, midspring, pink, fragrant, from New England to Virginia and barely tiptoeing into North Carolina's high mountains.

Plumleaf azalea (*R. prunifolium*), tall, summer, orange, not fragrant, a wonderful summer azalea, found in one row of counties on each side of the southern Alabama–Georgia state line.

Hammock-sweet azalea (*R. serrulatum*, often listed as *R. viscosum* var. *serrulatum*), tall, summer to late summer, white, very fragrant, spreading, coastal plain.

Pink-shell azalea (*R. vaseyi*) short, mid spring, pink, not fragrant, charming flowers unique in having seven stamens, endemic to mountains of North Carolina.

Swamp azalea (*R. viscosum*), medium to tall, very late summer, white, very fragrant, an adaptable plant for summer bloom, throughout the Southeast.

Rhododendron catawbiense

Rhododendron maximum ★★

Rose-bay, great-laurel

Ericaceae (heath family)

HABITAT & RANGE woods, slopes, and streambanks, mostly throughout the southern Appalachian Mountains and scattered in the adjacent Piedmont **ZONES** 6–8 **SOIL** moist, acidic, well drained **LIGHT** part shade to part sun.

DESCRIPTION Larger evergreen shrub to 18 ft. tall or taller, branches gnarled and spreading to form dense thickets. Leaves alternate, simple, very leathery, dark green above and rusty below, 4–8 in. long, tapering to a tip, older leaves turn yellow in autumn. Flowers very showy, white to deep pink with greenish spots, the five petals longer than the ten stamens, in dense clusters, mildly fragrant, blooming in early to mid summer. Fruits are fat, dry pods in erect clusters, maturing in autumn.

PROPAGATION seeds, without pretreatment; cuttings difficult; layering **LANDSCAPE USES** specimens, accents, borders, foundation plantings, naturalized in informal woodlands, kept as an understory shrub **EASE OF CULTIVATION** moderate

Rhododendron maximum

Rhododendron 'Maxecat'

Rhododendron minus

AVAILABILITY frequent.

NOTES Rose-bay is a ubiquitous shrub in the mountains, dominant in every moist forest and roadside below 3500 ft. elevation. It grows along streams and steep hillsides and usually forms dense, impenetrable, tangled "rhododendron hells." The soil in these sites is so acidic and the dense thickets so dark that almost nothing grows under the shrubs. It is beautiful in bloom and provides long-term summer displays along the Blue Ridge Parkway. Rose-bay is more heat tolerant than Catawba rhododendron and will grow in the cooler parts of Zone 8. It is widely used in hybridizing to impart heat tolerance. Two good hybrids are 'Judy Spillane' and 'Summer Snow'.

Catawba rhododendron (*R. catawbiense*) is world famous as a cold-hardy purple rhododendron forming vast stretches of early summer bloom on open balds and roadsides in the southern Appalachian Mountains above 3000 ft. It also grows into northeastern Alabama, with a few notable outliers in the Piedmont of North Carolina. Its flowers are lavender-purple, the leaves are smaller and distinctly whiter underneath, and the plant is shorter and denser than rose-bay. It is not heat tolerant, but has been used extensively in hybridizing beautiful cultivars which are adaptable in the cooler areas of Zone 8, especially 'Roseum Elegans' and 'English Roseum'.

One of the best hybrids for native gardens is a cross between *R. maximum* and *R. catawbiense* called 'Maxecat'. It is heat tolerant in Zone 8 and disease resistant. It blooms in late spring and forms a handsome 10-ft. tall lush shrub with reddish stems. This hybrid is a very good choice for the garden, better than either parent.

These large-leaved rhododendrons and their countless hybrids are tricky to grow in clay or sandy soils outside of their cool and moist mountains homes. They prefer loamy soil with a 2- to 3-in. mulch of bark or decomposed organic matter, and perhaps are best grown in raised beds. The extensive fine roots are very shallow, less than 6 in. deep. Watering must be carefully done to allow the plants to establish and then tolerate dry spells.

Rhododendron minus

Piedmont rhododendron

Ericaceae (heath family)

HABITAT & RANGE streambanks, wooded slopes, and rocky outcrops in the mountains and Piedmont from North Carolina to Alabama, with some notable coastal plain outliers **ZONES** 6–8 **SOIL** moist, acidic, well drained **LIGHT** shade to part sun.

DESCRIPTION Evergreen shrub to 12 ft. tall, usually shorter, producing an informal twiggy mounding plant. Leaves alternate, simply, leathery, ovate with pointed tips, 1½–2½ in. long, dark shiny green above and covered with brownish pitted scales below, seen best with a ten-times magnifying lens, older leaves turning red in autumn. Flowers very showy, white to deep pink, with five petals and ten stamens, 1½ in. wide, in few-flowered clusters, variably blooming in early to late spring. Fruits are fat pods in erect clusters, maturing in autumn.

PROPAGATION seeds, without pretreatment; layering **LANDSCAPE USES** specimens, accents, borders, foundation plantings, naturalized in informal woodlands, kept as an understory shrub **EASE OF CULTIVATION** moderate **AVAILABILITY** frequent.

NOTES Included here are plants with slightly larger leaves and flowers sometimes known as *R. carolinianum*. Piedmont rhododendron, also known as dwarf or Carolina rhododendron, makes an exceptional shrub for light shade or part sun, and if you want to try an evergreen rhododendron, this is the one. It is among our most beautiful species, growing into an informal shrub with very showy flowers, though they are not particularly fragrant. They make stunning flower displays in the more formal shrub garden or in an informal woodland setting. They are easier to grow than the large-leaved rhododendrons, more tolerant of various well-drained soils, but still needing mulch and regular watering. Some beautiful hybrid selections are available, including 'Dora Amateis' with white flowers and 'Llenroc' with pink. The 'PJM' hybrids are also very nice, have very good foliage, but tend to bloom in the fall in the Southeast.

Chapman's rhododendron (*R. chapmanii*) is distinctly different from Piedmont rhododendron, though closely related. It is very rare from central Panhandle Florida. The flowers are always pink, slightly more tubular, with 2-in. leaves that are distinctly wavy on the edges and rounded on the tips. I have grown the same plants for over forty years with no problems—the mass gets a bit larger every year, in light shade. One of the great early-blooming adaptable hybrids is 'Chapmanii Wonder'.

Rhus aromatica

Rhus aromatica ★★

Fragrant sumac

Anacardiaceae (cashew family)

HABITAT & RANGE rocky wooded slopes in less-acid soils, widely scattered, but primarily in the Piedmont throughout the Southeast into eastern Tennessee **ZONES** 3–9 **SOIL** moist to dryish, less acid, well drained **LIGHT** part shade to sun.
DESCRIPTION Deciduous, spreading shrub 1–6 ft. tall with sprawling, informal stems that are fragrant when scratched. Leaves not poisonous to touch, alternate, compound with three leaflets, 1–3 in. long, coarsely toothed, turning striking shades of red and yellow in autumn. Flowers tiny, bright yellow, opening from elongate catkin-like buds that are somewhat showy, blooming in late winter to early spring. Fruits are bright red, hairy, round berries (drupes) ¼ in. broad, maturing in autumn and persisting all winter.
PROPAGATION seeds, cold stratified; stem cuttings **LANDSCAPE USES** groundcover, upright informal shrub for specimens, borders, massing, woodland gardens **EASE OF CULTIVATION** moderate **AVAILABILITY** infrequent.
NOTES Fragrant sumac is an interesting plant in that it looks exactly like a diminutive poison oak, but is not poisonous. This supports the fact that sumacs with red berries are never poisonous, only those with white fruits (namely, poison ivy, poison oak, and poison sumac). Even in part sun, the fall color is outstanding, and it's fun to see the winter buds and scratch the twigs and smell the spicy fragrance.

Rhus glabra

The widely available cultivar 'Low Grow' forms a loose open shrub to 8 ft. tall and as wide, but does not spread except maybe a sucker or two. It has beautiful flowers and great fall color but is not low growing.

Smooth sumac (*R. glabra*) is a vigorous spreading coarse shrub of sunny old fields, roadsides, and waste places, growing commonly to 6 ft. tall, but sometimes much taller. It has compound leaves with eleven to twenty elongate-toothed leaflets that are not hairy, and striking plumes of small greenish yellowish flowers followed by 1 ft. clusters of flaming red berries that last well into winter. Birds eat the berries as emergency food; humans use the sour fruits to steep

Rhus typhina

a "sumac-ade." I love this beautiful shrub in fruit and fall color. Because it spreads, plant it only in sunny, dry areas well outside the prime garden. It is excellent for erosion control on roadbanks.

The same can be said for the gigantic staghorn sumac (*R. typhina*) found only in the mountains. Its fall color and red berries are fabulous, but it forms aggressive colonies with stems up to 10 ft. tall and 1 in. in diameter. There are no manageable cultivars of either species.

Rosa carolina ★★

Carolina rose

Rosaceae (rose family)

HABITAT & RANGE dry woods, rocky hillsides, and old fields and pastures, mostly throughout the Southeast **ZONES** 4–9 **SOIL** moist, well drained **LIGHT** sun.

DESCRIPTION Deciduous rhizomatous-spreading shrub to 4 ft. tall, forming dense colonies, stems slender, with sharp, straight prickles. Leaves alternate, compound, with five- or seven-toothed ovate leaflets, turning rusty red-orange in autumn. Flowers large, showy, fragrant, bright pink to lighter shades, with five symmetrical petals and numerous yellow stamens, to 2 in. wide, in clusters of one to three, blooming early summer and sporadically later. Fruits are bright red berrylike rose-hips about 1/4 in. wide, in loose clusters, maturing in late summer and persisting with good color into winter.

PROPAGATION seeds; cuttings **LANDSCAPE USES** massing in open natural areas **EASE OF CULTIVATION** easy **AVAILABILITY** frequent.

NOTES Carolina rose is almost indistinguishable from swamp rose (*R. palustris*), which grows in open wet meadows, has five to nine leaflets, and, diagnostically, curved prickles. These wild roses are charming in flower, but much too aggressive and prickly for the formal garden; they make excellent wild plants for adjacent sunny, dry or wetland natural areas. They are widely used by bees and butterflies in flower and birds in fruit.

Rubus occidentalis ★★

Black raspberry

Rosaceae (rose family)

HABITAT woodlands, clearings, and fencerows throughout the Southeast as far south as northern Georgia through Mississippi **ZONES** 6–8 **SOIL** moist, well drained **LIGHT** sun.

DESCRIPTION Deciduous prickly shrub with arching stems 4–6 ft. long, canes covered with curved prickles, arching and rooting at the tips to form open colonies, stems distinctly whitish with a waxy coating. Leaves alternate, compound, to 6 in. long, with five-toothed leaflets on first-year canes and three-toothed leaflets on second-year flowering branches, turning red in autumn. Flowers showy, white to pinkish, five-petaled with numerous white stamens, in clusters of twelve or more on the tips on flowering

Rosa carolina

Rubus occidentalis

branches, blooming in late spring. Fruits are clusters of tiny units, like "blackberries" except they remove freely from their support podiums (receptacle or torus), often whitish waxy, turning from red to black, maturing in late summer, eaten by birds.
PROPAGATION seeds, without pretreatment; rooted cane-tips **LANDSCAPE USES** informal shrub in gardens, perhaps on trellises, informal spreading masses in open woodland gardens or meadows **EASE OF CULTIVATION** easy **AVAILABILITY** frequent. **NOTES** Picking wild blackberries was a summer treat when I was growing up in the Southeast, although I rarely saw raspberries because they were cool-climate shrubs. This black raspberry is certainly more tolerant of southern heat than the red raspberry. Grow it in open woods, or for more flowering, in at least a half day of sun in woods edge or meadow. You will be delighted with the delicate flowers, fruits, whitish canes, and fall color—and the insects and birds they attract.

Many native animals utilize wild raspberries and blackberries in flower and fruit. While these plants may not be conducive to the formal garden, they may work in an informal border with trellises, or better naturalized in a meadow or woods edge. Treat them as ornamentals and the "thorns" will not hurt as much, as with roses.

Rubus cuneifolius with a zebra swallowtail butterfly

Several groupings of wild species of *Rubus* are available. The raspberries have whitish-waxy stems, often prickly and bristly, and berries with deep dimples when picked. The blackberries usually have coarser prickly green or red-purple stems with fruits that are solid. Only four species of blackberry are native in the Southeast: common blackberry (*R. pensilvanicus*) is the most frequently occurring species, the one seen on roadsides and meadows; sand blackberry (*R. cuneifolius*) is widespread in the coastal plain; and (almost) thornless blackberries (*R. allegheniensis* and *R. canadensis*) are plentiful in the higher mountains. Many other species are naturalized introductions. Blackberries are useful for stabilizing road banks and are valuable to many wildlife species. Many cultivated selections offer better fruit production, and all the blackberries turn from green to red to black as they ripen.

The creeping blackberries, often called dewberries, can make an interesting evergreen groundcover with cute little white flowers and tiny black fruits. They are easy to propagate from stem layering. Common species are northern dewberry (*R. flagellaris*) with several 1½ in. flowers on less-evergreen, hairless stems; southern dewberry (*R. trivialis*) with solitary 1½ in. flowers on hairy stems with evergreen leaves that turn reddish in winter; and swamp dewberry (*R. hispidus*) with smaller 1-in. flowers on slender, nearly thornless stems.

Rubus odoratus ★★★

Flowering raspberry

Rosaceae (rose family)

HABITAT & RANGE rocky woods, rocky borders, and rocky road banks in the Blue Ridge Mountains of North Carolina, Tennessee, and Georgia **ZONES** 5–8 **SOIL** moist, well drained **LIGHT** part sun to sun.

DESCRIPTION Deciduous upright shrub to 5 ft. tall, spreading by creeping rhizomes, forming lush colonies, stems brittle, unarmed (no prickles), not rooting at the tips. Leaves alternate, simple, five-lobed and toothed, large 4–12 in. across, handsome light green, rough textured but not leathery, turning yellow in autumn. Flowers very showy, petals deep rose to magenta, 1–2 in. wide, in loose, open clusters, unopened buds covered with attractive red hairs, with five symmetrical petals and numerous creamy stamens, blooming in summer. Fruits are bright red-purple mushy berries that can be removed from their little podiums, tart and tasty, maturing in summer as the flowers fade.

PROPAGATION seeds; division of rhizomes **LANDSCAPE USES** massing in half-sunny sites, informal borders along mowed lawn, naturalized in bright woodland gardens **EASE OF CULTIVATION** easy **AVAILABILITY** common.

NOTES Flowering raspberry is among our most distinctive and beautiful shrubs, with flowers reminiscent of rich raspberry sherbet. The shrub forms dense patches of striking foliage and flowers that can be controlled by periodic thinning, or mowing around the edges. It is best in bright light or a half-day of sun. It does not grow at all like the prickly cane-forming raspberries of commerce that have compound leaves and produce fruit only on two-year-old prickly stems.

Rubus odoratus

Sambucus canadensis ★★

Elderberry

Adoxaceae (moschatel family), formerly in Caprifoliaceae (honeysuckle family)

HABITAT & RANGE swamp forests, low woods, floodplains, pastures, and ditches throughout the Southeast **ZONES** 4–9 **SOIL** moist, well drained to wet **LIGHT** sun.

DESCRIPTION Deciduous coarse clump-forming or suckering shrub 5–12 ft. tall, with tan twigs distinctly pegged with lenticels (pores). Leaves opposite, compound, to 12 in. long overall, with five to seven leaflets, toothed, turning yellowish in autumn. Flowers small, white, fragrant, ¼ in. wide, very numerous arranged in flat-topped clusters to 8 in. broad, blooming in late spring through summer. Fruits are

Sambucus canadensis, flowers

Sambucus canadensis fruit with brown thrasher

purple, round juicy berries ¼ in. wide, in clusters that are usually abundant and hanging heavily, edible, maturing as the flowers bloom through late summer, relished by birds.
PROPAGATION seeds, warm stratified for two months, cold stratified for three; cuttings; suckers **LANDSCAPE USES** informal specimens, borders, screening, informal open woodlands or sunny natural areas, wet or dry **EASE OF CULTIVATION** easy **AVAILABILITY** common.
NOTES Elderberry is one of the commonest and most familiar shrubs in the Southeast. It grows in ditches and waste areas, and blooms and fruits all summer. It does become large and coarse, appearing overgrown. Humans use the flowers in fritters and salads, and the fruits to make wine, pies, and jams. Elderberry is one of the most important species for a wide variety of wildlife use, and for this reason could warrant a place in a back corner somewhere. While fresh and vibrant in spring, by late summer the plants become more overgrown and worn. They could be cut back hard each year, but keep some woody stems because they sprout forth the new stems that will flower, much like with hydrangeas.

Staphylea trifolia

Cultivar 'Aurea' has red fruits and looks good all seasons; 'Maxima' has much larger flower clusters. Planting two different selections will allow for greater fruit production.

Red elderberry (*S. racemosa*, syn. *S. pubens*) is a more refined shrub, the tan-white flowers are in elongated clusters, and the fruits are striking red. Unfortunately, it only grows well at high elevations (above 3500 ft.) and is not suitable south of a very cool Zone 7.

Staphylea trifolia ★★

Bladdernut

Staphyleaceae (bladdernut family)

HABITAT & RANGE wooded streambanks, swamp forests, and pond margins throughout the Southeast, much less common in the outer coastal plain, absent from Peninsular Florida **ZONES** 4–8 **SOIL** moist, well drained, tolerates flooding **LIGHT** part sun to sun.
DESCRIPTION Deciduous colony-forming dense shrub 8–15 ft. tall, with erect, straight green twigs marked by interesting longitudinal white stripes and with fat winter buds. Leaves opposite, 2–4 in. long, compound with three ovate toothed leaflets, turning yellowish in autumn. Flowers small, greenish white, ⅓ in. long in short pendulous spikes, blooming in spring. Fruits are curious inflated three-chambered pods 1–1½ in. long, turning yellowish green then brown, maturing in autumn.
PROPAGATION seeds, warm stratified, then cold stratified; cuttings root easily **LANDSCAPE USES** massing, naturalized in open woods and edges of wetlands **EASE OF CULTIVATION** easy **AVAILABILITY** frequent.
NOTES Bladdernut is not a flashy plant, but I like the subtle beauty of the white-striped stems, conspicuous green winter buds, cute creamy yellow flowers, attractive leaves, and unusual inflated seedpods. This shrub is good for naturalizing in wetland edges or open informal woodlands.

Symphoricarpos orbiculatus ★★ 

Coralberry, Indian-currant

Caprifoliaceae (honeysuckle family)

HABITAT & RANGE floodplain woods and creekbanks, in less-acidic soils, primarily throughout the Piedmont and mountains of the Southeast **ZONES** 6–9 **SOIL** moist, well drained **LIGHT** part sun to sun.
DESCRIPTION Deciduous or semievergreen shrub, much branched, 2–4 ft. tall, spreading by rhizomes to form a

scattered colony, with hairy twigs. Leaves opposite, simple, oval 1–1½ in. long. Flowers small, white ¼ in. long, clustered near the leaves, blooming in late summer. Fruits are very showy round berries ¼ in. in diameter, purplish red, densely clustered along the stems, persisting and colorful well into winter.

PROPAGATION seeds, warm stratified, then cold stratified; cuttings easy to root; suckers **LANDSCAPE USES** low groundcover in shade or sun, as stabilizer on steep banks, naturalized in informal woodland gardens **EASE OF CULTIVATION** easy **AVAILABILITY** frequent.

NOTES This small spreading shrub is so adaptable that it can be used to stabilize both roadbanks and streambanks. It is not as showy in bloom, but I am impressed with the abundant purple berries in winter on gracefully arching stems, brighter than the berries of any other species in the garden when grown in some sun. Birds eat the fruits as emergency food.

Symphoricarpos orbiculatus

Vaccinium corymbosum ★★★★

Highbush blueberry

Ericaceae (heath family)

HABITAT & RANGE thickets, streambanks, low woods, bogs, and balds throughout the Southeast **ZONES** 6–9 **SOIL** moist, acidic, well drained **LIGHT** part sun to sun.

DESCRIPTION Deciduous clump-forming shrub 6–12 ft. tall, with twiggy branches forming a dense spreading crown, twigs generally dull green with red winter buds. Leaves alternate, simple, ovate, 1–3 in. long, sometimes bluish green, turning brilliant red-purple in autumn. Flowers

Vaccinium corymbosum, fruit

Vaccinium corymbosum, flowers

Vaccinium elliotti

Vaccinium darrowii

showy, white, tubular (urceolate), 1/2 in. long, in dense clusters with the leaves, blooming in spring. Fruits are round, blue-purple berries, in tight clusters, maturing in late summer, edible, highly prized by birds.

PROPAGATION seeds, without pretreatment; cuttings in June
LANDSCAPE USES specimens, accents, groupings, borders, foundation plantings, naturalized in woodland gardens, as a food crop, kept as an understory shrub **EASE OF CULTIVATION** easy **AVAILABILITY** common.

NOTES Everyone recognizes and loves blueberries. They are the quintessential wild berry that we know we can eat without consequence. So many species exists that, like the deciduous azaleas, there is a blueberry for seemingly every habitat. Sometimes blueberries are difficult to identify, but a few distinctive types can be used in landscaping and are described here. The presence of the juicy multiseeded berries seems to hold the diverse group together. Bees and hummingbirds like the flowers, and birds like the fruits.

Highbush blueberry (*V. corymbosum*, syn. *V. ashei*) also known as rabbiteye blueberry, is the blueberry of commerce. It goes under many botanical names. It has typical deciduous leaves and juicy blueberry fruits, edible and tasty. There are many cultivars, mostly bred for edible fruits, but they can be used in the landscape like any other informal shrub. Grow several cultivars for cross-pollination and have a beautiful edible landscape. Be sure the soil is kept acidic—no lime from the lawn. Height varies but is mostly to about 8 ft. tall on a broadly spreading shrub with year-round interest.

Lowbush blueberries are rhizomatous spreading dwarf deciduous shrubs to 2 ft. tall. *Vaccinium tenellum* is from the mountains, *V. vacillans* from the coast. They are similar and work well as low attractive shrubby groundcovers in open woods.

Mayberry (*V. elliottii*) is similar to the common highbush blueberry, but has smaller shiny green leaves (1/2–11/2 in. long) and slender, arching or spreading smooth shiny green stems. This is my favorite tall blueberry to use as an ornamental shrub.

Creeping blueberry (*V. crassifolium*) is an evergreen trailing shrub with ovate leaves to 3/4 in. long that would make a great groundcover for sun or light shade in well-drained soil. It should be used more in the Southeast. A good cultivar is 'Well's Delight'.

Darrow's blueberry (*V. darrowii*) is a choice dwarf evergreen shrub to 2 ft. tall with tiny 1/2 in. leaves that can be very blue-green on contrasting pink stems. The cultivar 'John Blue' is a popular informal shrub for full sun. Don't confuse this selection with the highbush blueberry cultivar 'Darrow'.

Vaccinium stamineum

Viburnum dentatum

Viburnum nudum

Viburnum acerifolium

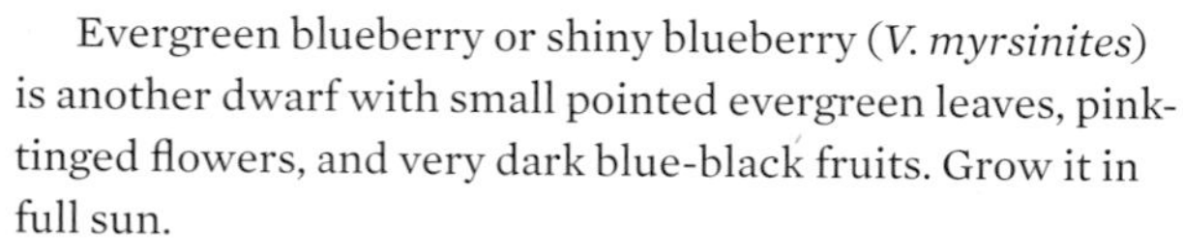

Evergreen blueberry or shiny blueberry (*V. myrsinites*) is another dwarf with small pointed evergreen leaves, pink-tinged flowers, and very dark blue-black fruits. Grow it in full sun.

Deerberry (*V. stamineum*) is a shorter species (to 3–4 ft. tall) in dry woods and has blue-green leaves with a white-waxy coating underneath, bowl-like open (not tubular) white flowers in loose masses, and hard bluish inedible greenish purple berries that hang in loose clusters.

Sparkleberry (*V. arboreum*) is the holy grail of the genus. It is a large clumping shrub or small tree, forming a few suckers over time, with gnarled trunks and branches. The rough reddish bark is exquisite on older trunks, vaguely resembling some of the western American madrones. The flowers are short-tubular, white, in loose clusters; the fruits

Viburnum prunifolium

are black and not juicy; the leaves are thick, leathery, semi-evergreen. The catch? The shrub is tricky to establish and grow, but worth the effort as the plants I have seen in cultivation have been captivating.

Viburnum dentatum

Arrow-wood

Adoxaceae (moschatel family), formerly in Caprifoliaceae (honeysuckle family)

HABITAT & RANGE dry woods, streambanks, bogs, and flood-plain forests throughout the Southeast **ZONES** 3–8 **SOIL** moist, well drained to wet **LIGHT** part sun to sun.
DESCRIPTION Deciduous clump-forming shrubs 6–12 ft. tall, the slender flexible stems often very straight, hence the common name. Leaves opposite, simple, ovate, coarsely toothed, 2–4 in. long, sometimes turning red-purple-yellow in autumn. Flowers showy, small, white, in flat-topped clusters 2–3 in. wide, blooming in late spring. Fruits are rounded, 1/4 in., blue to bluish black, maturing in late summer, eaten readily by birds.
PROPAGATION seeds are tricky; softwood cuttings in summer **LANDSCAPE USES** specimens, accents, groupings, naturalized in woodlands, kept as an understory shrub **EASE OF CULTIVATION** easy **AVAILABILITY** frequent.
NOTES Viburnums are staples of the shrub garden. They purport to have good fall color, but not all of them do. There are many selections, and these will give you a better chance of having good fall color than the species. A good choice for color in Southern gardens is 'Osceola'. A shorter shrub (4–5 ft.) is 'Christom' (trade name Blue Muffin) with glossy green summer leaves and abundant porcelain blue berries. The more sun, the better for viburnums.

Maple-leaved viburnum (*V. acerifolium*) is a short (4–5 ft.) spreading deciduous shrub with three-lobed leaves that look very much like those of a red maple and turn a good rusty red in autumn. The viburnum produces a few black berries but is highly desirable because of it low stature and good foliage.

Viburnum nudum

Smooth witherod, southern wild-raisin

Adoxaceae (moschatel family), formerly in Caprifoliaceae (honeysuckle family)

HABITAT & RANGE bogs, pocosins, savannas, and low woods throughout the Southeast **ZONES** 5–9 **SOIL** moist, well drained to wet **LIGHT** part sun to sun.
DESCRIPTION Deciduous clump-forming shrubs, sometimes suckering, 6–10 ft. tall, with arching branches. Leaves opposite, simple, ovate, coarsely toothed, glossy green, 2–5 in. long, turning red-purple in autumn. Flowers showy, small, white with yellow stamens, in flat-topped clusters 2–5 in. wide, blooming in late spring. Fruits are rounded, 1/4 in., in clusters, turning from green to pink, red, and blue over a long period, maturing in late summer, eaten readily by birds.
PROPAGATION seeds are tricky; cuttings root readily **LANDSCAPE USES** specimens, borders, groupings, informal woodlands, natural areas, edge of wetlands, kept as an understory shrub **EASE OF CULTIVATION** easy **AVAILABILITY** common.
NOTES Smooth witherod is a beautiful viburnum for its multicolored fruits, often with four colors at the same time. The leaves then turn great fall colors of rich red-purple. It is a plant for all seasons as its winter buds are large. Two good selections are 'Pink Beauty', with many fruits and stunning fall color, and 'Winterthur', one of the first selections, with abundant fruits and red fall leaves.

Withe-rod (*V. cassinoides*) is the mountain counterpart to the Piedmont and coastal *V. nudum*. It is very similar, with equally good fruits, but does not do as well south of Zone 7. It is exceptional on the Blue Ridge Parkway in the fall.

Black-haw (*V. prunifolium*) is common in low woods throughout the Upper Southeast. It is a tall shrub (18 ft.) with white early spring flowers, clusters of 1/2in. blue to black fruits, and good red-yellow fall color.

Small-leaf viburnum (*V. obovatum*) has leaves 1/2–2 in. long. It can be a large shrub, but several dwarf selections make it useful in landscaping. It is mostly evergreen, and thus it can be used as a foundation plant. Good 3-ft. cultivars are 'Mrs. Schiller's Delight', 'Reifler's Dwarf', and 'Whorled Class'. These selections are especially adapted to the warmer Southeast and make low-maintenance foundation shrubs for sun or light shade.

Xanthorhiza simplicissima

Yellowroot

Ranunculaceae (buttercup family)

HABITAT & RANGE shady streambanks and wet ledges, mostly throughout the Southeast, absent from southeastern South Carolina through Peninsular Florida **ZONES** 6–9 **SOIL** moist, acidic, well drained, tolerates wetness and flooding **LIGHT** shade to part sun.
DESCRIPTION Deciduous rhizomatous-spreading shrub to 4 ft. tall, with sparsely branched erect twigs, stems and woody roots show yellow when scraped (hence the name). Leaves alternate, compound, much divided with 1 in. deeply lobed and toothed leaflets, turning long-lasting yellow or purplish

Xanthorhiza sivmplicissima

Yucca filamentosa, leaves

Yucca filamentosa, flowers

(in sun) in autumn. Flowers small, purplish, 3⁄8 in. wide, in drooping branched clusters, blooming in spring. Fruits are yellowish pods in clusters, drooping, maturing in early summer.

PROPAGATION division of rhizomes **LANDSCAPE USES** delicate and lush groundcover for creekbanks and floodplains, naturalized in woodlands, kept as an understory shrub **EASE OF CULTIVATION** easy **AVAILABILITY** frequent.

NOTES Yellowroot is a multi-interest shrub, first for its ability to spread in average, moist to boggy soil and create a lush groundcover, and then for its early spring flowers which are unusual clusters of purplish stars. The fall color is charming. The bright yellow roots are a delightful surprise and are used as dye and herbal medicine. Even the naked stems that look like "sticks" with their pointy green buds are interesting in winter. It may become one of your favorite shrubs as it is mine; it was the first wild plant I ever transplanted as a young botany student (and it's still growing).

Yucca filamentosa ★★

Adam's-needle yucca, curlyleaf yucca

Agavaceae (agave family)

HABITAT & RANGE dry woods, roadsides, dunes, sandy disturbed areas, and railroad ballast throughout the Southeast **ZONES** 5–9 **SOIL** moist to dry, well drained **LIGHT** part sun to sun.

DESCRIPTION Evergreen shrub with a cluster of large basal leaves, 2 ft. tall from fleshy rootstock, with sharp points, trunk not evident. Leaves alternate, simple, 2 ft. long, 1½ in. wide, leathery, the edges fraying into long white threads or filaments. Flowers large, showy, white, fragrant, short-lived, pendulous, to 2 in. long, numerous, on erect, branched stalks rising to 8 ft. tall, blooming in late spring. Fruits are dry pods to 1 in. long, maturing in autumn.

PROPAGATION division of fleshy roots **LANDSCAPE USES** specimens, groupings, borders, informal woodlands, open areas **EASE OF CULTIVATION** easy **AVAILABILITY** frequent.

NOTES Yuccas are striking (and sticking) plants. Their long leaves are sharp-tipped and dangerous, so watch where you place them. In bloom, they are stunning, but short-lived considering how large they are. They are especially quaint on coastal sand dunes. Many insects visit the night-fragrant flowers.

Weakleaf yucca (*Y. flaccida*) is similar with smaller flowers and narrower leaves also with filamentous margins. It is native from Alabama and Tennessee, scattered east and west.

Spanish dagger (*Y. aloifolia*) has a trunk to 8 ft. tall. The leaves lack filaments but are finely notched or toothed, and the fruits are pendulous with plump seeds. The species is distributed mostly on coastal sand dunes from South Carolina to Mississippi.

Spanish bayonet (*Y. gloriosa*) is similar to the preceding. It has smooth leaf margins and pendulous fruits with flat seeds. It, too, is found on dunes.

Yucca gloriosa

Zenobia pulverulenta ★★★

Honey-cups

Ericaceae (heath family)

HABITAT & RANGE pocosins, bays, and bogs, mostly in coastal North and South Carolina **ZONES** 6–9 **SOIL** moist, acidic, well drained **LIGHT** part sun to sun.

DESCRIPTION Semievergreen clump-forming twiggy shrubs 2–5 ft. tall, new twigs usually red. Leaves alternate, simple, green or blue-green to white-waxy, some turning purple in autumn. Flowers showy, white, ½ in. wide, cuplike, pendulous, in loosely branched clusters on leafless stems, blooming in late spring. Fruits are round dry pods, ¼ in. in diameter, maturing in autumn.

PROPAGATION seeds; cuttings are difficult to root **LANDSCAPE USES** specimens, accents, borders **EASE OF CULTIVATION** moderate **AVAILABILITY** frequent.

NOTES Honey-cups can be unique and beautiful plants, with charming flowers and fabulous whitish blue-green leaves, or they can look a little disheveled if not grown in moist soil and full sun. Thus, they benefit from a little more cultural attention, especially if they are in a prime location. At their best, they are knockouts. The cultivar 'Woodlander's Blue' is superb with powder blue over purple fall foliage.

Zenobia pulverulenta

Conifers

There is a conifer for every habitat, from the longleaf pine of the southern sandhills or the Canada hemlock so characteristic of our beloved southern Appalachian Mountains to the bald cypress with its knees in the sultry swamps of the Deep South. Most conifers are evergreens, with upright structural trunks and needlelike leaves, and we see them as tough and tenacious. They typically dominate habitats that are too extreme for most broad-leaved trees: too wet, too dry, too rocky, too windy, and too cold. Their very presence tells us something about the land quality and its history. Conifers are usually pioneer species, coming into a disturbed landscape in the earliest stages and disappearing as the broad-leaved trees take over. Conifers persist only where the habitat remains harsh and inhospitable to flowering plants.

How, then, do conifers fit into cultivated landscapes? If your property comes with evergreen conifers, you might enjoy keeping them for their winter form and to provide a sense of place. If your property lacks conifers, you might want something more purposeful. Some conifers make distinctive specimen plants, fine-textured evergreen backdrops, or effective privacy screens. Additionally, you might try to mimic nature by creating a grove of loblolly pines that would provide light shade overhead and soft needles under foot.

Conifers are part of the larger group called Gymnosperms—vascular plants with naked seeds. This is in contrast to the plant group containing ferns, which have no seeds, only spores, and, on the other end of the spectrum, the Angiosperms—flowering plants whose seeds are inside of a fruit for protection and dispersal. The Angiosperms are the largest group by far.

Conifer seeds are not usually noticed by humans as they are mostly held inside large, protective woody cones until they are mature, and then they typically are released to fly on the wind with parchmentlike wings. That is not to say that conifer cones can't be ornamental, but most people don't grow conifers for their cones.

While the large to small woody cones seem to define the conifers, two species produce no cones, only solitary seeds surrounded by brightly colorful fleshy coverings: Florida yew and Florida torreya. Another strange conifer, the coontie, has large seeds inside a fleshy cone that are not wind dispersed. Yet an additional variation is the familiar red cedar, whose small fleshy, blue cones are attractive and relished by birds as if they were berries.

Like all Gymnosperms, conifers produce two types of cones as part of their sexual reproduction cycle. The smaller short-lived male cones, which can look like small dead scales or like purple or yellow fat caterpillars, shed pollen for a few days and then fall off like the catkins of wind-pollinated broad-leaved forest trees. The larger female cones must receive the wind-borne pollen for their eggs to become fertilized and develop into mature seeds with embryos. This latter process may take up to eighteen months in pines. We see tons of pine pollen produced every spring—it blows through the air and mostly lands on our cars, porches, and swimming pools. While it looks bad, it does not cause hay fever and is not toxic.

Conifers and broad-leaved flowering plants differ from each other in many ways, and yet, generalities can't be made. Both groups can have short- and long-lived species, occur in various stages of succession, and provide diverse products to wildlife and humans. Conifers are rarely eaten by humans, except for the nutritious seeds (pine nuts) of several Southwestern pines, but they do provide our most important building materials and paper fiber.

If one generality were to be made, it's that conifers thrive in harsh habitats where broad-leaved trees rarely dominate.

Male cones of longleaf pine (*Pinus palustris*). These will last a few days, then dry out, and shed.

Female cones of loblolly pine (*Pinus taeda*). Note the stout, sharp-prickle at the end of each scale.

Think, as examples, about the harshness of very high mountaintops and far northern subarctic climates; places where fires burn regularly (such as the long-leaf pine belt arcing through the Southeastern coastal pain); and rocky bluffs and crevices where hardy any soil or water is available. The largest (California redwood) and oldest (bristlecone pine) plants on earth are conifers, but they are restricted to specialized habitats in western United States.

If you look at the Southeast, we also have conifers that seem to defy the odds. The oldest plant species in the region is represented by a 1620-year-old bald cypress in eastern North Carolina. Table mountain pine can grow on the sheerest rock formations in the high mountains, holding its cones for decades until a fire causes them to open and release seeds to the barren rocky soil. Carolina hemlock grows only on the most rugged and barren ridges, avoiding competition by tolerating the effects of thin soil and severe drought. Bald cypress and pond cypress can survive in standing water in coastal ponds and swamps, tolerating mucky soil with little oxygen, making them good street trees for similar anaerobic soil conditions. Longleaf pine is the most fire tolerant of conifers, developing pure stands in the coastal zones where fires may burn every five years. They were the original source of sturdy lumber and naval stores (pitch, tar and turpentine) that made the early colonies successful. Lastly, the Florida torreya has persisted on the sandy bluffs of the Apalachicola River for eons, enduring ice age climate changes and continental alterations that few other species could survive.

Conifers are generally easy to grow. However, there can be the occasional problem due to cultural mistakes or attack by other organisms. For example, in the warmer parts of the South, species originally from the mountains, such as white pine and Canada hemlock, will need more shade. Trying to grow them in full sun results in stressed plants. Conversely, most conifers prefer sunny habitats and should not be planted in the shade or they will be weak and more susceptible to other problems. In general, conifers enjoy well-drained soil, with the exception of a very few species that naturally grow in wet areas—but even they can thrive in normal soils in cultivation. Ice build-up on evergreen needles can be a significant problem breaking brittle limbs on southern pines.

As for insect damage, watch out for and remove bagworms, the larvae of a moth that can defoliate and nearly kill smaller individuals of juniper, red cedar, arborvitae, and white cedar in just a few days. In the spring, you may notice orange gelatinous reproductive bodies of cedar-apple rust fungus that grows on and deforms branches and trunks, in some cases spreading throughout like a cancer and eventually killing the tree. Additionally, be vigilant for sawfly larvae, which are caterpillars that occur in large numbers aggregated together on pines (and oaks), and can devour the needles on whole branches in just a few days. Pine beetles can create local outbreaks that kill many trees, especially under drought stress. Wooly adelgids, related to aphids, have wrought great devastation by killing virtually all adult trees of Fraser fir and Canada hemlock throughout their southern ranges.

Lastly, a few words about three specific aspects of conifer culture. Unlike with Angiosperms, when you prune an older branch off a conifer, a new one will not grow back because conifers don't have dormant buds there. Second, conifers don't normally bend toward the light nearly to the extent that Angiosperms do. They simply don't have the hormonal and structural physiology to allow this. So, a spruce tree, for example, will still grow straight up even though it's planted under the edge of a large shade tree where strong light comes from only one direction. Third, if you propagate a conifer by rooting or grafting the growing tip of the central

Pond cypress (*Taxodium ascendens*) in standing water

LEAF OR NEEDLE TRAITS OF SOUTHEAST NATIVE CONIFERS

To use this key to identify an unknown tree, choose the best of the five lettered statements and then, if applicable, the best of the numbered statements. Then, compare pictures and descriptions in the text to find the correct species.

A. Needles 2–18 in. long, in tufts or bundles of two, three, or five, sharp pointed . . . . . . pines
B. Needless borne singly along the branch, short, less than 1 in. long, blunt or sharp
  1. Needles flat, blunt . . . . . . fir, hemlock
  2. Needles flat or needlelike, sharp pointed . . . . . . spruce, yew, torreya, ground juniper, juvenile red cedar
C. Needles borne singly on branchlets that fall off in autumn . . . . . . bald or pond cypress
D. Leaves compound, fernlike, thick and leathery, plant stemless . . . . . . coontie
E. Needles scalelike, in opposite pairs, closely placed on spraylike branches
  1. Branches of needles appearing flattened . . . . . . arborvitae
  2. Branches not appearing as flattened sprays . . . . . . Atlantic white cedar, red cedar

KEY TO SOUTHEAST NATIVE PINE (*PINUS*) SPECIES

To use this dichotomous key to identify an unknown plant, choose the best statement in each pair. Continue until you come to a plant name.

1. Needles in bundles of five . . . . . . *Pinus strobus*
1. Needles in bundles of two or three . . . . . . Go to 2
  2. Needles mostly in twos . . . . . . Go to 3
    3. Twigs smooth after needles fall . . . . . . Go to 4
      4. Bark of trunk scaly, flaky; needles not much twisted . . . . . . *Pinus clausa*
      4. Bark of trunk in tight plates; needles notably twisted . . . . . . *Pinus glabra*
    3. Twigs scaly, flaky, or roughened after needles fall . . . . . . Go to 5
      5. Needles stiff; prickles on cones very stout and cones > 3 in. wide (mountains only) . . . . . . *Pinus pungens*
      5. Needles pliable; prickles slender, weak; cones < 2 in. diameter . . . . . . Go to 6
        6. Needles twisted, less than 2½ in. long . . . . . . *Pinus virginiana*
        6. Needles not twisted, 3 in. or more long . . . . . . *Pinus echinata*
  2. Needles mostly in threes . . . . . . Go to 7
    7. Needles more than 10 in. long, and mature cones more than 6 in. long . . . . . . *Pinus palustris*
    7. Needles less than 10 in. long, and cones less than 6 in. long . . . . . . Go to 8
      8. Cones longer than broad . . . . . . Go to 9
        9. Cones reddish brown, glossy, needles green . . . . . . *Pinus elliottii*
        9. Cones brown, not glossy, needles yellowish green . . . . . . *Pinus taeda*
      8. Cones rounded, about as long as broad . . . . . . Go to 10
        10. Needles 6–8 in. long, flexible; coastal plain . . . . . . *Pinus serotina*
        10. Needles 3–5 in. long, stiff; Piedmont and mountains . . . . . . *Pinus rigida*

leader, it will continue to grow straight up. However, if you root the growing tip of a side branch, it will continue to grow horizontally just as the branch would. Only after several years, if ever, will a propagated side branch be able to form a new central leader. So, if you want an upright plant with a strong trunk, don't remove the central leader; and if you want to maintain a spreading or weeping form, don't let a central leader develop.

The descriptions that follow include information for the female cones (or fleshy seeds) only, as they are persistent and distinctive for each species. Width of younger actively growing trees can generally be considered about two-thirds of height, but as the trees mature, they lose their lower branches and width is a moot point.

Since there is a relatively small number of conifers in the Southeast (and only about six hundred in the whole world), it's possible to get to know them all and to understand their landscape uses; see "Leaf or Needle Traits of Southeast Native Conifers." Additionally, because the pines are a distinct group with only eleven species, I have included a key to their identification; see "Key to Southeast Native Pine (*Pinus*) Species."

Abies fraseri ★★

Fraser fir, she-balsam

Pinaceae (pine family)

HABITAT & RANGE Fraser fir forests, often mixing with red spruce, at high elevations in the southern Appalachian Mountains, generally above 4500 ft. to 6600 ft. **ZONES** 4–6 **SOIL** moist, well drained **LIGHT** part shade to sun.
DESCRIPTION Pyramidal evergreen tree to about 40 ft. tall or more. Bark somewhat smooth except for pustules of aromatic resin that can be milked (hence the name "she-balsam"). Needles very aromatic, flat, blunt, borne singly, crowded, swept upwards and forward, leaving the twig smooth when fallen. Cones purple turning brown at maturity, 1½–2½ in. long, erect, with parchmentlike bracts

Abies fraseri

Chamaecyparis thyoides

longer than individual thick cone scales (distinguishing this from the more northern balsam fir, *A. balsamea*, with shorter bracts), shattering at maturity and leaving upright naked axis.

PROPAGATION seeds, cold stratified for one month **LANDSCAPE USES** specimens, groupings, borders, pruned regularly to form a hedge **EASE OF CULTIVATION** easy (in cooler climates) **AVAILABILITY** common.

NOTES Fraser fir is a handsome, dark green formal tree, not very heat or drought tolerant in the Southeast and best planted only above 2500 ft. in sun or light shade. It makes the finest, most aromatic Christmas tree and is grown as such in extensive farms on steep slopes at mid-elevations in the mountains. By 1992, virtually all the wild Fraser firs in the southern Appalachian Mountains were killed by the balsam woolly adelgid insect, but most forests have made remarkable recovery with rapidly growing natural seedlings.

Chamaecyparis thyoides ★★

Atlantic white cedar

Cupressaceae (cypress family)

HABITAT & RANGE woodland, streambanks, moist pine flatwoods, wet ditches, and ponds, mostly in the outer coastal plain throughout the Southeast **ZONES** 4–9 **SOIL** moist, well drained **LIGHT** sun to part sun.

DESCRIPTION Columnar evergreen tree to about 50 ft. tall or more. Bark reddish brown, rough, irregular, ridged and furrowed, often twisting around the trunk. Needles flat, scale-like, opposite, successive pairs at right angles to each other, forming non-flattened spraylike branchlets. Cones borne on tips of branchlets, brown, woody, round, about ¼ in. diameter (similar cones of northern white cedar or arbor vitae, *Thuja occidentalis*, are distinctly elongate on flattened branchlets).

PROPAGATION seeds, cold stratified for one month; cuttings in winter **LANDSCAPE USES** groupings, borders, screening and hedges where it forms a dense crown without pruning **EASE OF CULTIVATION** moderate **AVAILABILITY** frequent.

NOTES This attractive, informal tree is heat, but not drought, tolerant. It is not accepting of shade and the lower branches eventually die out. Dozens of cultivars are known and have been evaluated at the University of Georgia, where Michael Dirr reported most succumbed to drought; therefore, few if any, are widely available. An attractive dwarf blue-green selection is 'Little Jamie'. More effort should be made to try this species in the Southeast, at least in reconstructed wetlands where moisture is assured.

Juniperus communis var. *depressa*

Juniperus communis var. *depressa* ★★★

Ground juniper

Cupressaceae (cypress family)

HABITAT & RANGE rock outcrops, shaded bluffs, and sandy soils in the mountains and coastal plain from Virginia to Georgia **ZONES** 5–8 **SOIL** well drained **LIGHT** shade to sun.

DESCRIPTION Prostrate evergreen shrub spreading to form dense mats, to about 2 ft. tall, horizontal branches rooting as it grows. Needles in threes around the stem, flat, sharp, angled forward, about ½ in. long, gray-green. Female cones fleshy, blue or black, waxy, about ⅜ in. in diameter, maturing in the second or third year.

PROPAGATION cuttings in winter **LANDSCAPE USES** groundcover **EASE OF CULTIVATION** easy **AVAILABILITY** rare.

NOTES This interesting dwarf conifer seems to be a southern segregate (relict from glacial times) from the widespread northern common juniper (*J. communis*) which does not grow well in the hot South. It is very worthwhile in the garden and is quite long-lived. At higher elevations, it may be grown in full sun, but in Zones 7 and 8 it needs light shade.

Juniperus virginiana ★★★

Eastern red cedar

Cupressaceae (cypress family)

HABITAT & RANGE old fields and upland woods throughout eastern North America **ZONES** 3–9 **SOIL** any well-drained type **LIGHT** part sun to sun.

Juniperus virginiana with cedar waxwing

DESCRIPTION Broadly pyramidal to narrowly columnar evergreen tree to about 50 ft. tall. Bark handsome brownish, shredding. Needles in pairs and of two types: juvenile foliage is sharp, needlelike, to 1⁄4 in. long, and is rare on mature trees unless pruned or broken and new growth results; adult foliage is scalelike, with opposite pairs of flattened scales arranged closely and overlapping. Crushed foliage smells like cedar shavings. The trees are unusual in being unisexual, with male and female cones on separate trees. The small persistent female cones are blue, round, and fleshy berrylike, about 1⁄4 in. wide. Nibble on a ripe one and you will get the taste of gin.

PROPAGATION seeds; cuttings in winter **LANDSCAPE USES** foundation plantings, specimens, accents, groupings, screenings, borders, hedges, windbreaks **EASE OF CULTIVATION** easy **AVAILABILITY** common.

NOTES The species is so common in many parts of the Southeast that it may be viewed as a roadside and pasture weed, but individual trees can be quite handsome, and they should

Picea rubens

be used more in landscaping, especially in full sun. The blue fleshy cones, which can be quite attractive as a mass in winter, are relished by cedar waxwings and other birds. Several cultivar forms are available for use in a variety of landscape situations. 'Burkii' is a handsome columnar cultivar with blue foliage, 'Grey Owl' grows 3 ft. tall by 6 ft. wide and has gray foliage, and other selections come in various shades of green. These cultivars can be tall and narrow, or wider, globose, or short and spreading. There is a form and color for every need.

Picea rubens ★★

Red spruce, he-balsam

Pinaceae (pine family)

HABITAT & RANGE high-elevation forests, mixed with Fraser fir, on the higher peaks of the southern Appalachian Mountains, generally above 3500 ft. **ZONES** 4–7 **SOIL** moist, well drained **LIGHT** shade to sun.

DESCRIPTION Pyramidal evergreen tree to about 50 ft. tall (or more). Bark scaly, reddish brown. Needles four-sided, about 1/2 in. long, sharp, neither flat nor aromatic like Fraser fir, borne singly, straight, crowded, leaving the twig with peg-like stubs when fallen. Cones on upper branches, reddish brown, 1–1 1/2 in. long, pendulous, falling intact.

PROPAGATION seeds, without pretreatment **LANDSCAPE USES** specimens or informal groupings **EASE OF CULTIVATION** moderate **AVAILABILITY** frequent.

NOTES Red spruce definitely prefers the cool, moist conditions of elevations above 2000 ft. I have grown it in the Piedmont of North Carolina in Zone 8 under high shade for forty-five years, and watched it struggle to become 10–15 ft. tall, whereas Fraser firs have simply died as the summers have become warmer. Red spruce is not recommended in zones warmer than Zone 7.

Pinus echinata ★★

Shortleaf pine

Pinaceae (pine family)

HABITAT & RANGE old fields and upland woods throughout the Southeast, rare in Peninsular Florida **ZONES** 6–9 **SOIL** moist to dryish, well drained **LIGHT** sun.

DESCRIPTION Evergreen tree to 50 ft. tall or more, pyramidal in youth but forming a large crown of crooked branches on a naked trunk. Bark broken into rectangular flat plates about 2–3 in. long. Needles two in a bundle, 3–5 in. long, not twisted. Cones oblong egg-shaped, 1 1/2–2 1/2 in. long, falling after a year or two.

Pinus virginiana and brown-headed nuthatch

PROPAGATION seeds, without pretreatment **LANDSCAPE USES** informal groupings or groves, screening **EASE OF CULTIVATION** easy **AVAILABILITY** frequent.

NOTES Shortleaf pine is very common throughout the Southeast and may be confused with large loblolly pine (*P. taeda*) from a distance. The latter may be recognized by its much straighter branches high in the canopy. Shortleaf pine is often planted in landscapes, but soon grows out of scale. Nevertheless, it is useful to create a canopy of high shade offering some advantage in understory cultivation of flowering broad-leaved evergreen shrubs such as azaleas, rhododendrons, and camellias.

Scrub pine (*P. virginiana*), also called old-field pine or Virginia pine, may be easily confused from a distance. It has shorter needles that are distinctly twisted, and its bark plates are usually half as large as those of shortleaf pine. Both species may hold their old cones for several years, but scrub pine typically holds more. The species is very common throughout the Southeast as it comes early in succession in old fields and poor, dry, rocky soils. It is often gnarled and stunted in the wild, making interesting specimens worth keeping.

Another two-needled pine of note is table mountain pine (*P. pungens*). It grows mostly in the mountains and is at its best on the harshest, barren rocky outcrops where it forms gnarled, windswept specimens. The very stiff needles are only 1–1½ in. long, and the dense rounded cones are 2–3½ in. long but with ¼ in. stout sharp spines. The cones can remain closed on the branches for many years waiting for a fire. This conifer is best in a sunny, well-drained site in Zones 6 and 7.

Pinus glabra ★★★★

Spruce pine

Pinaceae (pine family)

HABITAT & RANGE mixed in hardwoods forests, sandhills, and bluffs of the coastal plain from southeastern South Carolina to Florida, westward to Louisiana **ZONES** 7–9 **SOIL** moist, well drained **LIGHT** shade to part sun.

DESCRIPTION Narrow, pyramidal evergreen tree to 40 ft. tall

Pinus glabra

Pinus echinata

or more. Bark of trunk ridged and grooved, but the layers of bark tight, not breaking up into plates or flakes as with other pines. Needles in twos, 2–4 in. long, spirally twisted, dark green. Cones egg-shaped, 1–2½ in. long, held for three or four years.

PROPAGATION seeds **LANDSCAPE USES** specimens or informal groupings **EASE OF CULTIVATION** easy **AVAILABILITY** moderate. **NOTES** This species is a handsome dark-green fine-textured tree for mostly shaded landscapes. It should be used much more throughout the Southeast as it is heat and shade tolerant, holding its needles well on lower branches that may last for decades. Needles turn yellowish in strong winter sun, but recover in summer. A huge 90-ft. specimen at the Coker Arboretum in Chapel Hill, North Carolina, is impressive.

Sand pine (*P. clausa*) is similar but it holds its cones longer and the bark is flaky. While not widely cultivated, this small tree (to 40 ft. tall) is a rich green and may be useful as a windbreak and habitat cover in areas of deep, infertile sands. The occasional gnarled specimen may make an interesting landscape feature if left in place. I have seen beautiful large sand pines planted in open, sandy barren areas where nothing else would grow.

Pinus palustris ★★★

Long-leaf pine

Pinaceae (pine family)

HABITAT & RANGE dry sandy ridges and hills, and moist pine flatwoods, mostly in the coastal plain throughout the Southeast **ZONES** 7–10 **SOIL** well drained, moist to dry **LIGHT** sun. **DESCRIPTION** Tall, massive evergreen tree to 60 ft. tall or

Pinus palustris and *Quercus laevis* in the Carolina sandhills

more, forming a large crown of stout branches. Bark broken into large, flat plates. Needles three in a bundle, 10–18 in. long, not twisted. Cones brown, very large, 6–8 in. long, ovate, falling the next winter.
PROPAGATION seeds **LANDSCAPE USES** specimens, groupings or groves **EASE OF CULTIVATION** easy **AVAILABILITY** common.

Pinus palustris

NOTES This majestic tree is characteristic of the southeastern coastal plain and sandhills, where in forms pure stands in areas subject to regular fire that keeps out competition. Elsewhere it's grown as a curiosity and is subject to extensive breakage from ice and snow accumulating on the long needles. It is adapted to fire like no other species, forming dense tufts of long needles during its first few years, allowing fires to burn over the top while protecting the tender growth bud down inside the "nest." From this "grass stage," saplings grow rapidly to get above the fire's damage zone. Europeans first exploited this pine for naval stores and later as sturdy material for shipbuilding and hardwood flooring. Stumps are resinous and virtually rot-proof; I have several in my yard that are just as hard today as they were forty years ago. Long-leaf pine is somewhat paradoxical in that it can grow in both very dry and very wet places (but not in standing water). The species is characteristic of deep, sterile sands as well and those constantly moist pitcherplant bogs usually referred to as pine flatwoods.

Pinus rigida ★★

Pitch pine

Pinaceae (pine family)

HABITAT & RANGE rocky ridges, outcrops, and poor soil in the mountains and Piedmont of the Upper Southeast **ZONES** 4–7 **SOIL** moist, well drained, acidic **LIGHT** sun.
DESCRIPTION Evergreen tree to 40 ft. tall or more, irregular in growth and often becoming gnarled. Bark broken into small flat plates. Needles three in a bundle, 3–5 in. long, stiff, uniquely occurring also as tufts along the mature trunk. Cones 1–2 in. long, rounded.
PROPAGATION seeds, without pretreatment **LANDSCAPE USES** groupings **EASE OF CULTIVATION** easy **AVAILABILITY** infrequent.
NOTES Pitch pine grows from the wet bogs and peat lands of the New Jersey pine barrens to medium elevations in the southern Appalachian Mountains where it's often the common pine in exposed areas on dry ridges and poor soils. The tufts of needles on the trunk make for easy identification. This pine tree may not be ornamental enough for planting in the formal landscape, but it is worth keeping or planting in rough places on your mountain property.

Another three-needled pine of the Southeast is pond pine (*P. serotina*). It is typical of wet coastal scrubland and pocosins, and is the only pine that can grow in shallow standing water. The distinctly rounded cones are 1½–2 in. wide and they open only after fire to repopulate a dry burned-over habitat. This pine is best in coastal plain sandy soil.

Pinus strobus ★★★

Eastern white pine

Pinaceae (pine family)

HABITAT & RANGE dry to moist woods and old fields, mostly throughout the southern Appalachian Mountains **ZONES** 3–8 **SOIL** fertile, moist, well drained, acidic **LIGHT** light shade to sun.

DESCRIPTION Evergreen tree to 80 ft. tall or more, pyramidal in youth becoming a forest giant with erect upper branches and horizontal lower ones. It is of much finer texture than other pines, with branches in distinctive sets or whorls at each node along the trunk, one set per year. Needles five in a bundle, 3–5 in. long, soft to the touch, bluish to light green. Cones 4–7 in. long, slender and slightly curved, light brown often with white-encrusted resin, maturing and falling in autumn of the second year.

PROPAGATION seeds, cold stratified for three months; grafting

Pinus strobus

Pinus rigida

LANDSCAPE USES specimens, groupings, borders, groves, hedges **EASE OF CULTIVATION** easy **AVAILABILITY** common.
NOTES White pine is our most distinctive and graceful eastern pine, with five needles, slender cones, and whorled branches. It is among the few conifers that thrive in full shade and may volunteer as seedlings in moist woodlands, but is not invasive. It is a must-have for the woodland garden. In the Southeast it's weakened by heat stress and root rot, but still very reliable and worth the risk in a shady, well-drained site. It is subject to sudden death when grown in full sun in heavy clay soils.

There are many selections of Eastern white pine. Their cultivar names attest to their diverse shapes: 'Compacta', 'Pendula', 'Prostrata', 'Contorta', and 'Fastigiata'. South of Zone 6 they all benefit from protection from hot afternoon sun.

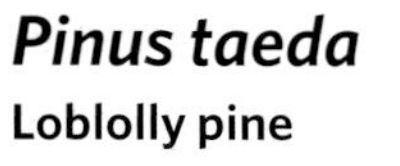

Pinus taeda ★★

Loblolly pine

Pinaceae (pine family)

HABITAT & RANGE lowland and upland old fields and woodlands, often wet, mixing with hardwoods, mostly Piedmont and coastal plain throughout the Southeast **ZONES** 6–9 **SOIL** moist to dry, well drained to poorly drained **LIGHT** sun.
DESCRIPTION Evergreen tree to 60 ft. tall or more, forming a massive trunk with large bark plates and a large crown of straight branches. Needles three in a bundle, 4–9 in. long, dark yellowish green. Cones elongate, 3–5 in. long, not stalked, yellowish brown to gray, held for three or four years.
PROPAGATION seeds, without pretreatment **LANDSCAPE USES** screens, groves, allées, pulpwood plantations **EASE OF CULTIVATION** easy **AVAILABILITY** common.
NOTES Loblolly pine is fast growing and adaptable. It can be quite attractive as its longer-than-average needles give it a look of fullness. It can be beautiful as groves on large estates, along highways, and in old fields; but is probably not suited as an isolated specimen in the residential landscape, looking out-of-place to my mind. Its older branches are much straighter than the similar shortleaf pine. It is very susceptible to breakage in ice and snow.

An almost identical species in the South is slash pine (*P. elliottii*), widely planted for its fast growth. Though inexpensive, it is even less suitable for landscape use, and easily breaks in snow and ice. To differentiate, loblolly pine consistently has three needles and dull-brown cones that are not stalked and are held for three or four years, while slash pine has mixed bundles of two or three needles and lustrous-brownish cones that are stalked and fall soon after maturing.

Pinus taeda

Taxodium ascendens

Taxodium ascendens ★★★★

Pond cypress

Cupressaceae (cypress family)

HABITAT & RANGE pond and lake margins, pocosins, savannas, and swamps in the coastal plain from southeastern Virginia throughout the Southeast **ZONES** 6–11 **SOIL** moist, well drained to shallow standing water **LIGHT** part sun to sun.
DESCRIPTION Deciduous tree to 50 ft. tall or more, pyramidal in youth and remaining so for many years, with a straight trunk becoming somewhat fluted at the base, forming occasional upright conical "knees" to 2 ft. tall. Needles narrow,

Taxodium distichum, fall color

½–¾ in. long, arranged in a flattened manner against short branchlets that are variable in size, 3–6 in. long such that they look like erect or diverging needles, bluish green in summer turning rich orange to reddish brown in autumn before falling. Cones globose, 1 in. in diameter, woody, brown at maturity, shattering to release the seeds in late autumn.

PROPAGATION seeds, cold stratified for three months; cuttings in winter **LANDSCAPE USES** specimens, informal groupings, allées

EASE OF CULTIVATION easy **AVAILABILITY** frequent.

NOTES Pond cypress is one of our greatest ornamental species. I can't say enough good things about its beauty. It is closely related to bald cypress (*T. distichum*), but is different. It does not become nearly as massive, and as such is much better suited to the smaller property. It also has needlelike leafy branchlets rather than feathery ones, shorter knees (rarely formed in cultivation), and is found in pocosins and around lakes rather than in river floodplains. It is sometimes confusing, but pond cypress does produce feathery branchlets, but only when young. It is fast growing and adaptable to wet or average soils, in sun or part sun. It is a relative newcomer to our plant palette; and while there are fewer named selections, it should become more widely available to consumers as a container plant. 'Prairie Sentinel' and 'Morris' (trade name Debonair) are excellent selections.

Taxodium distichum ★★★★

Bald cypress

Cupressaceae (cypress family)

HABITAT & RANGE riverbanks and floodplains, sloughs, backwaters, and pond and lake margins, mostly on the coastal plain throughout the Southeast **ZONES** 4–11 **SOIL** moist, well drained to shallow standing water **LIGHT** part sun to sun.

DESCRIPTION Deciduous tree to 80 ft. tall or more, pyramidal in youth, flat-topped in old specimens, with a stout straight trunk that becomes fluted or buttressed at the base, forming upright conical "knees" to 3 ft. tall. Needles flat, about ½–¾ in. long, arranged in a two-ranked featherlike (pinnate) pattern on short branchlets that are 3–6 in. long, soft green in summer turning rich orange to reddish brown

Taxodium distichum, foliage

in autumn before falling. Cones globose, 1 in. in diameter, woody, brown at maturity, shattering to release the seeds in late autumn.

PROPAGATION seeds, cold stratified for three months; cuttings in winter **LANDSCAPE USES** specimens, informal groupings, allées, street trees, in parks and estates **EASE OF CULTIVATION** easy **AVAILABILITY** common.

NOTES Bald cypress is a stately tree that grows rather rapidly. It is distinctive in every way: a conifer with shreddy bark, soft flat needles on deciduous branchlets, with good fall color, growing in standing water, producing knees, buttressed trunk, and round shattering cones. When draped with Spanish moss, bald cypress is fantastic. The fall color is striking, and this changeability adds to its character by offering an interesting winter twig texture. Knees are typically formed only where water levels fluctuate regularly. Make no mistake: this species is extremely adaptable as to soil type even though it's restricted to swamps in the wild. In uptown Charlotte, bald cypress is grown in small planting holes of compacted soil that undoubtedly lack water, air, and oxygen. It would make a good screen except it's deciduous—don't forget that.

Many cultivars are available, including dwarf, pyramidal, and weeping forms. One nice weeper with a central leader is 'Falling Waters'. The spectacular 'Cascade Falls' weeping form has to be one of the five most interesting garden plants you can grow.

Thuja occidentalis

Thuja occidentalis ★★★

Eastern arborvitae

Cupressaceae (cypress family)

HABITAT & RANGE rocky limestone cliffs and boulder fields in the mountains of Virginia into the Valley and Ridge of Tennessee **ZONES** 3–8 **SOIL** moist to dryish, well drained **LIGHT** part shade to sun.

DESCRIPTION Evergreen tree to 40 ft. tall or more, dense, pyramidal; branches formed into horizontal or vertical flattened sprays. Needles scalelike, paired, densely spaced on the branches, bright to dark green, may emit a mild spicy-tangy odor when bruised, especially in winter. Cones oblong, erect, brown, about ¾ in. long, formed of eight woody scales.

PROPAGATION cuttings in winter from the current year's woody growth **LANDSCAPE USES** foundation plantings, specimens, accents, hedges, screening, allées **EASE OF CULTIVATION** easy **AVAILABILITY** common.

NOTES Eastern arborvitae is a standard among landscape plants, with cultivars of many shapes, sizes, and shades of green or even yellowish. It is a more northern species, and as such may feel the stress of southern heat, so protect it from afternoon sun. One concern is that in winter sun, the foliage will bronze or turn yellowish brown, returning to green in spring. It would be best to choose one of the several cultivars.

I especially like the toughness and rounded columnar profile of the 15- to 20-ft.-tall 'DeGroot's Spire', which bronzes in full winter sun. Other cultivars of note are 'Golden Globe', 2–3 ft. tall, yellow; 'Malonyana', rigidly upright like a pole, 15 ft.; 'Lutea', pyramidal to 30 ft.; 'Nigra', pyramidal to 20 ft., staying green in winter; 'Pendula', open arching branches, 15 ft., staying green; 'Smaragd' (trade name Emerald), a compact pyramidal to 15 ft., staying green, good for hedges and becoming widely used; and 'Techny' (syn. 'Mission'), a broad pyramidal, to 15 ft., staying green, good for hedges.

The many cultivars of western arborvitae (*T. plicata*), such as 'Green Giant' and 'Zebrina', may be better choices for larger plants in full sun in the Southeast.

Torreya taxifolia ★★★

Florida torreya, stinking-cedar

Taxaceae (yew family)

HABITAT & RANGE wooded ravines and bluffs in the hills of the Apalachicola River in the Florida Panhandle and adjacent Georgia **ZONES** 7–9 **SOIL** moist, well drained **LIGHT** part shade to part sun.

DESCRIPTION Evergreen shrub, broadly pyramidal-rounded, to about 15 ft. tall, with irregular branching. Needles flat, dark green, about 1 in. long, arranged in a flat plane on the branch, stiff and sharp, strongly aromatic when crushed. Cone absent, instead a curious single large egg-shaped seed surrounded by a removable casing, about 1 in. long, purplish green when mature in early autumn.

PROPAGATION seeds **LANDSCAPE USES** specimens or groupings in semiformal or woodland gardens **EASE OF CULTIVATION** easy **AVAILABILITY** infrequent.

NOTES Florida torreya is a fascinating species, nearly extinct in the northern Florida wilds, whose closest relatives are in California, China, and Japan. Strangely, the plant is easy to grow in cultivation, is heat tolerant, and has made a handsome dark-green specimen at UNC Charlotte Botanical Gardens. As a dense background plant in light shade it's hard to beat. While difficult to obtain, it's well worth the effort to grow.

Torreya taxifolia

The somewhat related Florida yew (*Taxus floridana*), in my experience, has produced weak specimens that were not cold hardy. It should perform better in warmer districts of Zones 8–9, but it may produce little more than a curiosity. Yews in general are a favorite food of deer.

Tsuga canadensis ★★★

Canada hemlock, Eastern hemlock

Pinaceae (pine family)

HABITAT & RANGE moist to dry forests in mountainous eastern North America, much less common in the Piedmont, and with isolated populations in the coastal plain **ZONES** 3–8 **SOIL** moist, well drained, acidic **LIGHT** shade to part sun.

DESCRIPTION Evergreen tree to 50 ft. tall or more, broadly pyramidal, slightly descending branches with gracefully pendulous tips. Needles flat, ⅔ in. long, two-ranked, closely spaced along the branches in flattened sprays, dark green. Cones ovoid, about ¾ in. long, brown, pendulous, not persisting.

PROPAGATION seeds, cold stratified for three months **LANDSCAPE USES** specimens, groupings, groves, borders, screenings, and backgrounds, may be sheared as hedges or topiary, especially in the mountains, elsewhere best used in woodland gardens **EASE OF CULTIVATION** easy **AVAILABILITY** common.

NOTES Canada hemlock is a dominant tree in southern Appalachian mountain forest where it reaches enormous size and is our most massive native conifer, at least it once

Tsuga caroliniana

was until it was decimated by an aphidlike adelgid insect during the first decade of the twenty-first century. Its handsome gracefulness in youth and fragrant evergreen needles add a familiar and desirable "feel of the mountains" to Piedmont woodland gardens. This conifer is best not used in full sun south of Zone 7, as it's not suitable to prolonged heat or drought. Beware that it grows slowly into a large tree that produces dense year-round shade unsuitable for most other plants, and in this sense it takes up a good deal of space in the garden. The dwarf weeping forms, such as the cultivar 'Pendula' make excellent garden specimens.

A related species of great beauty and value in the garden is Carolina hemlock (*T. caroliniana*), and I highly recommend it. It is quite restricted in its wild habitat but it ranges down throughout the Blue Ridge Mountains. The species is characteristic of dry ridges, barren rock outcrops, steep slopes and rocky crags, mostly above 2000 ft. elevation.

Tsuga cones: (left) *T. caroliniana,* (right) *T. canadensis*

There are magnificent stands on Bluff Mountain in northwestern North Carolina. Carolina hemlock is much smaller in stature than Canada hemlock, growing half as fast and creating a dark green pyramidal form more in scale with other woodland garden plants. It has slightly larger needles to ¾ in. long that are formed uniformly around the branch to give a much fuller, non-flattened appearance, and the larger 1 in. ovoid cones that open fully to create an attractive display. It has performed well in Piedmont gardens when grown in high shade or part sun, in well-drained soil, and not overwatered. Even during our worst droughts, we try to avoid watering established Carolina hemlocks. They seem to be less affected by the adelgid insect, especially when grown outside the areas of infestation. A challenge to obtain and grow, but well worth the effort and should be more readily available.

Zamia floridana ★★

Coontie

Syn. *Zamia pumila*

Zamiaceae (zamia or cycad family)

HABITAT & RANGE coastal deciduous forests, palm hammocks, and scrubby pine flatwoods from southeastern Georgia and northern Peninsular Florida to the Keys and West Indies.

ZONES 8–11 **SOIL** moist, well drained, sandy **LIGHT** shade to part sun.

DESCRIPTION Low-growing palmlike evergreen, with a thickened underground fleshy-woody unbranched stem, forming clumps to 3 ft. high and 6 ft. wide. Leaves arise in an annual flush from the tip of the stem, pinnately compound like a coarse fern frond to 3 ft. long, and opening by uncoiling like a fern fiddlehead, with many narrow opposite segments (sometimes broad, sometimes narrow, depending on the individual specimen) arranged in a flat featherlike "blade." The species is unusual among Gymnosperms in having male and female cones on separate plants (though this is typical for all cycads), emerging in late summer. The brown

Zamia floridana

female cone arises on a stout stalk from the underground stem, is ovoid with many thick segments arranged in a compact 2-in. clublike "cone," each producing one or two orange fleshy seeds, maturing in December after a year of development. The brown male cone is also ornamental, though not long lived, and is cylindrical, 3–4 in. long. This species is the larval food source for the Florida atala butterfly.

PROPAGATION seeds, no pretreatment required, just be patient **LANDSCAPE USES** specimens as a curiosity, groupings, in dry ferneries **EASE OF CULTIVATION** moderate **AVAILABILITY** infrequent.

NOTES Coontie is quite tough and seems to grow well on its own in the warmer south. My grandmother had one for decades in eastern South Carolina, and I loved to see it poke its brown fuzzy cones up amid the stiffly arching evergreen "fronds" from beneath the pine needles mulch. I thought "coontie" was a great name (Seminole origin). Other local names are Florida arrowroot and Seminole bread, alluding to the starch that came from the underground stem. This species is protected, is making a comeback, and is readily grown from seeds. It is difficult to know how to use it in the landscape; perhaps to grow it more with palms than with ferns as it's quite drought tolerant, but representing a primitive evolutionary transition (it's a living fossil) from ferns to conifer-like seed plants (I know it's hard to believe). Coontie does become lusher in the Deep South.

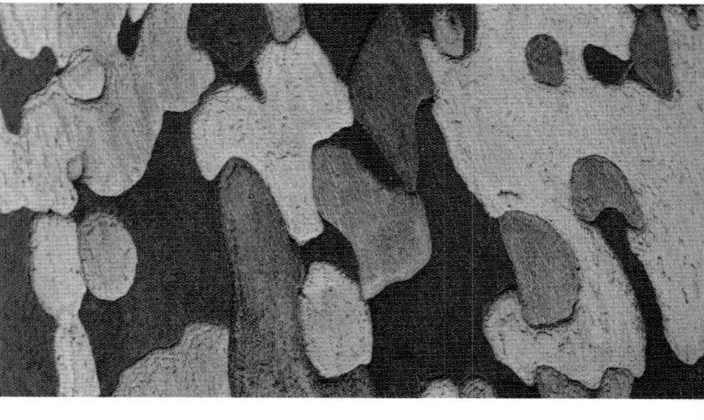

Trees

When our daughter, Suzanne, was very young she had a propensity to climb on things, including all the way to the top of a 25-ft. southern magnolia tree across the street from our house. Once we got over the initial shock, we realized the tree had sturdy, horizontal limbs that gave her a safe stairway to heaven, had enough small branches to break her fall (which never came), and provided a nesting place with interesting knots and branch formations. She loved that tree, and it was her sanctuary.

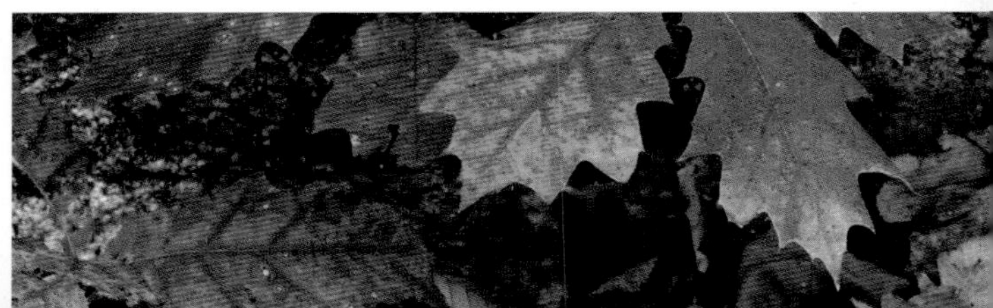

No matter where you grew up, you probably have a favorite tree from your youth, bigger than life, often associated with a magical awe-inspiring moment. Such memories tie us to nature. However, even as they can inspire us, we sometimes take trees for granted: they seem to have always been there, shading the play area, hosting a swing, compelling us to rake the leaves. But think of how much we receive from them: the cooling shade that improves our environment, the glorious sight of spring leaves and fall color, wood for everything from furniture and construction lumber to toothpicks and baseball bats, fruits and nuts for us and the critters, fuel for warmth, fiber for writing paper, and the superstructure of a pleasant habitat in which to live.

Trees as a growth form are familiar to us all. But what are trees really? And what are they trying to accomplish?

Like any other living organism, trees want to survive, make offspring, and carry on into the next generation. They have found a place in nature, becoming large woody plants built of tough cellulose and lignin material, added cell-by-cell, layer-by-layer, year-after-year, to create a sturdy scaffold. Such structure is necessary to support an enormous crop of sun-loving leaves to carry on photosynthesis, bear much weight in fruits or nuts to reproduce themselves, and to stand up against the various forces of nature. It is hard to realize that each oak tree, for example, producing tons of acorns throughout its long life, needs merely to replace itself

The quintessential southern tree, live oak (*Quercus virginiana*), typically seen with Spanish moss festooned on the branches.

once. Only one seedling on average needs to survive to maturity to keep the species represented in a normal oak-hickory woods.

Trees grow in height only at the tips of their branches, where overwintering buds expand in spring to push out the new growth for the season. They produce a crop of fresh leaves and flowers, bear the fruits until dispersed, drop their leaves on schedule in autumn, and return to a dormant state for the winter. Trees grow in girth evenly every spring, from the tiniest twig to the massive trunk, by adding a continuous ring of new growth produced by the vascular cambium.

It is important to understand that the new leaves and flowers are produced only on new tip growth each year. If it appears that an older twig (more than one year old) has leaves, it's due to small side branches (called spurs) that form and grow just enough each year to produce a few leaves and maybe flowers and fruits. Since tiny embryonic cells for next spring's blooms are preformed in the buds made the previous summer, the weather conditions during the summer growth period determine the quality of flowering next spring. It is then reasonable to give your flowering trees extra care in cultivation so they produce the finest floral and leaf displays. A little extra water (if needed) and fertilizer during late spring or early summer can help, but no later. Also, don't be too quick to prune off branches in the fall as you may be removing buds that would be blooming the following spring.

When we look at the large shade trees in our neighborhoods and cities, we realize their benefit even though we did not plant them. In some cases, these are the forest trees that were left when homes were built; in other cases, they were planted five or more decades ago. In any case, virtually all

Spectacular fall color along the Blue Ridge Parkway in North Carolina.

of the best shade trees in the Southeast are native canopy trees from our common forests. Most of the trees described in this chapter are deciduous; the few evergreen exceptions are noted.

Landscapes, as well as forests, benefit from more than just tall canopy specimens. A number of smaller understory trees, like redbud and dogwood, are grown for their showy flowers. These species are adapted to survive in the shady understory, but we often plant them in full sun where they produce more flowers and leaves. Sometimes they suffer from high heat and light, and it may explain why we may have trouble growing certain small trees in the sunny landscape for as long a time as they are known to survive in their natural shady habitats.

Since most forest trees don't produce showy flowers, at least not that we can readily see and enjoy, one of the delights we receive from them when they are used as shade trees is their fall color. We look forward to color changes perhaps as much as we anticipate spring flowers. Nowhere is fall color more exceptional than in the Upper Southeast—the Blue Ridge Mountains and the cooler regions of hardiness Zones 7 and 8. The diversity of tree species there is greater than in all of Europe. Only in Japan, Korea, and China would fall color be as spectacular, so it's no wonder that most of our nonnative cultivated plants hail from temperate China and Japan.

Good fall color displays are a function of three factors: the tree's inherent ability to turn colors, the qualities of the sites on which they grow, and the weather in the autumn that allows the colors to develop. First, not every tree species is capable of turning all the possible colors of the autumn rainbow. Trees that turn good fall color produce either striking shades of yellow and orange, or various hues of red, creating amazing displays of crimson and scarlet. The yellows pigments were there are along, aiding in photosynthesis, but were masked by the more abundant green chlorophyll. Only certain species of trees can produce red pigments, known as anythocyanins, in the fall as their leaves age. These red colors can also meld with the yellows, oranges, and even greens to make patterns of purples and deep oranges along with the rich yellows, sometimes creating a wondrous mosaic pattern.

Site also influences fall color. Most trees that naturally grow in floodplains and wetlands tend not to have good fall color. Neither red nor yellow, they just slowly fade to yellowish brown. This may be due to the higher nutritional levels found in these sites from regular flooding. We don't know why this affects colors. Those species with especially bright reds grow in sites that are more nutrient-poor, such as drier woods and upland slopes and ridges. This might lead you to conclude that red and yellow color changes have something to do with nutrition levels in the soil, and you would probably be correct; but the conclusive scientific proof is lacking. Nevertheless, because this is what we observe, it would be wise to avoid excess fertilizing and watering of established cultivated trees after July 1 less the extra nutrients late in the season may interfere with good fall color production.

The third factor, local weather conditions, is also very important. Leaves turn colors for two reasons. When green chlorophyll breaks down and is reabsorbed, it's due to cooler night temperatures and hormonal changes as the days get shorter in late summer, then the yellow and orange pigments show through. The red pigments are produced by converting sugars into anthocyanin pigments when the autumn leaves are exposed to direct sunlight. So, the brighter the sun, the stronger the red colors. This is why clear, sunny days and cool nights lead to the formation of strong reds, purples, and oranges. Gloomy, rainy weather decreases the exposure to bright sun and thus less intense red colors. As further evidence, you will notice the south sides of trees will turn red earlier than the north sides, and leaves in the shade turn only after exposure to sun after canopy leaves drop.

So how do you choose a tree for good fall color in the landscape? First, pick a species, especially a selected cultivar, known to produce good color. Then, place it where strong sunlight can shine on it in the late summer and into fall, if it can tolerate full sun, avoiding overwatering or overfertilizing after July, and pray for clear fall weather. And there you have it, simple as that.

There may be an art to planting a tree, but there are also scientific principles that play a role in achieving success. When planting a tree, dig the planting hole much wider than deep (twice as wide or more); keep the soil beneath the root ball solid so it will not sink when the tree root ball is set in place. Spread the new roots as far wide as you can, gently pulling them apart if they are pot-bound. Remove or straighten any larger roots that may be closely encircling, or girdling, the trunk of the tree, and remove all rope or wire ties that could strangle the trunk later as it grows in diameter. Create a soil that has a crumbly texture, not too sandy and not heavy red clay. It should form a ball in your hand when you squeeze it, but should loosen up easily when teased apart. Incorporate the new soil with the existing soil and roots—use your hands. Pack the new soil tightly against the root ball as you refill the hole. Avoid adding peat moss to any garden planting soil because it acidifies (possibly too much) and holds more water than you really want. If you have alkaline soil and want to grow sourwood trees or other acid-loving species, build a raised bed and use only acidic clay-based soil or composted leaves as a growing medium.

If the tree is top-heavy with branches or taller than 5–6 ft., you may want to stake it up when planting. Prune the young tree minimally to keep the branches evenly spaced, but do prune out any dead, broken, or crossing branches. In addition, it's best to have a single trunk, or leader, unless you have purposely selected a multitrunked tree branching from ground level. When you prune branches greater than 1 in. in diameter from an older tree, make your cuts just outside the branch collar, not flush with the trunk. The collar is the thickened bark surrounding the juncture of branch with trunk. Don't apply wound dressing or tree paint to such cut surfaces, despite conventional wisdom. It actually hampers the self-healing process of the tree by killing the regenerative cells around the wound.

Be aware that the active major roots of your tree are just a foot or so underground, and the actively growing roots are mostly at the drip line, an imaginary ring where the edge of the leafy canopy would come down and touch the ground. Don't disturb this zone. Remember there is a root for every leaf, and the more roots you disturb, the more leaves you lose. If you want to feed or water your tree, do it mostly at the drip line; there are few feeder roots near the trunk of a large shade tree.

Trees, of course, can tolerate some disturbance and regrow their small branches and roots readily, but the greatest injustice we do to trees is to rake up and dispose of their leaves each fall. These leaves are meant to rot under the tree, forming the leaf litter, slowly decomposing and producing the recycled nutrients the tree needs for next year's growth. If we put it in the compost pile, or send it off to the municipal dump, where does the tree gets its food? Probably from lawn fertilizer, at best. No wonder older street and lawn trees in urban settings decline after so many years. They are slowly starved by raking, deprived of water by paving, subjected to compacted soil by parking, and hit by cars while driving. Give your shade trees a little more loving care, and they will live even longer to cool your house, hold your hammocks, and support your swings, not to mention brighten your sky with the signs of seasonal changes.

People often ask about the difference between a tree and a shrub. I use the concept that a tree is usually single trunked and more than 20 ft. tall, while a shrub is more naturally multitrunked and less than 20 ft. tall. Some large trees are multitrunked if they have regrown as several stump sprouts after logging or natural damage. There are exceptions to these rules, and I have made choices here that seem useful in home landscapes.

Acer barbatum ★★★★

Southern sugar maple

Syn. *Acer floridanum*

Sapindaceae (soapberry family), formerly in Aceraceae (maple family)

HABITAT & RANGE streambanks and woods in the Piedmont and coastal plain throughout the Southeast excluding Tennessee **ZONES** 6–9 **SOIL** moist, well drained **LIGHT** part sun to sun.

DESCRIPTION Deciduous tree 40–60 ft. tall. Leaves opposite, simple, palmately three- or five-lobed with rounded sinuses and shallow rounded teeth, the blade to 4 in. long, often wider than long, the underside hairless and waxy-whitish (glaucous), turning beautiful shades of red to orange in autumn. Flowers greenish yellow in stringy pendulous clusters, blooming in midspring. Fruit a detachable pair of one-seeded winged samaras 1 in. long, maturing in autumn.

PROPAGATION seeds, cold stratified for three months **LANDSCAPE USES** shade tree, specimen, grouping, kept as an understory tree **EASE OF CULTIVATION** easy **AVAILABILITY** frequent.

NOTES This species is one of the most beautiful in the Southeast. It makes a great shade tree over time because it is slow growing, sturdy, and long-lived. Southern sugar maple mostly replaces the closely related larger northern sugar maple (*A. saccharum*) south of Tennessee and east of the Blue Ridge Mountains. It is smaller in all ways and more heat tolerant, and should be used more widely in Zones 8 and 9. Some of the cultivars of northern sugar maple will do well through Zone 7, but heat stress is becoming more of a negative factor for them. I know of no named cultivars of southern sugar maple.

Acer leucoderme 

Chalk maple

Sapindaceae (soapberry family), formerly in Aceraceae (maple family)

HABITAT & RANGE moist to dry woods above streams, in typically less-acidic soil, mostly in the Piedmont from central North Carolina to Louisiana **ZONES** 6–9 **SOIL** moist to dryish, well drained **LIGHT** part sun to sun.

DESCRIPTION Small deciduous tree to 30 ft., often multitrunked, older bark lighter in color. Leaves opposite, simple three- or five-lobed with rounded sinuses and shallow rounded teeth, the blade 1½–3 in. long, light green and hairy beneath, turning exquisite bright red-orange in late autumn. Flowers greenish yellow in stringy pendulous clusters, blooming in midspring. Fruit a detachable pair of one-seeded winged samaras, maturing in autumn.

PROPAGATION seeds, cold stratified for three months **LANDSCAPE USES** specimen, grouping, kept as an understory tree **EASE OF CULTIVATION** easy **AVAILABILITY** infrequent.

NOTES Chalk maple is certainly related to southern sugar maple, and sometimes considered merely a variety of it, but

Acer barbatum

Acer leucoderme

I have treated it separately here to give it recognition, as it's among the finest small native trees I have grown. The best ones have several crooked trunks, develop a crown that is rounded to upright, and are twiggy but open. They are unrivaled for fall color of brilliant dark red and orange, among the very last trees to turn in late autumn (usually peaking in early November). In many ways, chalk maple grows like a Japanese maple and would be a suitable native substitute. It is easy to tell from southern sugar maple as the leaf is light green and hairy underneath, and the middle leaf lobe is usually (not always) wider at the base than at the tip, whereas southern sugar maple is whitish underneath and the middle lobe is actually parallel-sided or narrower than the tip toward the base of the lobe. I give chalk maple my highest recommendation and suggest it become more available.

Acer pensylvanicum, flowers and foliage

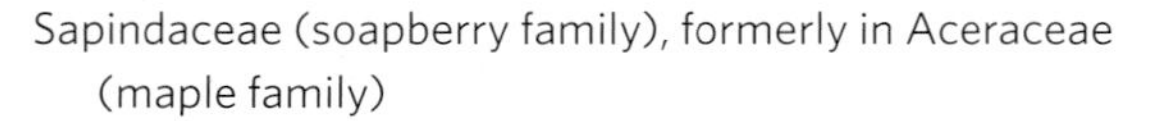

Acer pensylvanicum ★★★

Striped maple

Sapindaceae (soapberry family), formerly in Aceraceae (maple family)

HABITAT & RANGE dry to moist forests at mid elevations in the southern Appalachian Mountains from Virginia to Georgia **ZONES** 5–7 **SOIL** cool, moist, well drained **LIGHT** shade to part shade.

DESCRIPTION Deciduous small tree or multitrunked shrub to 40 ft., usually much shorter, the bark greenish with conspicuous white stripes when young. Leaves opposite, simple, three-lobed, finely toothed, 5–6 in. long, turning a mellow yellow in autumn. Flowers are noticeable, greenish, several on simple pendulous stalks under the leaves, male and female on separate plants, blooming in early spring. Fruits are paired greenish winged seeds (samaras) with 1 in. long, widely divergent wings, in pendulous clusters, maturing in late summer to autumn.

PROPAGATION seeds, cold stratified for three months **LANDSCAPE USES** specimen in a woodland garden, kept as an understory tree **EASE OF CULTIVATION** moderate **AVAILABILITY** frequent.

NOTES Striped maple, called moosewood up north, is a tree for all seasons. The leaves are handsome, the flowers and fruits are interesting, the winter twigs are noticeable, and the striped bark is outstanding. The tree struggles with the heat south of Zone 7, but is worth trying in a woodland setting with extra water during droughts.

Acer pensylvanicum, bark

gray-brown and less peely with age. Leaves alternate, simple, finely toothed, triangular-ovate and squared-off at base, turning a poor gold-yellowish in autumn. Flowers tiny, in attractive pendulous male catkins and small erect female spikes, blooming in earliest spring. Fruits are tiny winged seeds in conelike clusters, shedding in early summer.
PROPAGATION seeds, without pretreatment **LANDSCAPE USES** specimen, shade tree, grouping in an estate lawn or park, street tree **EASE OF CULTIVATION** easy **AVAILABILITY** common.
NOTES River birch is a widely used tree in many situations. Fast growing and sturdy, it's most attractive as a multitrunked specimen (with odd number of trunks), showing its pendulous lower branches and attractive bark (its greatest asset). It does not tolerate drought and will begin to shed leaves in prolonged dry weather. It is one of the fastest-growing shade trees that minimize negative traits; another is red maple.

Betula nigra

Carpinus caroliniana ★★

Ironwood, musclewood, American hornbeam

Betulaceae (birch family)

HABITAT & RANGE streambanks and low, rich woods throughout the Southeast **ZONES** 5–9 **SOIL** moist, well drained **LIGHT** shade to part sun.
DESCRIPTION Small deciduous tree to 35 ft. tall, with irregularly fluted smooth gray bark. Leaves alternate, simple, finely toothed, barely rounded base, hairless below, 1½–3 in. long, turning somewhat yellow in autumn. Flowers not showy, in short pendulous spikes (catkins), blooming in early spring. Fruits dry, nutlike, with a three-lobed leaflike bract attached for wind dispersal, maturing in late summer.
PROPAGATION seeds, warm stratified for three months, then cold stratified for three months **LANDSCAPE USES** specimen in open woodland or streamside, kept as an understory tree **EASE OF CULTIVATION** easy **AVAILABILITY** common.
NOTES The great attraction for ironwood is its fabulous smooth musclelike bark that becomes more fluted with age. The wood itself is very hard. This tree does not like to be planted in full sun, especially in dry soil. It is a very informal tree and can have distinctive character (unusual branching patterns) as a specimen. Ironwood is often called simply "hornbeam," which can be confusing, because many people feel the latter name is better reserved for hop hornbeam (*Ostrya virginiana*, which see). Both species have similar leaves. Ironwood differs in having bark that is smooth rather than flaky, leaves that are darker bluish green and not hairy below, two-tone brown winter buds, and dry fruits with leaflike "wings."

Carpinus caroliniana

Carya glabra ★★

Pignut hickory

Juglandaceae (walnut family)

HABITAT & RANGE moist to dry upland forests throughout the Southeast **ZONES** 6–9 **SOIL** moist, well drained **LIGHT** part shade to sun.

DESCRIPTION Deciduous tree 70–80 ft. tall with compact upright crown and tight fishnet-patterned (reticulate) bark. Leaves alternate, compound like a feather, 8–12 in. long, with five- or seven-toothed leaflets 3–6 in. long, the tip three times larger, hairless, turning good yellow, or sometimes not, in autumn. Flowers not showy, in pendulous male catkins, blooming in late spring. Fruit a spherical nut, covered with a thin husk, a small swollen knob on the end, splitting away, seed sweet, maturing in autumn.

PROPAGATION seeds, cold stratified for one month **LANDSCAPE USES** shade tree, kept as a forest tree **EASE OF CULTIVATION** easy **AVAILABILITY** rare.

NOTES Pignut hickory is easy to recognize by its small, thin-husked (to 1⁄8 in. thick) "piggy nipple" nuts with knobby ends. It is slow growing and not readily transplanted, but makes a great shade tree if you already have one.

Shagbark hickory (*C. ovata*) has distinctive bark that falls off in long narrow plates, resulting in a shaggy appearance. The nuts have a very thick (3⁄8 in.) husk.

Mockernut hickory (*C. tomentosa*) has tight reticulate bark but its large leaves and stout twigs are very hairy, and the nuts also have thick (3⁄8 in.) husks.

These three hickories, along with several others, have edible nuts, but they are messy under the trees. They are sturdy and long-lived. All can have spectacular orange-yellow, long-lasting fall color. For this reason alone, they should be saved in the landscape when possible.

Carya illinoinensis ★★

Pecan

Juglandaceae (walnut family)

HABITAT & RANGE rich, moist soils, barely crossing the Mississippi River eastward into the Southeast **ZONES** 4–9 **SOIL** moist, well drained, tolerates flooding **LIGHT** part shade to sun.

DESCRIPTION Large deciduous tree 60–80 ft. tall and as wide or wider, with fissured bark. Leaves alternate, compound like a feather, 12–20 in. long, with seven to eleven lance-shaped, sharply toothed leaflets 2–6 in. long, distinctly curved, the terminal three not larger than the others.

Carya illinoinensis

Carya glabra

Flowers not showy, in pendulous male catkins, separate female flowers clustered at twig tip, blooming in late spring. Fruit an elongate nut with a thin husk splitting away, seed edible and choice, maturing in late summer. GRAFTS

PROPAGATION seeds, cold stratified for one month **LANDSCAPE USES** nut tree **EASE OF CULTIVATION** easy **AVAILABILITY** common.

NOTES This massive, often impressive, tree is relatively fast growing and long-lived, but lacks good fall color. Like all hickories, pecan trees are slow to establish during their first two or three years. The nuts and leaves can be messy. While graceful with their arching branches, the trees become much too large for the small home landscape and generally lack ornamental value. Plant cultivars that are selected for better quality nuts.

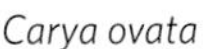

Carya ovata

Carya tomentosa

Castanea dentata ★★

American chestnut

Fagaceae (beech family)

HABITAT & RANGE rocky upland forests scattered throughout the Piedmont and mountains of the Southeast **ZONES** 6–8 **SOIL** moist, well drained **LIGHT** part sun to sun.
DESCRIPTION Formerly a massive deciduous tree, but introduced chestnut blight killed virtually all of the mature trees by 1950. Now it exists as fast-growing stump sprouts, rarely reaching 20 ft. tall before the blight hits again and the sprouts die down. Leaves alternate, simple, elliptical with a long tapering tip and many coarse sharp teeth, not hairy beneath, 6–11 in. long, turning yellowish brown in autumn. Flowers very showy, males small but arranged on long erect spikes (catkins), females in clusters of two or three at the base of the male catkins, blooming in early summer. Fruit a large nut, 1 in. across, flattened on one side, arranged two or three inside a spiny bur 2–3 in. across, maturing in late summer, edible and choice.
PROPAGATION seeds, cold stratified for two months **LANDSCAPE USES** specimen, nut tree, kept as an understory tree **EASE OF CULTIVATION** moderate **AVAILABILITY** rare.
NOTES Losing the American chestnut was an ecological and social disaster: it was a dominant tree in the forest and had so many uses (food, barrel wood, rail fences, house siding, leather tanning). Today it has a chance of making a comeback in cultivation with the development of 98 percent pure blight-resistant hybrids with Chinese chestnut, the

work of the American Chestnut Foundation (www.acf.org). Almost every chestnut that someone finds today as a mature nut-bearing tree is an escaped or planted blight-resistant Chinese chestnut (*C. mollissima*) that has shorter, wider leaves that are distinctively soft-hairy underneath.

The much smaller chinquapin (*C. pumila*) is a compact tree to 25 ft. or a large multitrunked shrub. The leaves are alternate, simple, broadly elliptical with small but distinct teeth, 3–6 in. long, and hairy underneath. The flowers are very similar to American chestnut, but the nut is smaller, about 1/2 in. wide, and the 1 in. bur holds only one nut. Chinquapin makes an interesting specimen, is readily commercially available, and is widespread as an understory tree in dry forests throughout the Southeast.

Castanea dentata

Catalpa bignonioides ★

Southern catalpa, Indian-cigar

Bignoniaceae (bignonia family)

HABITAT & RANGE streambanks, fields, and roadsides from south-central Mississippi east just into Georgia, now spread throughout the Southeast **ZONES** 7–9 **SOIL** moist, well drained **LIGHT** sun.
DESCRIPTION Deciduous tree to 60 ft. tall, with massive trunk for its height, and large brittle branches. Leaves in whorls of three (a unique condition for our native trees), simple, heart-shaped, large, to 11 in. long and 8 in. wide, light green. Flowers large and very showy, but short-lived, to 2 in. long, broadly tubular, white with wavy-margined petals and purple and yellow-orange blotches and stripes in the throat, in branching clusters in late spring. Fruit a pencil-thin dark-brown pod 6–18 in. long, splitting open to release numerous curious flat-winged seeds in autumn, persisting all winter.
PROPAGATION seeds, without pretreatment **LANDSCAPE USES** specimen, curiosity **EASE OF CULTIVATION** easy **AVAILABILITY** frequent.
NOTES Southern catalpa can look pretty broken down and ratty, and the foliage looks pale, but the large leaves and beautiful flowers can make you smile when you see them. Some folks might grow this tree just as a breeding site for Catawba worms, the large green and black-spotted caterpillars of the catalpa sphinx moth, used as fish bait.

The northern catalpa (*C. speciosa*) is practically identical except the seedpods are much thicker.

Castanea pumila

Catalpa bignonioides

Celtis laevigata ★

Southern hackberry, sugarberry

Ulmaceae (elm family)

HABITAT & RANGE floodplains and upland woods in limestone regions throughout the Southeast, except absent in the mountains **ZONES** 6–10 **SOIL** moist to dry, tolerates flooding **LIGHT** part sun to sun.

DESCRIPTION Deciduous tree to 80 ft. tall with an open, spreading crown, smooth gray bark with distinctive corky outgrowths. Leaves alternate, simple, heart-shaped to elliptic, the bases asymmetrical, inconsistently toothed, turning weakly yellow in autumn. Flowers small, inconspicuous, in loose clusters, blooming in spring. Fruit a hard, dryish orange berry, 1/4 in. in diameter, slightly sweet, maturing in autumn.

PROPAGATION seeds, cold stratified for three months **LANDSCAPE USES** specimen for bark **EASE OF CULTIVATION** easy **AVAILABILITY** frequent.

NOTES There is hardly anything to recommend planting a hackberry. They are weedy, unkempt trees with little of beauty, but the bark, oh the bark. The trunk has those enigmatic corky ridges that form as if portions of a normal covering of bark had been etched away by erosion, leaving islands and groupings of layered bark rising from the surface. You can even see the annual growth rings. Fascinating. And birds love the fruits.

Celtis laevigata

Cercis canadensis ★★★★

Eastern redbud

Fabaceae (legume family)

HABITAT & RANGE woodlands and open disturbed roadsides throughout the Southeast, except absent in the Atlantic Coastal Plain **ZONES** 6–9 **SOIL** moist to dry, well drained **LIGHT** part sun to sun.

DESCRIPTION Small deciduous tree to 40 ft. tall, often multitrunked. Leaves alternate, simple, heart-shaped, toothless, 3–5 in. wide, turning a nice yellow in autumn. Flower buds are black-purple, conspicuous all winter at nodes along the trunks, branches and twigs (a unique feature), producing dense clusters of fragrant (and tasty) purple-pink two-lipped flowers about 1/2 in. long, blooming in early spring (usually the first showy native tree to bloom). Fruits are dry, brown, flat bean-pods, 3–4 in. long, maturing in autumn and usually persisting into winter, inedible.

PROPAGATION seeds, cold stratified for two months, nicking the hard seed coat helps germination **LANDSCAPE USES**

Cercis canadensis, flowers

Cercis canadensis, habit

specimen, grouping, street tree, kept as an understory tree **EASE OF CULTIVATION** easy **AVAILABILITY** common.
NOTES Eastern redbud is one of our best-known and most beautiful small trees. It is widely adaptable to various light and soil conditions, is drought tolerant, recovers from storm damage by sprouting from the base, and is really a harbinger of spring. The tree is especially floriferous along the sunny edges of deciduous woods. Redbud is our best fast-growing small tree for quick shade on a small scale, being effective after one or two full growing seasons from a 3- to 5-ft. tall single stem seedling under good conditions. There are many attractive cultivars involving flower and leaf color, including 'Royal White', 'Tennessee Pink', and 'Silver Cloud', which has white leaf blotches. 'Forest Pansy' has purple leaves, but does not keep the color well in hot sun. 'Traveler' has a wonderful weeping habit with dark green glossy foliage.

Chionanthus viginicus

Chionanthus virginicus ★★★★

Fringe tree, grancy gray-beard

Oleaceae (olive family)

HABITAT & RANGE dry woods, slopes around rock outcrops, glades, barrens, and pocosins throughout the Southeast, except rare in Tennessee, and absent in the high mountains **ZONES** 7–9 **SOIL** moist to dry, well drained **LIGHT** part sun to sun.

DESCRIPTION Deciduous small tree or large shrub, often multitrunked, to 30 ft. tall. Leaves opposite, simple, not toothed, oblong-oval 3–6 in. long, turning a nice clear yellow in autumn. Flowers pure white, fragrant, with four narrow petals, about 1 in. long, in loose hanging clusters displayed in profusion in spring. Fruits on female trees are oval bluish black drupes (like an olive) about ¾ in. long, ripening in late summer.

PROPAGATION seeds difficult, warm stratified for three

months, then cold stratified for one month **LANDSCAPE USES** specimen, grouping, naturalized in a woodland garden, kept as an understory tree **EASE OF CULTIVATION** easy **AVAILABILITY** common.

NOTES A massive flurry of wispy white flowers on a fringe tree in full bloom in a yard is a typical spring scene in the rural Southeast. The small tree grows everywhere in dryish woods and in open areas. At its best, it's a multitrunked clump in mostly full sun, to about 10 ft. tall. Although a bit slow growing, it's worth the space for the lush spring flowers and tranquil fall colors. Makes you feel at home, connected to the natural landscape.

Cladrastis kentukea ★★★★

Yellowwood

Syn. *Cladrastis lutea*
Fabaceae (legume family)

HABITAT & RANGE rich woods, very widely scattered from western North Carolina through Tennessee into Missouri, and in Alabama and Kentucky **ZONES** 6–8 **SOIL** moist, well drained **LIGHT** part shade to sun.

DESCRIPTION Deciduous tree to 50 ft. tall, with a crown of upward-spreading branches, sometimes wider than tall, bark smooth, gray. Leaves alternate, compound with seven to eleven alternating toothless leaflets that are ovate, 2–3 in. long, turning a beautiful golden yellow in autumn. Flowers are like wisteria (pea flowers), in pendulous clusters, white with a yellow throat, fragrant, blooming in spring. Fruits are small dry bean-pods in hanging clusters, maturing in autumn and persisting into winter.

PROPAGATION seeds, cold stratified for three months, nicking the hard seed coat helps germination **LANDSCAPE USES** specimen, grouping, street tree, in parks **EASE OF CULTIVATION** easy **AVAILABILITY** common.

NOTES There is no finer tree for the landscape; yellowwood stands out in all seasons. It is one of the most celebrated ornamental trees from the mid-South. It has showy spring flowers, excellent foliage, spectacular warm-yellow fall color, appealing smooth bark, and interesting branching structure. The compound leaves are unique in that the leaflets alternate with each other and the enlarged base of the leaf stalk (petiole) covers the winter bud completely. It may help to train (by selective pruning) young trees to ensure a symmetrical growth pattern. This could be said of many trees and shrubs.

Cladrastis kentukea

Cornus alternifolia 

Alternate-leaf dogwood, pagoda dogwood

Cornaceae (dogwood family)

HABITAT & RANGE rich woods, shrub balds, and streambanks, mostly in the mountains, but widely scattered in the Piedmont of North Carolina to the upper coastal plain of Mississippi **ZONES** 6–9 **SOIL** moist, well drained **LIGHT** shade to sun.

DESCRIPTION Deciduous small tree or large shrub to 25 ft. tall, usually smaller, often multitrunked, very young twigs smooth, shiny green becoming red in winter sun, developing white stripes in trunks as they grow. Leaves alternate (unique in our dogwoods), simple, not toothed, the leaf veins arching towards the tip, turning an intriguing array of red-orange-yellow-purple in autumn. Flowers small, white, four-petaled, in flat-topped clusters 2–4 in. wide, on the tips of leafy branches, blooming in spring. Fruits are fleshy (drupes), blue-black, about 1/4 in. in diameter, on bright red stalks, maturing in late summer.

PROPAGATION seeds, warm stratified for three months, then cold stratified for three months **LANDSCAPE USES** specimen, grouping, kept as an understory tree **EASE OF CULTIVATION** easy **AVAILABILITY** infrequent.

NOTES This wonderful native dogwood has ornamental traits that have not been fully appreciated, and it should be more widely planted. It can be trained as a single- or multitrunked tree or shrub. It has excellent garden appeal in all seasons, not the least of which is the kaleidoscope of fall colors that might ensue, depending on the strength of the sunlight that might fall on the foliage during color changing time. In full shade, the leaves are purplish yellow. The fruits are readily eaten by birds in late summer.

Cornus alternifolia

Cornus florida ★★★★

Flowering dogwood

Cornaceae (dogwood family)

HABITAT & RANGE woodlands of all kinds throughout the Southeast **ZONES** 6–9 **SOIL** moist, well drained **LIGHT** part sun to sun.

DESCRIPTION Small deciduous tree to 30 ft. tall, with distinctive layered branches forming a flat-topped crown with age, bark thick and platy, winter twigs tipped with a single large onion-shaped flower bud. Leaves opposite, simple, ovate, bases tapering, not toothed, the veins arching towards the tip, 3–6 in. long, turning rich red in autumn. Flowers small, hardly noticeable, four-petaled, in tight clusters surrounded by four large showy leaflike bracts, blooming in early spring. Fruits are bright red fleshy drupes, in clusters on the tips of short branches, ripening in late summer, relished by birds.

PROPAGATION seeds, cleaned of red covering and cold stratified for three months **LANDSCAPE USES** specimen, grouping, street tree, in parks, naturalized in informal woodlands, kept as an understory tree **EASE OF CULTIVATION** easy **AVAILABILITY** common.

NOTES This species needs no introduction as it's the overall most beloved and planted deciduous tree of the South. It has utmost appeal in all seasons: flower, foliage, form, fruit, and bark. (So it lacks floral fragrance! Nobody is perfect.) Wild flowering dogwood grows in shady woods. Therefore, planting it in a site that gets full sun is akin to light-skinned humans getting a rich suntan: it makes them look great, but it's unhealthy. I am seeing more (virtually all in Charlotte) open-grown older dogwoods succumb to a fungal-heat-stress syndrome that is exacerbated by high temperatures and drought. Young trees seem to do fine, and grow rapidly when replanted to replace the old. Try to protect flowering dogwoods from hot afternoon sun; you can still have a beautiful specimen. The fatal dogwood anthracnose disease seems to affect trees only in cooler, wetter climates, mostly in the mountains. Many cultivars based on flower and leaf variations are listed in Dirr (2009), with 'Cloud 9' and 'Cherokee Princess' being two of the best for the Southeast. For a pink-flowering dogwood, select 'Cherokee Chief' or 'Cherokee Sunset'.

Cornus florida, flowers

Cornus florida, fruit with wood thrush

Cotinus obovatus

Crataegus viridis

Cotinus obovatus ★★★★

American smoketree

Anacardiaceae (cashew family)

HABITAT & RANGE dry limestone woods, widely scattered across the Upper South from northwestern Georgia to Missouri **ZONES** 4–8 **SOIL** dry to moist, well drained **LIGHT** part sun to sun.

DESCRIPTION Deciduous small tree or large shrub to 25 ft. tall, often multitrunked, with wide-spreading branches, inner wood is yellow, cut twigs and leaves exude a milky sap that may be irritating. Leaves alternate, simple, ovate blades 2–6 in. long on long leaf stalks (petioles), dark green hairy beneath, sometimes having a whitish waxy cast, turning wonderful mixed colors of purple-red-orange-yellow in autumn. Flowers miniscule but borne in unique pinkish smokelike plumes of feathery branchlets, giving an overall billowy look, blooming in spring. Fruits are tiny nutlike drupes 1/4 in. long, on detachable pieces of the fluffy plume for dispersal, maturing in summer.

PROPAGATION seeds, treated briefly with acid to soften the seed coat, then stratified for three months **LANDSCAPE USES** specimen, grouping, kept as an understory tree **EASE OF CULTIVATION** easy **AVAILABILITY** rare.

NOTES Smoketree, with its unique puffy plumes of tiny flowers, is limited in nature to rare limestone bluffs and ravines of the mid-West. It grows readily and rapidly in cultivation, creating exciting small trees with magnificent fall color when grown in sun. 'Grace', its hybrid with the common smoketree (*C. coggygria*) of China, is becoming more widely grown and makes an excellent compact small tree. The species is still rare in cultivation, but that should change.

Crataegus viridis ★★

Green hawthorn

Rosaceae (rose family)

HABITAT & RANGE swamps and wet woods throughout the Southeast **ZONES** 7–9 **SOIL** moist, well drained **LIGHT** sun.

DESCRIPTION Small deciduous tree 25–30 ft. tall, with a dense spreading crown, bark flaking off with age to expose orange-brown inner bark, twigs notorious for thorns up to 1½ in. long. Leaves alternate, simple, toothed and often lobed as well, slightly narrowed at base, 2–4 in. long, turning red or yellow in autumn. Flowers showy, white with five petals in rounded clusters 2 in. wide, blooming in spring. Fruits small, about 1/4 in. in diameter, red, applelike, in clusters, ripening in autumn and often persisting into winter if not eaten by birds.

PROPAGATION seeds, cold stratified for three months **LANDSCAPE USES** specimen, grouping, street tree, in parks **EASE OF CULTIVATION** easy **AVAILABILITY** frequent.

NOTES The most famous green hawthorn is the cultivar 'Winter King', which makes a wonderful street tree, heavily laden with colorful "berries" in the winter, and does better in the Southeast than most hawthorns.

There are many species of hawthorns in the Southeast, and they are notoriously difficult to identify. Some, called May haws, are locally known for making jams and jellies from the ripe fruits, likening to crab apples. Hawthorns are so variable in their appearance and performance, you never know when you will find just the right one for your conditions. It's a gamble, but the multiseason beauty may be worth it.

May haw (*C. aestivalis*) does well in the Southeast, grows in swampy woods in the coastal plain, and flowers in very

Cyrilla racemiflora, flowers

Cyrilla racemiflora, fall color

early spring with beautiful one- to three-flowered clusters and leaves that taper narrowly to the base. The one-flower hawthorn (*C. uniflora*) is a smaller shrub that grows in very dry sites throughout the region. Parsley hawthorn (*C. marshallii*) grows in a variety of wet to dry habitats and is distinctive in having deeply cut leaves and small flowers and fruits. Littlehip hawthorn (*C. spathulata*) is widespread and has beautiful flaking bark and distinctly three-lobed leaves long tapering to the base.

Cyrilla racemiflora ★★★★

Titi, swamp cyrilla

Cyrillaceae (titi family)

HABITAT & RANGE shallow swamps, pocosins, pond margins, flatwoods, streambanks, and ditchbanks in the coastal plain throughout the Southeast **ZONES** 6–9 **SOIL** moist, well drained, tolerates seasonal flooding **LIGHT** part sun to sun. **DESCRIPTION** Semievergreen shrub or small tree to 25 ft. tall, usually smaller, with an informal crown and widely spreading branches often arching upwards, gently bending twigs are beautiful red-brown with slight ridges. Leaves evergreen, alternate, simple, narrow, toothless, shiny green, 2–4 in. long, a few turning yellow in autumn. Flowers small, white, densely aggregated into very showy elongate spikes borne in clusters at the base of new growth, blooming in early summer (and attracting the most interesting insects). Fruits are tiny dried round tan pods persisting into winter. **PROPAGATION** seeds, without pretreatment; stem cuttings in summer; root cuttings in fall–winter **LANDSCAPE USES** specimen, grouping, screening, foundation plant, kept as an understory tree **EASE OF CULTIVATION** easy **AVAILABILITY** frequent. **NOTES** Titi (pronounced "tie-tie") is an underutilized Southeastern native. It is both drought tolerant and accepting of wet feet. While it can grow into a very large shrub or small tree, I have kept one pruned as a foundation plant next to our bedroom window for thirty years. I love to see the unusual flowers and the plant is open enough to create a see-through shrub. It grows informally, allowing me to keep it selectively pruned, which is how I like it without worrying about what is "the right way." I think more people should try them.

Two other *Cyrilla* species that I like are not always recognized as different, but can be obtained and grown as distinctive plants. Dwarf titi (*C. parviflora*) is smaller in every way, with leaves to just 1½ in. long. Scrub titi (*C. arida*), according to Bob McCartney (pers. comm.) of Woodlanders Nursery, "may now be extinct in the wild [from sandy habitats in Florida] but is well established in cultivation and proving surprisingly hardy and a very worthy garden ornamental."

A related species, buckwheat tree (*Cliftonia monophylla*), is a large thicket-forming shrub to 18 ft., with many branches and crooked trunks. It can form beautiful evergreen masses in the wild, usually in wet areas. The pinkish white flowers are very showy against the bluish green leaves. It does best in the warmer Deep South.

Diospyros virginiana, bark

Diospyros virginiana, fruit

Diospyros virginiana ★★

Persimmon

Ebenaceae (ebony family)

HABITAT & RANGE dry woods, floodplains, fencerows, and disturbed places throughout the Southeast **ZONES** 6–10 **SOIL** moist to dry, well drained **LIGHT** sun.
DESCRIPTION Deciduous tree to 70 ft. tall, usually smaller, straight trunked or dwarf and gnarled depending on the site, twigs with rounded tip buds often quirked to one side, bark distinctly formed into square, blocklike plates. Leaves alternate, simple, not toothed, dark green above and lighter below, with an abrupt tip, 2–6 in. long, turning striking red-orange in autumn. Flowers white, ½ in. with four thick petals, not very conspicuous among the leaves, male or female, usually on separate trees (you'll need two to get fruits), blooming in early summer. Fruit a very large pulpy orange berry to 1½ in. in diameter, with four persistent sepals usually attached, very astringent when unripe, containing several large brown flattened seeds, ripening in autumn.
PROPAGATION seeds, cold stratified for three months **LANDSCAPE USES** specimen, grouping, in woodland or on open property **EASE OF CULTIVATION** easy **AVAILABILITY** common.
NOTES Possums love persimmons, at least that what everybody thinks. You will love them too, ripe ones, in puddings and pies (don't eat an unripe one, but suggest to someone else to try it). But you will not love them if you plant a female next to your sidewalk. Boy, the fruits when they fall are messy, squishy, ooey, gooey, yucky! But the spectacular fall color—words can't describe the intensity and richness—combined with the fabulous bark, makes all else forgiven. You may plant a persimmon tree for its tasty fruit, or you may already have one in a good location—away from the walk. The wood is one of the heaviest and hardest in North America.

Fagus grandifolia ★★

American beech

Fagaceae (beech family)

HABITAT & RANGE rich hardwoods throughout the Southeast **ZONES** 5–9 **SOIL** moist, well drained **LIGHT** shade to sun.
DESCRIPTION Very large deciduous tree to 100 ft. tall, slow growing with spreading crown, maintaining distinctive smooth gray bark throughout the life of the tree (usually developing carved initials), twigs with distinctive long, spindle-shaped, sharp-pointed buds in winter. Leaves alternate, simple, ovate, 3–5 in. long, with each principal side vein ending in a small tooth, turning a rich yellow-brown in fall and persisting as dead leaves on lower branches until spring. Flowers not showy, males clustered into drooping heads, females single on 1-in. stalks with the leaves, blooming in spring. Fruit a triangular nut enclosed in a slightly prickly bur maturing in autumn, relished by wildlife.
PROPAGATION seeds, cold stratified for three months **LANDSCAPE USES** specimen, grouping in parks and estates, kept as a forest tree **EASE OF CULTIVATION** easy **AVAILABILITY** infrequent.
NOTES American beech can grow to a massive size and dominate the forest and the landscape. It is majestic and stately, and a joy to watch grow from a sapling. The tree is easy to spot with its huge light gray trunks and tan dry leaves

Fagus grandifolia

Frangula caroliniana

hanging on in winter. You may see the pale-purple-flowered parasitic plant called beech-drops (*Epifagus virginiana*) growing about 8 in. tall from beech roots—they just naturally go together. A gardening friend once gave me a yellow lady's-slipper orchid (*Cypripedium*) and said, "Be sure to plant it under a beech tree." I did, but did not understand why. Later, I realized it was because beech is the last tree to leaf out in spring, affording the plants under it a few extra precious days of light before the canopy shade closes in.

Frangula caroliniana ★★

Carolina buckthorn

Syn. *Rhamnus caroliniana*
Rhamnaceae (buckthorn family)

HABITAT & RANGE moist, less-acidic woods throughout the Southeast, but rare in the Atlantic Coastal Plain **ZONES** 6–9 **SOIL** moist, well drained **LIGHT** shade to part sun.
DESCRIPTION Small deciduous tree barely to 20 ft. tall, slender with horizontal branches. Leaves alternate, simple, toothed, 2–5 in. long, with distinctive regular veining pattern, turning a clear yellow (sometimes more red in the sun) in autumn. Flowers small, not showy, blooming in late spring. Fruit a noticeable small ¼-in. round "berry" turning from red to black, maturing in autumn, and eaten by birds.
PROPAGATION seeds, cold stratified for three months **LANDSCAPE USES** specimen in some sun, naturalized in informal woodlands, kept as an understory tree **EASE OF CULTIVATION** easy **AVAILABILITY** infrequent.
NOTES Carolina buckthorn is a slight tree in all respects. It is but a wisp in the woods, but you notice it for its distinctive leaves and fruits when you pass by. With more sun it may become more dense and shrubby, and more noticeable. It may seed around, but is by no means as aggressive as its invasive exotic relatives further north.

Franklinia alatamaha ★★★★

Franklinia, Franklin-tree

Syn. *Gordonia alatamaha*
Theaceae (tea family)

HABITAT & RANGE floodplain in northeastern Georgia, may now be extinct in the wild **ZONES** 6–9 **SOIL** moist, well drained **LIGHT** part sun to sun.
DESCRIPTION Small deciduous tree to perhaps 20 ft. tall, somewhat upright growing, with smooth bark. Leaves alternate, simple, irregularly toothed, shiny dark green, about 6 in. long, wider near the tip, turning a beautiful purple-red to orange-yellow in autumn. Flowers very showy, large, with five crinkle-edged petals, 3 in. wide, fragrant, white with conspicuous yellow stamens, clustered at the tips of the branches but opening one at a time, blooming in mid to late summer. Fruit a dry round pod with numerous flat-winged seeds, maturing in autumn.
PROPAGATION seeds, collected when ripe and kept moist, cold stratified for one month; cuttings in summer **LANDSCAPE USES** specimen, accent **EASE OF CULTIVATION** moderate **AVAILABILITY** common.
NOTES Franklinia is one of our most celebrated Southeastern species, right up there with Oconee bells and Venus flytrap, and it's certainly one of the showiest native woody plants. It is of historical importance because it disappeared over two hundred years ago, thirty-five years after it was discovered by John and William Bartram on the banks of the Altamaha River in northeastern Georgia. It has not been seen since 1803 and is presumed extinct in the wild, but thrives

Franklinia alatamaha

Fraxinus pennsylvanica

Gymnocladus dioica

Halesia diptera

in cultivation. It is tricky to establish because it likes moist but well-drained soil. It is infamous for being easier to grow in a pot than in the ground. It did best for me when planted in very sandy soil and kept watered during droughts. Grow it somehow, even keeping it containerized.

A very similar close relative is loblolly bay (*Gordonia lasianthus*), an evergreen tree to 25 ft. tall with flowers almost identical to those of franklinia. It grows in wet areas of the coastal plain and is worth trying in well-drained (sandy) but moist soil. Nice ones border the parking lot at Kalmia Gardens in Hartsville, South Carolina.

An exciting new hybrid between *Gordonia* and *Franklinia*, called *Gordlinia grandiflora*, has been developed by Tom Ranney at North Carolina State University's Mountain Research Center. It is fast growing, more forgiving in culture, and very floriferous over the summer. Look for the cultivar 'Sweet Tea'; it's the perfect Southern plant.

Fraxinus americana ★★★

White ash

Oleaceae (olive family)

HABITAT & RANGE rich upland soils and old fields throughout the Southeast, but rare in the Atlantic Coastal Plain **ZONES** 6–9 **SOIL** moist, well drained **LIGHT** shade to sun.
DESCRIPTION Large deciduous tree often 100 ft. tall, with fishnet-patterned (reticulate) bark. Leaves opposite, compound like a feather, leaflets usually seven, faintly toothed if at all, ovate, green above and whitish beneath, each 3–4 in. long, with outstanding fall color of rich purple-red-yellow. Flowers not showy, male and female on separate trees, in spreading clusters on last year's twigs, blooming in spring. Fruits are winged, one-seeded, flat, paddlelike, about 1 in. long, maturing in late summer in hanging clusters.
PROPAGATION seeds, warm stratified, then cold stratified for two months **LANDSCAPE USES** shade tree, specimen, in parks, street tree, kept as a forest tree **EASE OF CULTIVATION** easy **AVAILABILITY** infrequent.
NOTES White ash is slow-growing and one of the best sturdy and long-lived shade trees. It is a handsome tree with one of the best fall color displays. The fruits are a bit messy, but like the fruits of maples, are scattered by the wind and are not as bad as the heavy fruits of walnuts or persimmons. The dense wood of white ash is great for baseball bats and tool handles.

Green ash (*F. pennsylvanica*) is inferior to white ash but grows faster and thus is more used as a shade tree. It grows in floodplains and has finely fuzzy twigs and leaves, and poor fall color.

Gymnocladus dioica ★★★

Kentucky coffeetree

Fabaceae (legume family)

HABITAT & RANGE bottomlands, forest, and pastures in central and western Tennessee and perhaps northern Alabama, but widely scattered from cultivation **ZONES** 3–8 **SOIL** moist, well drained **LIGHT** part sun to sun.
DESCRIPTION Deciduous tree to 100 ft. tall, usually smaller, with stout twigs and rounded crown, bark exquisite with ridges and broad vertical swaths. Leaves alternate, two-times compound, very large, 1–2 ft. long, the leaflets ovate 1–2 in. long, turning yellow in autumn. Flowers somewhat showy, in elongate clusters, greenish white, to about ¾ in. long, male and female on separate trees, blooming in early summer. Fruits are large, very thick, dark brown flattened bean-pods 4–10 in. long, with three to five very hard round seeds in a sticky bittersweet pulp, persisting into winter.
PROPAGATION seeds, soaked in acid for four hours, no further pretreatment necessary **LANDSCAPE USES** specimen, grouping, shade tree for parks and estates, allées **EASE OF CULTIVATION** moderate **AVAILABILITY** frequent.
NOTES This beautiful, robust species has fine-textured foliage even though the compound leaves themselves are about the largest of any tree. The bark is incomparable and among my favorites, looking like the long-swirled icing on a tall layer cake. This sturdy tree is majestic and should be planted more widely. See newly establish ones at the North Carolina Arboretum in Asheville. Kentucky coffeetree barely gets into the Southeast in its upper Midwest native range, but it has been widely planted and persists from cultivation.

Halesia diptera ★★★★

Two-wing silverbell

Styracaceae (storax family)

HABITAT & RANGE bottomland forests in the coastal plain from southeastern South Carolina to East Texas **ZONES** 5–9 **SOIL** moist, well drained **LIGHT** part sun to sun.
DESCRIPTION Deciduous tree to about 30 ft. tall, broadly pyramid-shaped. Leaves alternate, simple, ovate 3–6 in. long, turning yellow in autumn. Flowers showy, white, bell-shaped, with four petals deeply split forming a very short tube, up to about 1 in. long, blooming in early spring. Fruits dry, one-seeded, nutlike with two thin wings or flanges along the length, maturing in late summer.
PROPAGATION seeds, warm stratified for three months, then

cold stratified for three months **LANDSCAPE USES** specimen, grouping, kept as an understory tree in native forests **EASE OF CULTIVATION** easy **AVAILABILITY** frequent.

NOTES Two-wing silverbell is a wonderful semiformal tree when grown in full sun, somewhat pyramidal and covered in white flowers. The bell-shaped flowers are delightful, and with the tree's growth form and fall color, make it one of the best small flowering trees for the Southeast. It absolutely should be much more widely grown, even though propagation is somewhat difficult.

The large-flowered two-winged silverbell (*H. diptera* var. *magniflora*) is endemic to southwestern Georgia and the Florida Panhandle. It has slightly larger flowers on a more informal gracefully low-sweeping small tree that is very worthwhile and widely available.

While different people pronounce the genus name "hal-eeze-ya" or "hal-eese-see-yah," I like to say "hale zia" to commemorate Stephen Hales, the man for whom it was named.

Halesia tetraptera

Ilex opaca

Halesia tetraptera ★★★

Mountain silverbell

Syn. *Halesia carolina, H. monticola*
Styracaceae (storax family)

HABITAT & RANGE moist slopes, coves, creekbanks, and bottomlands, widespread throughout the region, especially in the mountains, but absent from Florida and rare in Mississippi **ZONES** 4–8 **SOIL** moist, well drained, tolerates flooding **LIGHT** part sun to sun.

DESCRIPTION Deciduous tree 50–60 ft. tall, young bark green striped, older bark ridged and scaly. Leaves alternate, simple, ovate, finely toothed, 3–6 in. long, turning a rich yellow in autumn. Flowers showy, white, bell-shaped, with four petals slightly split forming a short tube, about 1 in. long, blooming in early spring. Fruits dry, one-seeded, nutlike with four thin wings or flanges along the length, maturing in late summer.

PROPAGATION seeds, warm stratified for three months, then cold stratified for three months; cuttings with strong rooting hormone **LANDSCAPE USES** specimen, grouping, kept as an understory tree **EASE OF CULTIVATION** easy **AVAILABILITY** frequent.

NOTES Mountain silverbell is considered one of the finest flowering trees, with charming bell-shaped flowers in abundance, sometimes tinged with pink. They are spectacular in spring bloom throughout the mountains. The flowering period is brief, but the bark is interesting. I consider the fruits a nuisance, but some people might like them. I think the two-winged silverbell is a better choice in the Southeast.

This *Halesia* species is not to be confused with little silverbell, a smaller species correctly named *H. carolina* (syn. *H. parviflora*). Little silverbell occurs from South Carolina to western Florida and Mississippi and is actually more like two-wing silverbell. It is not as vigorous in cultivation.

Ilex opaca ★★★

American holly

Aquifoliaceae (holly family)

HABITAT & RANGE dry, moist to wet woodlands throughout the Southeast **ZONES** 5–9 **SOIL** moist, well drained, tolerates flooding **LIGHT** shade to sun.

DESCRIPTION Evergreen tree to 70 ft. tall, bark gray, smooth even in old age. Leaves alternate, simple, thick, very spiny toothed, 2–4 in. long, a few older leaves turning yellow in autumn. Flowers small, white, not conspicuous, male and female on separate plants (so you need both sexes to

Juglans nigra

get fruits), blooming in late spring. Fruits are the familiar round, bright red berries ¼ in. in diameter, long lasting and usually eaten by birds in late winter.

PROPAGATION seeds difficult, requiring alternating warm and cold stratification; cuttings root well just after summer growth hardens **LANDSCAPE USES** specimen, grouping, in lawns of parks and estates, kept as a tree in natural forests **EASE OF CULTIVATION** easy **AVAILABILITY** frequent.

NOTES American holly is a beautiful and tough tree, surviving urban disturbance. It is very cold hardy and our only widespread broad-leaved evergreen forest tree. When grown in full sun, a female plant will become covered with red berries, creating a cheerful sight it winter. It is then startling to see a flock of cedar waxwings literally denude the tree in mid-March. Several good, named selections (both male and female) are listed by Dirr (2009). It may be difficult to find a pure American holly at nonspecialist nurseries because there are so many hybrids and exotic ones that grow faster.

Yaupon holly (*Ilex vomitoria*), from the outer coastal plain, is a widely grown evergreen that is more well known in the landscape as a 4-ft. dwarf (cultivar 'Nana') or a 12-ft. weeping shrub (*I. vomitoria* forma *pendula*). The species has shiny dark green evergreen leaves about 1 in. long that are merely shallowly toothed, not spiny. It has many forms and can have abundant red, yellow, or orange berries. It is a very worthwhile shrub for many landscape uses because it can endure extremes of heat and drought better than most other species.

Juglans nigra ★★

Black walnut

Juglandaceae (walnut family)

HABITAT & RANGE moist hardwood forests and bottomlands throughout the Southeast, except absent to rare in the outer coastal plain **ZONES** 5–9 **SOIL** moist, well drained **LIGHT** part sun to sun.

DESCRIPTION Deciduous tree to 100 ft. tall, bark regularly ridged in a fishnet (reticulate) pattern, twigs somewhat stout with partitions in the cut open core (pith). Leaves alternate,

compound, with nine to twenty-one leaflets arranged like a feather, each leaflet 2–3 in. long, hairy. Flowers tiny, males in fetching tight green pendulous catkins, blooming in late spring. Fruits are large edible oval nuts 1½ in. long with very rough coats, surrounded by a light green nonsplitting shell that has a distinctive lemony smell, eventually crumbling off and staining everything, maturing in autumn.
PROPAGATION seeds, cold stratified for three months **LANDSCAPE USES** specimen, nut tree **EASE OF CULTIVATION** easy **AVAILABILITY** common.
NOTES We perhaps have a love-hate relationship with black walnuts. They produce a highly prized nut that has been used by humans (and squirrels) forever. The heartwood of very old trees is invaluable for furniture. But that is the end of the love part. The trees are quite widespread and seem to find their way into almost any disturbed urban space (placed there by these same squirrels). They grow rapidly and make distinctive, even admirably majestic trees when given the space, but they are messy. They constantly drop leaves, especially in dry weather, and tennis-ball size fruits that stain the pavement and our clothes. Worse, they exude a chemical (juglone) that leeches into the soil and retards the growth of some desirable plants, so they don't make good companion trees.

Liquidambar styraciflua ★★

Sweetgum

Altingiaceae (sweet gum family), formerly in Hamamelidaceae (witchhazel family)

HABITAT & RANGE floodplains, bottomlands, old fields, disturbed areas, and roadsides throughout the Southeast **ZONES** 5–9 **SOIL** moist, well drained **LIGHT** part sun to sun.
DESCRIPTION Deciduous tree 80–100 ft. tall with an upright crown, wounds ooze a golden gummy sap (hence the origin of the genus name from 'liquid amber"), twigs typically developing corky ridges when young. Leaves alternate, simple, glossy green, 3–6 in. long, deeply five lobed (star-shaped), the lobes toothed, with a pleasant resinous fragrance when crushed, turning a most gorgeous array of rich purples, red, oranges, and yellows in autumn. Flowers not showy, male and female on the same tree, in tight balls, blooming in early spring. Fruits are sharp-beaked little pods, aggregated into the infamous hard and spiny "gum balls," maturing in autumn, falling off in midwinter, seeds eaten by many different birds.
PROPAGATION seed, without pretreatment **LANDSCAPE USES** shade tree, specimen **EASE OF CULTIVATION** easy **AVAILABILITY** common.

Liquidambar styraciflua with prairie warbler

Liquidambar styraciflua, fall color

NOTES No native tree has a finer fall color display than sweetgum. It rivals white ash for a long and beautiful show. That does not mean you should plant one, and if you do, it should certainly not be near a path or walk. There is no nuisance worse than having to walk on hard sweetgum balls, barefooted or not. So, what is the best thing to do if your only shade trees are sweetgums? Collect the balls, spray them gold and silver, and use them as holiday decorations. The species is very adaptable as to soils and situations. It is weedy and volunteers readily in disturbed areas. There are several selections of sweetgum, including variegated ones and even some nice columnar ones. A fruitless cultivar, 'Rotundiloba', tends to be disease-prone and is not worthwhile to plant in the Southeast.

Liriodendron tulipifera ★★

Tulip-poplar, yellow-poplar

Magnoliaceae (magnolia family)

HABITAT & RANGE moist forests, stream bottoms, and cove hardwoods throughout the Southeast **ZONES** 5–9 **SOIL** moist, well drained **LIGHT** part sun to sun.
DESCRIPTION Very large deciduous tree more than 100 ft. tall and 3–4 ft. in diameter, crown upright due to self-pruning of older limbs. Leaves alternate, simple, shallowly four- or six-lobed such that each vertical half is a perfect mirror image of the other half, turning yellow in autumn. Flowers showy, cup-shaped with six greenish yellow petals with orange markings, 2 in. wide, blooming in late spring. Fruits are conelike structures made up of dozens of nutlike fruits, each with a narrow 1-in.-long wing, shedding in autumn but some persisting into winter, creating interesting pronglike structures.
PROPAGATION seeds, cold stratified for one month **LANDSCAPE USES** shade tree, grouping for parks and estates **EASE OF CULTIVATION** easy **AVAILABILITY** common.
NOTES Tulip-poplar becomes our largest eastern tree, well over 100 ft. tall, growing 3 ft. a year. This magnificent and beautiful tree is not for the small home landscape, and at the very least grows so fast that it quickly becomes out of scale with the house. The tree grows continuously throughout the season, stopping only when frost hits. It characteristically develops a very tall and very straight trunk with no lower limbs. There are smaller cultivars, including 'Ardis', 'Fastigiatum', and 'JFS-Oz' (trade name Emerald City).

Liriodendron tulipifera

Maclura pomifera ★★

Osage-orange

Moraceae (mulberry family)

HABITAT & RANGE woodlands and prairie edges, originally from southwestern Arkansas, Oklahoma, and Texas, but now naturalized throughout the Southeast **ZONES** 6–9 **SOIL** moist, well drained **LIGHT** part sun to sun.
DESCRIPTION Deciduous, massive stocky tree to 50 ft. tall, trunk short with irregular rounded crown, inner bark of trunk and roots orange, twigs with stout thorns ½ in. long. Leaves alternate, simple, smooth, turning yellow in autumn. Flowers minute, male and female on separate plants (you need both to have fruits), females in a spherical ball 1 in. across, blooming in early summer. Fruits one-seeded, aggregated into very large, bizarre, light green, lemony fragrant spherical balls the size of grapefruits with the surface texture of brain-coral, with milky sap inside, ripening in early autumn, not edible to humans.

Maclura pomifera

PROPAGATION seeds, cold stratified for one month **LANDSCAPE USES** specimen, exclusion hedge, curiosity **EASE OF CULTIVATION** easy **AVAILABILITY** frequent.
NOTES Osage-orange trees are very tenacious and long-lived. I have dated large specimens to two hundred years old in Charlotte and have seen other individuals continue to survive decades after a fire has killed 90 percent of the trunk. The trees are not particularly ornamental, but if you have room near the back of your property, the fruits make wonderfully fragrant autumn decorations and are a conversation piece. Just don't plant them near your car or patio.

Magnolia grandiflora, flower

Magnolia grandiflora, fruit

Magnolia grandiflora ★★★

Southern magnolia

Magnoliaceae (magnolia family)

HABITAT & RANGE moist forests in the coastal plain from South Carolina to Texas **ZONES** 7–9 **SOIL** moist, well drained to heavy clay **LIGHT** part sun to sun.

DESCRIPTION Spreading evergreen tree 30–50 ft. tall, bark smooth, twigs stout with large hairy buds, limbs sturdy, lower ones often resting on the ground in age and rooting to form a "grove" of sprouts, making the tree appear even more massive. Leaves alternate, simple, very thick, glossy green above, brown-hairy below, 6–12 in. long, lasting two years, a few turning yellow in autumn. Flowers very large, 7–8 in. wide, with six to nine waxy white petals that last for two days, producing a very strong lemony-sweet fragrance the first day, blooming all summer. Fruits are small pods (follicles) aggregated into plump oblong conelike structures 3–5 in. long, turning blush red as they begin to ripen, opening to release bright red seeds dangling on a thin stretchable thread that is immediately devoured by a bird, ripening in early autumn, the spent cones shedding as they dry.

PROPAGATION seeds, cold stratified for three months; cuttings are difficult **LANDSCAPE USES** specimen, grouping, especially for parks and estates **EASE OF CULTIVATION** easy **AVAILABILITY** common.

NOTES Southern magnolia is an aristocrat of southern trees, at its best providing majestic dark evergreen specimens and classic sweet summer fragrance. It has come to symbolize the warmth and friendliness of the South. But be mindful: it takes up a great deal of room as it can attain a width of more than 40 ft., and it drops old leaves and fruit cones constantly, which don't decompose very fast. A not-so-handsome specimen can be an eyesore forever, so plant a named selection, not an inexpensive seedling.

There are many variations of southern magnolia. Of the ninety cultivars listed by Dirr (2009), 'Alta' is a very columnar form suitable for screening, 'Bracken's Brown Beauty' is overall one of the best with exquisite dark green foliage that is uniformly rusty-brown underneath, 'D.D. Blanchard' is more pyramidal, and 'Little Gem' is dwarf and early flowering.

Magnolia macrophylla ★★★★

Bigleaf magnolia

Magnoliaceae (magnolia family)

HABITAT & RANGE moist wooded ravines and stream bottoms, scattered from central North Carolina southwestward to western Mississippi, but absent in the mountains **ZONES** 5–9 **SOIL** moist, well drained **LIGHT** shade to sun.

DESCRIPTION Small deciduous tree 40–50 ft. tall, with well-spaced branches, twigs stout with large white-hairy buds, the leaves produced in a symmetrical cluster near the twig tips such that they look like an umbrella. Leaves alternate, simple, light green above and whitish beneath, unlobed except for two 1-in. lobes at the base of the leaf, up to 36 in. long, the largest simple leaf in a temperate-zone tree, lacking significant fall color except for the silvery undersides, looking like so much limp white laundry hanging out on shrubs when they fall. Flowers are the largest of any temperate plant, open wide-spreading 10–12 in. wide, white with a purple blotch at the base of each petal, sweetly fragrant, lasting two days, blooming in late spring. Fruit is a

Magnolia macrophylla, flower

Magnolia macrophylla, fruit

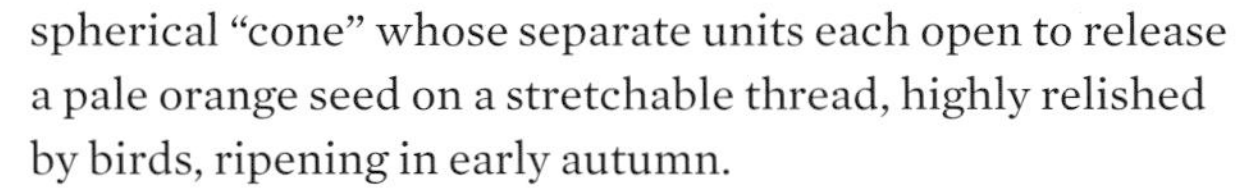

spherical “cone” whose separate units each open to release a pale orange seed on a stretchable thread, highly relished by birds, ripening in early autumn.

PROPAGATION seeds, cleaned and then stratified for one month **LANDSCAPE USES** specimen, grove, parks, in lawns and estates, kept as an understory tree **EASE OF CULTIVATION** easy **AVAILABILITY** common.

NOTES Bigleaf magnolia is truly a striking plant; words fail to convey the feeling of awe one experiences upon seeing it for the first time, or standing under a mature specimen and noting the gigantic leaves. It is widespread in rich bottomland forests along streams and seems to appreciate the wind protection the surrounding forest affords. It is well adapted as its seedlings are among the most prodigious self-sowers of all the native magnolias we grow in Charlotte. It is very easy to cultivate and should be more widely grown. When grown in full shade, it’s a spreading graceful tree that flowers quite well; when grown in full sun, the branches and leaves are held more vertically, giving a more compact appearance. The huge dried leaves make interesting decorations, and when they fall you can’t rake them, you just go out and gather them up.

A similar species, sometimes considered a variety, is Ashe magnolia (*M. ashei*) from one small area in the western Florida Panhandle. It is smaller in all ways but begins producing flowers at the much younger age of three to five years rather than ten years, and is a delight for the garden.

Often confused with bigleaf magnolia is Fraser magnolia (*M. fraseri*), which has a much smaller leaf that is not at all hairy, but still with the ear lobes at the base of the blade. The early spring (before the leaves) flowers are much smaller and not so fragrant. The bright red fruiting “cones” are narrower. Fraser magnolia does not tolerate heat well and is not recommended in areas warmer than Zone 7.

Magnolia macrophylla, fallen leaves

Magnolia fraseri, flower

Magnolia fraseri, fruit

Magnolia virginiana, flower

Magnolia tripetala

Magnolia virginiana, fruit

The pyramid magnolia (*M. pyramidata*) is sometimes considered a smaller variety of *M. fraseri* and is more heat tolerant.

Umbrella magnolia (*M. tripetala*), a deciduous tree, has large hairless leaves to 20 in. long that are not lobed but gradually taper to the base. It is easy to grow with crooked arching trunks to 30 ft. tall. Its large flowers have a fetid odor.

The leaves of deciduous cucumber magnolia (*M. cordata*) are more ovate with slightly rounded bases, 6–10 in. long, and the flowers are more cup-shaped and less conspicuous in the tree. Cucumber magnolia is not as widely grown. The flowers are a yellow-green, but some nice pure yellow specimens have been found and used in hybridizing with the early flowering Chinese yulan magnolia (*M. denudate*) to make the beautiful yellow cultivars 'Butterflies' and 'Elizabeth'.

Magnolia virginiana ★★★★

Sweet-bay magnolia

Magnoliaceae (magnolia family)

HABITAT & RANGE woodlands, floodplains, swamp forests, and pocosins in the Piedmont and coastal plain throughout the Southeast **ZONES** 5–9 **SOIL** moist, well drained **LIGHT** part sun to sun.

DESCRIPTION Evergreen large shrub or medium tree to 50 ft. tall, usually smaller, multitrunked or single-trunked. Leaves alternate, simple, shiny green above and very waxy whitish beneath, the leaves may drop in a very cold winter. Flowers single on twig tips, creamy white, 2–3 in. across, very fragrant, blooming over a long period from spring to midsummer. Fruits in conelike structures about 2 in. long, seeds bright red-scarlet, maturing late summer into autumn.

Malus angustifolia

Morus rubra

PROPAGATION seeds, cleaned and then cold stratified for one month **LANDSCAPE USES** specimen, foundation planting, screening, grouping, patio container, parks and estates, kept as an understory tree
NOTES Sweet-bay magnolia makes a wonderful small semi-evergreen shrub that has many uses. The flowers are so fragrant, and the form is pyramidal but not as dense as that of an evergreen holly. Many cultivar selections are available. Sweet-bay is better kept as a multitrunked shrub than a tall tree so you can see the flowers. Cut down the trunks and they will readily resprout.

There are many cultivars: 'Henry Hicks' and 'Moonglow' are more reliably evergreen, and 'Santa Rosa' has larger, handsome leaves.

Malus angustifolia ★★

Southern crab apple

Rosacea (rose family)

HABITAT & RANGE along streams, rich woodlands, barrens, and old fields, in a patchy distribution pattern throughout the Southeast **ZONES** 6–9 **SOIL** moist, well drained **LIGHT** part shade to sun.
DESCRIPTION Deciduous thicket-forming shrub or small tree to 30 ft. tall, twigs with occasional stout thorns 1 in. long. Leaves alternate, simple and toothed or lobed and toothed, to 2½ in. long, turning unreliably reddish purple in autumn. Flowers showy, various shades of pink, 1 in. across, in clusters on short shoots along the branch, very fragrant, blooming in spring. Fruits are a yellowish green berry (pome) about 1 in. in diameter, very sour, maturing in late summer, eaten by various wildlife.
PROPAGATION seeds, cold stratified for two months **LANDSCAPE USES** informal specimen, in a thicket in natural woodland **EASE OF CULTIVATION** easy **AVAILABILITY** frequent.
NOTES Our native crab apples (also including *M. coronaria*) are interesting, but not prized as flowering and fruiting ornamentals the way hybrid crab apples are in the North. The flowers are subtle, beautiful, and very fragrant, and the fruits are greenish. The fall color quality depends on the specimen and conditions. The bark can be rough, flaky, and noteworthy. The tree may be more attractive architecturally than ornamentally for its informal gnarled appearance.

Morus rubra ★

Red mulberry

Moraceae (mulberry family)

HABITAT & RANGE bottomlands and upland woods throughout the Southeast **ZONES** 4–9 **SOIL** moist, well drained **LIGHT** part sun to sun.
DESCRIPTION Deciduous tree to 30 ft. tall, with spreading brittle branches. Leaves alternate, simple, coarsely toothed, 3–8 in. long or more, rough-hairy on both surfaces, broadly ovate with rounded base and long-pointed tip, often leaves

with none or up to two large lobes on the same branch (as seen in sassafras), the juvenile foliage more consistently lobed, turning somewhat yellow in the autumn. Flowers not showy, in small clusters, male and female on separate plants, blooming with the leaves in spring. Fruits are conspicuous reddish black false "berries" (like blackberries), cylindrical to 1 in. long, thickly sweet and juicy, relished by birds, ripening in early summer, and very messy.
PROPAGATION seeds, cold stratified for two months **LANDSCAPE USES** limited, as a specimen in a mixed woodland setting, fruit tree
NOTES Red mulberry is a coarse and determined species that find its way into any disturbed woodland or managed setting, where it grows fast and tends to take over. It can have bold leaves, which provide an attractive backdrop. Stems sprout back readily after cutting. Some folks do like the fruits, which seem to have a laxative effect on birds and humans, so plant a cultivar selected for improved flavor. Mulberry can be distinguished (from basswood especially) by the milky sap that exudes from the picked leaf.

Nyssa sylvatica

Nyssa sylvatica ★★★

Black gum, sour gum

Nyssaceae (tupelo family)

HABITAT & RANGE streambanks, bottomlands, and upland wood, ubiquitous throughout the Southeast **ZONES** 6–9 **SOIL** moist, well drained **LIGHT** part sun to sun.
DESCRIPTION Deciduous tree 90–100 ft. tall, with blocky bark. Leaves alternate, simple, shiny dark green, untoothed except for occasional shallow lobelike teeth, 2–6 in. long, turning magnificent dark to bright crimson or orange-scarlet in autumn. Flowers greenish, nonshowy, male and female on separate plants, in spherical clusters on 1- to 2-in.-long stalks, blooming in late spring. Fruit a blue-black fleshy drupe ½ in. long, maturing in late summer, relished by birds.
PROPAGATION seeds, cold stratified for two months **LANDSCAPE USES** specimen, as a shade tree for parks and estates, kept as an understory tree **EASE OF CULTIVATION** easy **AVAILABILITY** frequent.
NOTES Black gum is one of the best native trees for fall color, consistently producing outstanding displays under most circumstances. The form shows a straight main trunk and branches almost horizontal, well worth noticing in winter. The blocky bark is handsome as well. The fruits are little noticed, but good for wildlife. When young, black gum is similar in appearance to persimmon. It has leaves with occasional teeth, which are not whitish beneath, and tip

Ostrya virginiana

buds that are distinctly pointed (not rounded). Both persimmon and black gum can have similar blocky bark and red-orange fall color. The common name "tupelo" is ambiguous and is applied to all species in the genus. Tupelo honey is made from the nectar of one or more species of *Nyssa* in the Deep South.

Black gum (*N. biflora*), water tupelo (*N. aquatica*) and Ogeechee tupelo (*N. ogeche*) are wetland species that can have colorful leaves or very large leaves, and can be ornamental in the right setting.

Ostrya virginiana ★★

Eastern hop hornbeam

Betulaceae (birch family)

HABITAT & RANGE moist to dry upland slopes throughout the Southeast, except absent in the Atlantic Coastal Plain south of Virginia **ZONES** 4–9 **SOIL** moist, well drained to dryish **LIGHT** shade to sun.

DESCRIPTION Deciduous tree to 35 ft. tall, with longitudinally flaky bark. Leaves alternate, simple, finely toothed, 2–5 in. long, finely hairy underneath. Flowers not showy, in short pendulous spikes (catkins), blooming in early spring. Fruits dry, nutlike about 1/4 in. long, enclosed in papery leaf scales such that the whole fruiting spike resembles a loose head of hops, maturing in late summer.

PROPAGATION seeds difficult, warm stratified for three months, then cold stratified for three months **LANDSCAPE USES** specimen, grouping, in parks, street tree, kept as an understory tree **EASE OF CULTIVATION** easy **AVAILABILITY** common.

NOTES The bark of this species, one of my favorites, looks like a cat has scratched it, giving it a finely shredded appearance, with the pieces of bark curling and easily picked off. The bark is the tree's main ornamental feature, but the clusters of tan-colored fruits have some appeal. Unfortunately the common names "ironwood" and hornbeam" are often used interchangeably for *Ostrya* and *Carpinus*. There is no confusing the plants, just their names, so you never know what you might get in mail order.

Oxydendrum arboreum ★★★

Sourwood, sorrel-tree

Ericaceae (heath family)

HABITAT & RANGE mixed hardwoods, fields, and roadsides throughout the Southeast, except absent in the coastal plain from South Carolina through Georgia **ZONES** 5–9 **SOIL** moist, acidic, well drained **LIGHT** part sun to sun.

DESCRIPTION Deciduous tree to 50 ft. tall, narrow pyramid shape when grown in full sun, rarely growing straight up in the forest, bark deeply furrowed, not blocky like black gum and persimmon. Leaves alternate, simple, elliptical, to 7 in. long, finely toothed, sour to taste, turning rich vibrant red in autumn, Flowers showy, white, pendulous, 3/8 in. long, bell-shaped like lily-of-the-valley, arranged on conspicuous long branched stalks (racemes) from the tips of the leafy branches, blooming in early summer. Fruits are numerous yellowish tan erect pods about 3/8 in. long, maturing in autumn and striking when seen with the red leaves, persisting into winter as drooping clusters.

Oxydendrum arboreum

PROPAGATION seeds, without pretreatment **LANDSCAPE USES** specimen, grouping, kept as an understory tree **EASE OF CULTIVATION** moderate **AVAILABILITY** common.

NOTES Found only in southeastern United States, sourwood is a splendid tree for all seasons: outstanding in flowers, form, foliage, fall color, and bark. If you are in doubt about its identity (the leaves can resemble black cherry and silverbell), just nibble a leaf and detect the sour taste. It is best grown in full sun to achieve characteristic form, abundant flowering, and wonderful fall color. The tree is a bit difficult to transplant and establish, but worth several tries. It is also susceptible to some disfigurement by fall webworm.

Persea borbonia

Persea borbonia ★★

Red bay

Lauraceae (laurel family)

HABITAT & RANGE dry woodlands and dunes, maritime forests, and pocosins in the coastal plain throughout the Southeast **ZONES** 7–9 **SOIL** moist, well drained **LIGHT** part sun to sun.

DESCRIPTION Small evergreen tree to 60 ft. tall, with irregular crown. Leaves alternate, simple, leathery, lustrous, medium green above and covered with golden hairs beneath, 2–8 in. long, aromatic when crushed, often infected with characteristic galls. Flowers small, not showy, yellowish, in small clusters on long stalks, blooming in early summer. Fruits are dark blue leathery drupes, interesting to notice, maturing in late summer, eaten by birds.

PROPAGATION seeds, cold stratified **LANDSCAPE USES** specimen, grouping, screening, kept as an understory tree **EASE OF CULTIVATION** easy **AVAILABILITY** rare.

NOTES This tree has some useful ornamental value, most especially adding a nice evergreen flair to the winter landscape. The leaves can be harvested, dried, and used as a substitute for commercial bay leaves in cooking, as it's closely related. Unfortunately, a laurel wilt is spreading in the wild and killing trees. I like to see red bay naturalizing in a woodland setting, where I can crush and smell the fragrant leaves. It is also salt tolerant and can be used (or kept) in coastal landscaping. It is the host plant for palamedes swallowtail butterfly.

Pinckneya bracteata

Pinckneya bracteata ★★★

Fevertree, pinckneya

Syn. *Pinckneya pubens*
Rubiaceae (madder family)

HABITAT & RANGE swamp margins and poorly drained sites, mostly limited to the coastal plain in southeastern Georgia, but from extreme southeastern South Carolina to the central Florida Panhandle **ZONES** 7–9 **SOIL** moist, well drained **LIGHT** part sun to sun.

DESCRIPTION Deciduous large shrub or small tree to 20 ft. tall. Leaves opposite, simple, softly hairy, ovate, 4–8 in. long, turning yellowish in autumn. Flowers tubular, pinkish, with five petals, about 1 in. long, certain sepals are greatly enlarged (a unique trait) to form showy pink "flags" 2–4 in. long surrounding the flower cluster. Fruits are dry, rounded pods about ½ in. across, maturing in late summer, persisting into winter.

PROPAGATION seed, without pretreatment, cuttings from young plants root easily **LANDSCAPE USES** specimen, grouping, border **EASE OF CULTIVATION** moderate **AVAILABILITY** frequent.

NOTES This fabulous small tree is surprisingly quite showy and easy to grow in average conditions of soil and sun. It is tricky to establish and needs attention to watering the first summer, but then it's tough and persists despite adverse conditions, as long as it's adequately moist. The pink bract-like sepals are very showy and long effective. Fevertree (once used to treat malaria) should be planted much more widely.

Platanus occidentalis

Prunus caroliniana

Platanus occidentalis ★

Sycamore

Platanaceae (sycamore family)

HABITAT & RANGE streambanks and bottomlands throughout the Southeast **ZONES** 5–9 **SOIL** moist, well drained, tolerates flooding **LIGHT** part sun to sun.

DESCRIPTION Massive deciduous tree to 100 ft. tall, with an open crown and striking gray-white bark. Leaves alternate, simple, broadly lobed and large-toothed, coarse, 4–8 in. long and wide (much larger on stump sprouts). Flowers inconspicuous, in round clusters about ½ in. across, blooming in spring. Fruits are tiny, nutlike, tightly arranged in hanging balls about 1 in. across, disassociating to release the individual fruits in a fluffy mass in late autumn.

PROPAGATION seeds, cold stratified for two months **LANDSCAPE USES** specimen for a wetland edge in a large park or estate, kept as a native tree in wetland habitats **EASE OF CULTIVATION** easy **AVAILABILITY** common.

NOTES Sycamore is a paradoxical tree. It is widely known for its striking whitish bark that is exposed in a coarse jigsaw puzzle pattern. This is its only claim to ornamental fame. It is one of our fastest growing native trees (3 ft. or more a year) and is thus widely planted as a street tree and around parking lots. However, it quickly becomes too large, dropping brittle limbs, and losing many leaves during the mildest droughts. Don't plant a sycamore outside of a permanently moist place; enjoy it in the wild.

Prunus caroliniana ★★★

Carolina cherry-laurel

Rosaceae (rose family)

HABITAT & RANGE thickets, low woods, old fields, and maritime forests in the coastal plain throughout the Southeast **ZONES** 7–9 **SOIL** moist, well drained, tolerates seasonal flooding **LIGHT** part sun to sun.

DESCRIPTION Small evergreen tree to 40 ft. tall, often having a dense upright form. Leaves alternate, simple, glossy, dark green, elongate, 2–5 in. long. Flowers showy, white, about ½ in. wide, in many-flowered clusters with the leaves, blooming in early spring. Fruits are noticeable, oval, ½ in. long, shiny blue-black somewhat fleshy drupes, maturing abundantly in autumn and persisting into winter, sparingly eaten by birds.

PROPAGATION seeds, cold stratified one month; cuttings in summer and winter **LANDSCAPE USES** specimen, screening, kept as an understory tree **EASE OF CULTIVATION** easy **AVAILABILITY** frequent.

NOTES Carolina cherry-laurel is related to the more widely planted English cherry-laurel (*P. laurocerasus*) and could be a substitute. Ours is a good evergreen and is widely adaptable. It is one of the most common volunteer seedlings in shady thickets where it's spread by birds. At its best, in good sun, it forms a handsome small tree with beautiful glossy evergreen foliage. I would like to see it used more widely and to have more interesting selections to choose from since it tolerates the southern heat and sun.

Prunus serotina with summer tanager

Prunus serotina ★★

Black cherry

Rosaceae (rose family)

HABITAT & RANGE forests, old fields, bottomlands, and roadsides throughout the Southeast **ZONES** 4–9 **SOIL** moist, well drained **LIGHT** part sun to sun.

DESCRIPTION Deciduous tree to 80 ft. tall, mature bark dark reddish black, forming thin plates, young twigs with the odor of bitter almond. Leaves alternate, simple, sharply toothed, 2–6 in. long, hairless except for characteristic strip of dense brown hairs by midvein underneath, also with characteristic tiny glands on the leaf stalk (petiole), turning beautiful shades of red in autumn. Flowers small but showy, white, ½ in. wide, arranged on pendulous racemes, blooming in spring. Fruits are dark red-purple, very juicy ¼-in. "cherries," ripening in summer, and relished by birds who get drunk on the fermenting fruits.

PROPAGATION seeds, cold stratified for three months **LANDSCAPE USES** shade tree for parks and estates, kept as an understory tree **EASE OF CULTIVATION** easy **AVAILABILITY** infrequent.

NOTES Black cherry is ubiquitous in the Southeast and can appear in almost any habitat except the wettest. It is not a species that you would plant, but if you have a tree in an advantageous location, you can enjoy the interesting summer flowers, outstanding fall color, and distinctive "potato chip" bark. The fruits, while messy, are relished by birds and that alone may be a good reason to keep a tree on your property. It is also a host plant for many beautiful butterflies. Older trees are less vigorous than oaks, but the reddish orange wood is prized for furniture. The tree is susceptible to disfiguring by tent caterpillars in spring.

Quercus alba

White oak

Fagaceae (oak family)

HABITAT & RANGE deep soils in oak-hickory forests throughout the Southeast, except absent from Peninsular Florida **ZONES** 3–9 **SOIL** moist, well drained **LIGHT** part sun to sun.

DESCRIPTION Long-lived deciduous tree 80–100 ft. tall, bark loosely splitting into longitudinal plates. Leaves alternate, simple, 5–9 in. long, deeply cut into nine or ten rounded lobes of varying depth, turning reddish in autumn. Flowers tiny, separate male and female on the same tree, males in hanging catkins, females clustered separately near the leaves, blooming in early spring. Fruits are nuts called acorns, bullet-shaped up to 1 in. long, the tight-fitting cap of warty scales covers the lower fourth of the acorn, maturing in autumn, the shape, size and texture of the caps and nuts helps tell oaks apart.

Quercus alba, trunk

Quercus alba, leaves and fruits

Quercus lyrata

PROPAGATION seed, no pretreatment required for immediate fall germination **LANDSCAPE USES** specimen, grove, shade tree for streets, parks, and estates, kept as a desirable tree in forest **EASE OF CULTIVATION** easy **AVAILABILITY** frequent.
NOTES White oak is the classic oak, widely distributed and known by anyone who can recognize an oak. The leaves can be quite variable, depending on whether exposed to shade or sun; in sun the leaves have narrower lobes. Some years the nuts can cover the ground, other years there are none.

Approximately forty species of oaks are native in the Southeast, many of which have become our best shade trees. They are not generally prized for fall color, but exceptions are noted. Most oaks are slow growing, taking thirty to fifty years to approach mature size, and they are usually planted too close together as youngsters. They are certainly favored to be planted on larger properties and as street trees or groupings. With some attention as to placement and pruning early on, oaks can lead long and effective lives.

Overcup oak (*Q. lyrata*) is 60–80 ft. tall and of floodplains and bottomlands. It has irregularly cut seven-lobed leaves 6–8 in. long and unique acorns that are mostly enclosed by their caps and never separate. The tree can have good reddish fall color. Because of the upswept lower branches on an upright growth habit, it could fit in a narrow space. It should be grown more in the Southeast.

Mossy-cup oak (*Q. macrocarpa*), also called bur oak, is an impressive tree, large (to 80 ft. tall) and broad spreading, with rough bark. The leaves have five to nine lobes and are markedly broader toward the tip. The acorns can be huge, our largest by far, up to 2 in. long and wide, sitting in a shallow cup with long-shaggy bristles, very conspicuous and decorative. I highly recommend the big-cupped forms.

Post oak (*Q. stellata*) is common in drier woodlands throughout the region and is drought tolerant. It grows to 70 ft. tall, and is broad spreading with horizontal branches and somewhat rough bark. The shiny dark green leaves are three- or five-lobed such that the overall blade is cross-shaped.

Quercus michauxii ★★★

Swamp chestnut oak, basket oak

Fagaceae (oak family)

HABITAT & RANGE bottomlands and floodplains throughout the Southeast, except absent in the mountains of eastern Tennessee **ZONES** 5–9 **SOIL** moist, well drained **LIGHT** part sun to sun.

DESCRIPTION Massive deciduous tree 70–90 ft. tall and as wide or wider, with a rounded-spreading crown, bark somewhat loose as with white oak. Leaves alternate, simple, 4–9 in. long, with regular, rounded teeth up to twelve, smooth above and hairy below, turning reddish brown to bronzy yellow in autumn. Flowers tiny, males in hanging catkins, females clustered separately near the leaves, blooming in early spring. Fruits are acorns, slightly elongate-rounded to 1¼ in. long, broadest at the base, cup loosely warty-scaly with a small fringe enclosing ½–⅓ of the nut, maturing in autumn.

Quercus michauxii

PROPAGATION seed, no pretreatment required for immediate fall germination **LANDSCAPE USES** specimen, shade tree for parks and estates, kept as a tree in natural forests **EASE OF CULTIVATION** easy **AVAILABILITY** infrequent.

NOTES Swamp chestnut oak can become an impressive spreading tree over time. The leaves can be broader or narrower, mostly hairy underneath, so similar to chestnut oak that you would need acorns for accurate identification (always pick one up, with its cup). A huge champion specimen in Charlotte at Historic Rosedale Plantation is over one hundred twenty-five years old.

Chestnut oak (*Q. montana*, syn. *Q. prinus*) is similar to and easily confused with swamp chestnut oak, but it's usually much more upright and always in drier upland sites. The leaves are usually narrower and less hairy. Chestnut oak can be distinguished by its bark that is quite tight with distinct nonscaly ridges, and by its acorns, which are consistently more elongate (bullet-shaped) with a cap sloped inside like a martini glass. The fall color is reddish yellowish.

Quercus montana

Quercus phellos, flowers

Quercus phellos ★★★

Willow oak

Fagaceae (oak family)

HABITAT & RANGE bottomlands, floodplains, wooded slopes, and old fields throughout the Southeast, except not in the mountains, also absent in southeastern Georgia and Peninsular Florida **ZONES** 5–9 **SOIL** moist, well drained **LIGHT** part sun to sun.

DESCRIPTION Large deciduous tree to 100 ft. tall, forming an upright, spreading crown, relatively fast growing. Leaves alternate, simple, nonlobed, very narrow 2–5 in. long and less than 1 in. wide, with a small bristly tip, turning reddish yellowish in autumn. Flowers tiny, males in hanging catkins, females clustered singly near the leaves, blooming in early spring. Fruits are acorns, almost round, 3⁄8 in. in diameter,

Quercus phellos

barely sitting in a small cup, maturing in autumn of the second year after forming.

PROPAGATION seeds, cold stratified for one month **LANDSCAPE USES** specimen, grouping, street tree, parks and estates **EASE OF CULTIVATION** easy **AVAILABILITY** common.

NOTES Willow oak is widely used as a street tree, for which it has excellent, upright spreading growth form, good branch structure, and a tolerance of city conditions. It somewhat resembles the upright vase form of the majestic American elm, but it branches at a higher level, forming a taller canopy and creating a cathedral effect over neighborhood streets. Its narrow leaf is fine-textured and beneficial in the landscape as mulch, being easier to rake and looking less messy than coarser oak leaves. The leaves are also more likely to blow over into the neighbor's yard. It is widely misnamed "pin oak" by the public, I suppose because the narrow leaf resembles a quill pen point (and true pin oak, *Q. palustris*, has deeply lobed leaves). The tiny acorns (eaten by blue jays) are less troublesome on sidewalks, and crush easily when you walk on them. It is perhaps the fastest-growing oak and would be the best choice to see it mature in your lifetime.

Quercus rubra, bark

Laurel oak (*Q. laurifolia*) is often confused with willow oak. The leaves are similar, but blunt without a bristle-tip, and usually wider in the middle. It makes a great open-grown tree for an expansive lawn. Laurel oak loses its leaves in mid-winter, making it distinctive from willow oak that is deciduous in autumn.

Sand laurel oak (*Q. hemispherica*), or Darlington oak as it's sometimes known in the nursery trade, is often confused with laurel oak and willow oak. It differs in holding its leaves all winter until the new leaves grow, being effectively evergreen. All three of these oaks are worthy of planting and could appear similar in the landscape when young, differing only in the timing of their leaf drop.

Water oak (*Q. nigra*) has small acorns and narrow leaves with three shallow lobes at the slightly broader tip. The leaves are tardily deciduous and a little thicker and darker green than the others mentioned here. The tree, which is acorn-messy in cultivation, but good for wildlife, is short-lived and is weaker than most oaks in snow and ice, keeping dead limbs in the canopy. Keep one "for the birds" if it's in a suitable site, but don't plant one as a shade tree.

Quercus rubra ★★★

Northern red oak

Fagaceae (oak family)

HABITAT & RANGE average hardwood oak-hickory forests in the Piedmont and mountains throughout the Southeast **ZONES** 3–8 **SOIL** moist, well drained **LIGHT** part sun to sun.

DESCRIPTION Deciduous tree 90–100 ft. tall, straight trunk, tight bark with characteristic long gray streaks about 1 in. wide. Leaves alternate, simple, 4–9 in. long, with seven to eleven shallow lobes having smaller lobes or teeth, bristle tips, turning red in autumn. Flowers tiny, males in hanging catkins, females clustered separately near the leaves, blooming in early spring. Fruits are acorns, slightly oblong-rounded to 1¼ in. long, broadest at the base, cup loosely scaly covering 20–25 percent of the nut, maturing in autumn of the second year, germinating the following spring.

PROPAGATION seeds, cold stratified for one month **LANDSCAPE USES** specimen, grouping, street tree, parks and estates, kept as a tree in native forest **EASE OF CULTIVATION** easy **AVAILABILITY** frequent.

NOTES Red oak is a handsome and popular oak, with a symmetrical upright growth form. It is a northern species that comes well south, and it would be best to use trees of

Quercus shumardii

Quercus rubra, fall color

Quercus texana

southern origin to ensure heat tolerance. The gray streaks in the bark are diagnostic for this and scarlet oak.

Closely related species in the "red oak" group have bristle-tipped leaf lobes and acorns that mature the second fall and germinate the following spring. Scarlet oak (*Q. coccinea*) has leaves with fewer lobes (7–9) and the lobes are cut more than halfway to the midrib. Its rounded sinuses form perfect C's, and the leaves consistently turn a beautiful bright scarlet red in autumn. For fall color alone, this oak should be planted more often. It grows throughout the Piedmont and mountains of our region.

Southern red oak (*Q. falcata*) grows into a 100 ft. tall tree with a rounded crown. It can be found in dryer, less fertile fields and forests. The leaves are deeply lobed, almost to the midrib, with three to five lobes, but the central lobe is much longer and deeply cut than the side lobes, and the leaves are noticeably hairy underneath. Fall color is poor, perhaps a little red-brown. The acorns are about ½ in. long, with the cup covering a third of the nut. A nice enough heat- and drought-tolerant tree if you have one, but not ornamental enough to plant.

Cherry-bark oak (*Q. pagoda*) can grow to 100 ft. tall with a straight trunk. Its leaves have five to seven broader, shallower lobes and more rounded sinuses than southern red oak. The young bark is smooth and somewhat shiny; the older bark is scaly like black cherry. This oak is more handsome overall than southern red oak and is more often sold by nurseries.

Shumard oak (*Q. shumardii*) is a large, rounded, spreading oak 80–90 ft. tall with deeply furrowed dark bark. Its shiny leaves are regularly deeply seven- to eleven-lobed, with many secondary lobes such that there could be as many as thirty bristle-tips. A distinguishing feature is the consistent production of distinct tufts of hairs at the junctions of the main veins on the underside of the leaf. I would highly recommend Shumard oak because it's a sturdy, symmetrical tree, is readily available, is heat tolerant, and has good red fall color.

Nuttall oak (*Q. texana*, syn. *Q. nuttallii*) is a tree to 80 ft. tall from central Alabama and westward. In the past, you would not have heard much about it, but it has become a popular replacement for pin oak and Shumard oak in the hot Deep South. It transplants better, grows faster, is more adaptable, has better fall color, and exhibits clean leaf drop in autumn (pin oak holds its dead leaves). The leaves are deeply cut, but the lobes are asymmetrical and angle forward.

Quercus virginiana ★★★

Live oak

Fagaceae (oak family)

HABITAT & RANGE dry sandy woods, edges of salt marshes, and maritime forests in the coastal plain throughout the Southeast **ZONES** 8–10 **SOIL** moist, well drained **LIGHT** part sun to sun.

DESCRIPTION Huge evergreen tree to 60 ft. tall and more than 80 ft. wide, with a massive trunk and widely spreading limbs. Leaves alternate, simple, thickened, leathery, 2–4 in. long and up to 1½ in. wide, sometimes wavy or with a few teeth, shiny green, pointed but not ending with a bristle-tip, often with edges slightly rolled under (revolute), all old leaves dropping in spring. Flowers tiny, males in catkins, females in separate clusters near leaves, blooming in spring. Fruits are attractive acorns, oblong to 1 in. long, shiny dark brown, maturing in autumn on current year's twig.

PROPAGATION seeds, no pretreatment required for immediate

Quercus virginiana

fall germination **LANDSCAPE USES** specimen, groves, street tree if properly pruned, parks and estates **EASE OF CULTIVATION** easy **AVAILABILITY** common.

NOTES Live oak is the quintessential southern tree, conspicuous in coastal areas with old homes and plantation mansions, typically with Spanish moss festooned on the branches and resurrection fern growing on the rough-barked trunks. Usually the older trees are much wider than tall, often with lowest limbs (large than most trees' trunks) resting on the ground and then growing upwards again. This is especially striking in the famous Angel Oak in Charleston, South Carolina, that is perhaps four hundred years old and has a trunk 9 ft. in diameter and a crown spread of 180 ft. If grown as a street tree or in the home landscape, live oak needs to be carefully pruned when young to create an uncrowded horizontal limb structure capable of clearing street traffic and withstanding snow loads on the evergreen leaves. The tree does create dense shade, unsuitable for most ornamental understory plants. The shiny acorns are very attractive against the dark green leaves. Many birds and mammals use the evergreen canopy for cover, and the acorns for food, and it's the larval host plant for the rare white-M hairstreak butterfly and consular oak moth. This species is drought and salt spray tolerant.

Rhapidophyllum hystrix ★ ★ ★

Needle palm

Arecaceae (palm family)

HABITAT & RANGE swamps and stream margins in the coastal plain from the extreme southeastern corner of South Carolina (Spring Island) to Peninsular Florida and west to Mississippi **ZONES** 6–10 **SOIL** moist, well drained **LIGHT** shade to sun.

DESCRIPTION Evergreen shrub to 12 ft. tall, branching near the base to form a big clump, with a rounded crown of "palmate" leaves. Leaves alternate, 4–7 ft. long including leaf stalk, simple but appearing fan-shaped with segments splitting towards the center. Flowers tiny, creamy brown, male and female on separate plants, in tight bunches near the base of the trunk, surrounded by abundant 6- to 20-in.-long needles, blooming in late spring. Fruits are hard, brown drupes, 1 in. in diameter, covered by curious tan woolly hair, maturing in autumn, described by some as having a sweet odor and by others as being vile, eaten by black bears.

PROPAGATION seed, without pretreatment, slow to germinate **LANDSCAPE USES** specimen, grouping, foundation planting, kept in natural forests **EASE OF CULTIVATION** easy **AVAILABILITY** common.

NOTES Needle palm is a good garden plant for light shade (or even some sun), and I highly recommend it. It develops a short trunk over time and the leaves are less massive than *Sabal palmetto*. The leaves are always a nice green, even in the snow. Be careful of the unique needles, which are very sharp—presumably to keep the bears away until the seeds are ripe. This palm is good as a single specimen or mixes in well with other shade-loving shrubs (but give it room to spread its leaves). It is extremely cold hardy, at least to -10°F. Needle palm is exciting because it can grow far north of its range and look great in a garden where palms are not expected.

Sabal palmetto 

Sabal palmetto, cabbage palm

Arecaceae (palm family)

HABITAT & RANGE margins of brackish marshes and in maritime forests along the extreme outer coastal fringe near the ocean from South Carolina to Peninsular Florida, where it spreads inland **ZONES** 7–10 **SOIL** moist, well drained **LIGHT** sun.

DESCRIPTION Evergreen tree to 60 ft. tall, trunk columnar, mostly uniform in size up to 1 ft. in diameter, unbranched

Rhapidophyllum hystrix

with persistent leaf bases acting as "bark," and a crown of very large "palmate" leaves at the top. Leaves alternate, 4–7 ft. long including leaf stalk (petiole), simple but appearing fan-shaped due to the blade splitting into narrow corrugated segments. Flowers small, yellowish white ½ in. wide, in conspicuous massive inflorescences, blooming in summer. Fruits are hard, shiny black drupes, ½ in. long, maturing in autumn, eaten by many birds and small mammals. **PROPAGATION** seeds, without pretreatment **LANDSCAPE USES** specimen, grouping, street tree in coastal areas **EASE OF CULTIVATION** moderate **AVAILABILITY** frequent.

NOTES Palms make a striking statement in the landscape, especially among traditional trees and shrubs. Sabal palmetto is the only native palm that makes a trunked tree outside of Florida. It is now being successfully planted far north of its range. It can survive single digit temperatures if water does not freeze in the crown. There are two secrets to successful transplantation. First, plant in pure sand, not too deeply, in late spring; and then, most importantly, water it daily and abundantly into the top of the crown the first

Serenoa repens

Sabal palmetto at Brookgreen Gardens, SC

summer (at least), or during any period of very hot, dry weather. It may take two or more years to begin to form full-sized leaves after the first summer. It will eventually become stunning. Water, water, water—the crown.

Dwarf palmetto (*Sabal minor*) does not have an aboveground trunk but forms a large clump of palm leaves like sabal palmetto but without a distinct midrib; the leaf segments arise from an extension of the end of the leaf stalk (petiole). Saw palmetto (*Serenoa repens*) in similar to dwarf palmetto, but is a bit more compact and has spiny teeth along the leaf stalk. Either can take light shade and add a new dimension to Southeastern landscape gardening.

Salix nigra ★

Black willow

Salicaceae (willow family)

HABITAT & RANGE sunny wetlands, streambanks, and ditches throughout the Southeast **ZONES** 5–9 **SOIL** moist to wet **LIGHT** sun.

DESCRIPTION Deciduous tree to 80–100 ft. tall, bark rough and gray, twigs limber and wispy. Leaves alternate, simple, regularly toothed, narrow, 2–4 in. long, turning yellowish in autumn. Flowers somewhat showy, male and female on separate trees, in elongated erect spikes (catkins), fuzzy like a pussy willow, flowering in early spring. Fruits are small dry pods that split open to release seeds with fluffy hairs that parachute them away, maturing in summer.

PROPAGATION seeds, without pretreatment; cuttings root readily **LANDSCAPE USES** naturalized along streams and wetlands **EASE OF CULTIVATION** easy **AVAILABILITY** common.

NOTES Willows have ornamental value in their pussy-willow flowers and graceful branches. The huge trees along streams and rivers can be impressive in their size. They sprout back readily if cut or broken. Willows are widely used for erosion control in wetlands as well as in streambank restoration and stabilization, short pieces of mature stems often simply being simply stuck in the wet ground where they root over winter (as live stakes). If you have a suitable wetland, plant a black willow and keep it cut back so it provides abundant young branches and flowers at eye level. The leaves are used by many butterflies as larval food, especially the beautiful mourning cloak.

Sassafras albidum ★★★

Sassafras

Lauraceae (laurel family)

HABITAT & RANGE old fields, roadsides, and forests throughout the Southeast **ZONES** 4–9 **SOIL** moist, well drained to dry **LIGHT** part sun to sun.

DESCRIPTION Deciduous tree 40–50 ft. tall, most often forming a colony from root sprouts, twigs smooth, green, aromatic when scraped. Leaves alternate, simple, occurring as unlobed, one-lobed, and two-lobed, all on the same branch, 3–6 in. long, turning spectacular colors of red-orange-yellow-purple in autumn. Flowers small, yellow, in loose spherical clusters at the tips of twigs before the leaves emerge in early spring, male and female on separate plants. Fruits are blue-black leathery drupes, ½ in. long, each sits

Salix nigra

Sassafras albidum

in a shallow cup on the swollen tip of a bright red stalk in small clusters 2–3 in. long, maturing in late summer. **PROPAGATION** seeds difficult, cold stratified for four months; roots sprouts dug up and replanted **LANDSCAPE USES** specimen, grouping, kept as a tree in meadow or forest edge **EASE OF CULTIVATION** moderate **AVAILABILITY** infrequent.
NOTES Sassafras is a beloved tree and often the first one that a child learns when shown the leaves that are like mittens or gloves, depending on the number of lobes. The tree has incomparable fall color, and it is awesome to see a large clone along a roadside in all its fall finery. The trees start out in old fields and grow up with the forest to persist as isolated trees. If you have this situation, keep the trunks cut down in sunny areas to regrow to exhibit their beautiful leaves as a low shrublike form. Be aware that they can be a nuisance in the sense that the root system will resprout readily if you are trying to remove them selectively, and using herbicide on one trunk may affect others that are connected. The attractive fruits are rarely seen because they are high in the treetops, and birds eat them as quickly as they turn red. Like other members of the laurel family, sassafras is the larval food of the peculiar tiger swallowtail caterpillar, the one with the big fake eye behind the head to fool (or scare) predators. Sassafras roots were once used to make root beer, and the very deep, blocky bark is gorgeous on large specimens.

Stewartia malacodendron

Stewartia ovata

Stewartia malacodendron ★★★

Silky camellia

Theaceae (tea family)

HABITAT & RANGE moist forests, often on limestone, widely scattered and rare in the Piedmont, more frequent but still rare in the coastal plain throughout the Southeast **ZONES** 7–9 **SOIL** moist, well drained, not as acidic **LIGHT** shade to part sun.
DESCRIPTION Deciduous large shrub or small tree to 20 ft. tall, sometimes multitrunked, with spreading spraylike leafy branches. Leaves alternate, simple, ovate, 2–4 in. long, finely toothed, turning rich yellow in autumn. Flowers showy, large at 2–3 in. across, the numerous stamens characteristically purple with blue pollen, not fragrant, blooming in late spring. Fruits are hard, dry, rounded pods about ½ in. wide, splitting into five chambers, maturing in autumn.
PROPAGATION seeds, very difficult, warm stratified for three months, then cold stratified for three months; cuttings difficult **LANDSCAPE USES** specimen, grouping, well placed in naturalized woodland garden, kept as an understory tree **EASE OF CULTIVATION** moderate **AVAILABILITY** infrequent.
NOTES Silky camellia is one of the most wonderful year-round trees that I have grown. It's also one of the most difficult. I had to move it three times as a young plant before it "took" and grew into a wonderful specimen. Grow it moist but well drained, never letting it dry out until well established. Active growth of juveniles occurs in late summer. Fertilize only sparingly. In bloom, it's breathtaking, with dozens of gorgeous white flowers well displayed on the sweeping leafy branches. Part of its appeal is that it does not present itself like any other tree. Alas, the flowers are short-lived (two days) and the flowering period is brief (about two weeks), but it's still an interesting tree for its form, fall color, winter features, and anticipation of next spring.

Of equal beauty is mountain camellia (*S. ovata*), an elusive tree of the southern Blue Ridge where four states come together. It grows in rich woods along streams, flowering in summer. There is no more charming flower on a delightful small tree. It blooms a little longer than silky camellia, with fewer flowers at once, and is distinguished by its yellow stamens. Difficult to obtain, but worthy of the effort.

Styrax americanus

Styrax americanus ★★

American snowbell

Styracaceae (storax family)

HABITAT & RANGE moist to wet hardwood forests in the Piedmont and coastal plain throughout the Southeast **ZONES** 6–9 **SOIL** moist, well drained **LIGHT** part sun to sun.

DESCRIPTION Small deciduous tree to 18 ft. tall, often with a single trunk that is somewhat striped. Leaves alternate, simple, weakly toothed if at all, 1–3 in. long, turning yellowish in autumn. Flowers modest, white, somewhat fragrant, in small clusters with the leaves, the five petals flaring back to reveal the yellow stamens, blooming in late spring. Fruits are unassuming hard, rounded, tan pods about ½ in. long, maturing in late summer.

PROPAGATION seeds, cold stratified for three months; cuttings root easily in summer **LANDSCAPE USES** specimen, grouping in woodland settings, kept as an understory tree **EASE OF CULTIVATION** easy **AVAILABILITY** frequent.

NOTES American snowbell is not as showy as its Japanese counterpart (*S. japonicus*), yet it makes a decent small tree for the mixed collection. Flowering is sudden and brief, but the flowers are interesting and the tree should be placed where the flowers can be seen.

Its relative, bigleaf snowbell (*S. grandifolius*), is definitely more often a suckering shrub forming a thicket, even though stems can reach 18 ft. tall. Its much larger leaves (to 6 in. long) and eye-catching flowers are worth noting, but the show is brief. I would be careful not to plant it where it could spread into specimens that are more valuable. It is best in light shade in an informal woodland.

Tilia americana

Tilia americana ★★★

American basswood, linden

Tiliaceae (linden family)

HABITAT & RANGE rich or rocky woods and stream bottoms in the Piedmont and mountains of the Southeast, rare in the coastal plain **ZONES** 4–8 **SOIL** moist, well drained **LIGHT** part sun to sun.

DESCRIPTION Deciduous tree 60–100 ft. tall, often massive and characteristically multitrunked, twigs slender, tip buds rounded, flattened, and uniquely wider than the twig. Leaves alternate, simple, toothed, heart-shaped with asymmetrical sides at their bases, to 4–7 in. long, turning yellow in autumn. Flowers yellowish, ½ in. wide, fragrant, in small clusters on 2- to 4-in.-long stalks, attached to narrow leaflike bracts, blooming in early summer. Fruits are round pea-sized nuts, the stalks detaching when dried and using the unique leafy bracts as dispersal propellers to twirl in the wind (great fun to play with), maturing in late summer.

PROPAGATION seeds, no good method known; cuttings difficult **LANDSCAPE USES** specimen, groves, shade tree for parks and estates, kept as an understory tree **EASE OF CULTIVATION** easy **AVAILABILITY** infrequent.

NOTES Linden trees are underutilized, probably due to their difficulty in propagation and the lack of good selections. They make handsome open-grown specimens, and their opening leaves and flowers are charming and graceful. One can make linden tea from the fragrant dried flowers, and bees can make a good basswood honey. Sometimes you will see native basswoods listed as *T. heterophylla*, with hairy leaves and hairless twigs, or *T. caroliniana*, with hairy leaves and hairy twigs. All are good.

Ulmus americana ★

American elm

Ulmaceae (elm family)

HABITAT AND RANGE floodplains, rich forests, and dry upland woods throughout the Southeast **ZONES** 3–9 **SOIL** moist, well drained **LIGHT** part sun to sun.

DESCRIPTION Deciduous tree more than 100 ft. tall, usually branching near the base to create an incomparable gracefully upward-arching crown of wide-spreading branches. Leaves alternate, simple, doubly toothed (each tooth has a smaller tooth), ovate, to 6 in. long, rounded at the base, characteristically uneven on each side, surfaces rough-hairy, turning yellowish in autumn. Flowers not showy, abundantly produced in small clusters, blooming in mid to late winter, long before the leaves appear. Fruits are flattened and nutlike (samaras), totally surrounded by a papery wing about ⅜ in. wide with a notch at the tip, ripening in early spring before the leaves mature.

PROPAGATION seeds, cold stratified for three months **LANDSCAPE USES** shade tree, especially for parks and estates **EASE OF CULTIVATION** easy **AVAILABILITY** rare.

NOTES The glory days of the American elm, once the most beautiful and graceful tree for northern streets and lawns, are gone. Most of the trees were wiped out by Dutch elm disease. In the Southeast, the trees were never used as much for that purpose, and the disease is less rampant. You could still have a big beautiful elm in your backyard, but it could die anytime (and it is no good as firewood). If you have one, enjoy it. But don't let volunteer seedlings grow up.

Slippery elm (*U. rubra*) is practically identical to American elm, with coarser leaves and hairy buds.

Winged elm (*U. alata*) has smaller leaves and attractive winged bark. It can make an interesting specimen with yellow fall color. I would not plant one, but I would enjoy one that I had. The young fruits are quite hairy.

Ulmus alata

Zanthoxylum clava-herculis ★★

Hercules'-club

Rutaceae (citrus family)

HABITAT & RANGE sand dunes, maritime forests, and river bluffs along the extreme outer coastal fringe on the Atlantic coast, then somewhat more inland in Peninsular Florida and along the Gulf Coast **ZONES** 7–9 **SOIL** moist to dry, well drained **LIGHT** sun.

DESCRIPTION Small deciduous tree to 25 ft. tall, stocky, the trunk smooth and gray but boasting striking sharp prickles that grow in size from year to year as the tree increases in girth, reaching enormous proportions (for a prickle) of 1 in. long or longer in older specimens where they become pyramidlike with sculptured faces. Leaves semievergreen or tardily deciduous, alternate, once-pinnately compound (like a feather), with ovate leaflets to 2 in. long, and with small spines on the leaf stalks, citrusy odor when crushed, turning yellowish in autumn. Flowers tiny, greenish yellow in branched clusters, blooming in early spring. Fruits are small reddish brown pods about ¼ in. long, opening to reveal a shiny black seed hanging by a thread, maturing in early summer.

PROPAGATION seeds, cold stratification helps **LANDSCAPE USES** specimen, curiosity **EASE OF CULTIVATION** moderate **AVAILABILITY** infrequent.

NOTES Hercules'-club is bizarre for a temperate tree with its ever-growing trunk spines, but then it belongs to the citrus family with mainly tropical affinities. Try growing it in full sun and certainly in sandy, well-drained soil. It will take many years to form the old, rugged bark with large prickles that you can see on the sand dunes of Cumberland Island, Georgia, where it grows with the wild ponies and cabbage palmettos.

Zanthoxylum clava-herculis

Plants for Special Situations and Purposes

Plants that Tolerate Wet Soil in Sun

Note: Soil can be wet but not continuously inundated.

FERNS

Dryopteris ×australis (Dixie wood-fern)
Matteuccia struthiopteris (ostrich fern)
Osmunda cinnamomea (cinnamon fern)
Osmunda regalis (royal fern)

GRASSES

Andropogon glomeratus (bushy broomsedge)
Arundinaria gigantea (river cane)
Carex glaucescens (wax sedge)
Cyperus strigosus (umbrella sedge)
Panicum virgatum (switch grass)
Saccharum giganteum (sugarcane plumegrass)
Scirpus cyperinus (wool-grass)

WILDFLOWERS

Asclepias incarnata (swamp milkweed)
Chelone glabra (turtlehead)
Eutrochium fistulosum (Joe-pye weed)
Helianthus angustifolius (swamp sunflower)
Helenium autumnale (autumn sneezeweed)
Hibiscus coccineus (scarlet hibiscus)
Hibiscus moscheutos (marsh mallow)
Lobelia cardinalis (cardinal flower)
Monarda didyma (beebalm)
Physostegia virginiana (obedient plant)
Stenanthium gramineum (featherbells)
Thalictrum polygamum (tall meadowrue)
Vernonia noveboracensis (ironweed)

VINES

Ampelaster carolinianus (climbing aster)
Apios americana (groundnut)
Clematis virginiana (virgin's-bower)
Wisteria frutescens (American wisteria)

SHRUBS

Aesculus pavia (red buckeye)
Aronia arbutifolia (red chokeberry)
Cephalanthus occidentalis (buttonbush)
Clethra alnifolia (sweet-pepperbush)
Cornus amomum (silky dogwood)
Ilex verticillata (winterberry holly)
Itea virginica (Virginia-willow)
Lyonia lucida (shining fetterbush)
Rosa palustris (swamp rose)
Viburnum nudum (wild raisin)

CONIFERS

Chamaecyparis thyoides (Atlantic white cedar)
Pinus taeda (loblolly pine)
Pinus serotina (pond pine)
Taxodium ascendens (pond cypress)
Taxodium distichum (bald cypress)

TREES

Asimina triloba (pawpaw)
Fraxinus pennsylvanica (green ash)
Magnolia virginiana (sweet-bay magnolia)
Platanus occidentalis (sycamore)
Quercus michauxii (swamp chestnut oak)
Quercus shumardii (Shumard oak)
Salix nigra (black willow)

Plants that Tolerate Wet Soil in Shade

Note: Soil can be wet but not continuously inundated.

FERNS

Dryopteris ×australis (Dixie wood-fern)
Matteuccia struthiopteris (ostrich fern)
Onoclea sensibilis (sensitive fern)

Osmunda cinnamomea (cinnamon fern)
Osmunda regalis (royal fern)
Woodwardia areolata (netted chain fern)

GRASSES

Arundinaria gigantea (river cane)
Carex crinita (fringed sedge)
Chasmanthium latifolium (river oats)

AQUATICS

Peltandra virginica (arrow-arum)

WILDFLOWERS

Amsonia tabernaemontana (blue-star)
Impatiens capensis (jewelweed)
Mertensia virginica (Virginia bluebells)
Packera aurea (golden ragwort)
Zephyranthes atamasco (Atamasco-lily)
Zizia aurea (golden Alexanders)

VINES

Berchemia scandens (American rattan vine)
Decumaria barbara (climbing hydrangea)
Menispermum canadense (moonseed))
Schisandra glabra (magnolia-vine)
Smilax smallii (Jackson-vine)

SHRUBS

Alnus serrulata (tag alder)
Eubotrys racemosa (fetterbush)
Itea virginica (Virginia-willow)
Sabal minor (dwarf palmetto)
Staphylea trifolia (bladdernut)
Rhapidophyllum hystrix (needle palm)

CONIFERS

Chamaecyparis thyoides (Atlantic white cedar)
Taxodium distichum (bald cypress)
Xanthorhiza simplicissima (yellowroot)

TREES

Asimina triloba (Pawpaw)
Celtis laevigata (hackberry)
Cornus amomum (silky dogwood)
Fraxinus pennsylvanica (green ash)
Ilex opaca (American holly)

Plants That Tolerate Dry Soil in Sun

GRASSES

Andropogon gerardii (big bluestem)
Andropogon ternarius (split-bear bluestem)
Aristida stricta (wiregrass)
Eragrostis spectabilis (purple love-grass)
Muhlenbergia capillaris (muhly grass)
Panicum virgatum (switch grass)

WILDFLOWERS

Asclepias tuberosa (butterfly-weed)
Lupinus perennis (sundial lupine)
Salvia coccinea (scarlet sage)
Symphyotrichum ericoides (heath aster)
Symphyotrichum georgianum (Georgia aster)
Rudbeckia hirta (black-eyed Susan)
Tephrosia virginiana (goat's-rue)

VINES

Gelsemium sempervirens (yellow jessamine)
Vitis rotundifolia (muscadine)

SHRUBS

Amorpha canescens (indigo-bush)
Ceanothus americanus (New Jersey tea)
Chionanthus virginicus (fringe tree)
Clinopodium coccinea (scarlet calamint)
Myrica cerifera (wax-myrtle)
Rhus glabra (smooth sumac)
Yucca filamentosa (curlyleaf yucca)

CONIFERS

Juniperus virginiana (red cedar)
Pinus clausa (sand pine)
Pinus echinata (shortleaf pine)
Pinus palustris (longleaf pine)
Pinus taeda (loblolly pine)
Pinus virginiana (scrub pine)

TREES

Aralia spinosa (devil's-walking-stick)
Cotinus obovatus (American smoketree)
Quercus falcata (southern red oak)
Quercus macrocarpa (mossy-cup oak)
Sassafras albidum (sassafras)

Plants That Tolerate Dry Soil in Shade

FERNS

Asplenium platyneuron (ebony spleenwort)
Cheilanthes lanosa (hairy lipfern)
Dennstaedtia punctilobula (hayscented fern)
Thelypteris kunthii (southern shield fern)

GRASSES

Carex pensylvanica (Pennsylvania sedge)
Elymus hystrix (bottlebrush grass)

WILDFLOWERS

Chrysogonum virginianum (green-and-gold)
Coreopsis latifolia (broadleaf coreopsis)
Eurybia divaricatus (wood aster)
Hepatica americana (round-lobed liverwort)
Heuchera americana (alumroot)
Sedum ternatum (sedum)
Silene virginica (fire pink)
Solidago caesia (axillary goldenrod)

SHRUBS

Callicarpa americana (beautyberry)
Chionanthus virginicus (fringe tree)
Croton alabamensis (Alabama croton)
Hamamelis virginiana (witchhazel)
Vaccinium stamineum (deerberry)

CONIFERS

Juniperus communis (creeping juniper)
Tsuga caroliniana (Carolina hemlock)

TREES

Frangula caroliniana (buckthorn)
Ilex opaca (American holly)
Ostrya virginiana (hop hornbeam)
Oxydendrum arboretum (sourwood)
Viburnum prunifolium (black haw)

Plants That Attract Butterflies

Aesculus parviflora (bottlebrush buckeye)
Asclepias species (milkweeds)
Coreopsis species (tickseeds)
Echinacea species (coneflowers)
Erigeron pulchellus (robin's-plantain)
Eupatorium perfoliatum (boneset)
Eutrochium species (Joe-pye-weeds)
Gaillardia pulchella (blanket-flower)
Helianthus species (sunflowers)
Hibiscus species (hibiscus)
Hymenocallis species, attract hawkmoths (spider-lily)
Kosteletzkya pentacarpos (seashore mallow)
Liatris species (gayfeathers)
Marshallia graminifolia (grass-leaved Barbara's–buttons)
Mertensia virginica (Virginia bluebells)
Phlox species (phlox)
Physostegia virginiana (obedient plant)
Platanthera ciliaris (yellow-fringed orchid)
Pontederia cordata (pickerelweed)
Rhododendron species, especially *R. prunifolium* (native azaleas)
Rudbeckia species (black-eyed Susans)
Silene species (pinks)
Silphium perfoliatum (cup-plant)
Solidago species (goldenrods)
Symphyotrichum species (asters)
Vernonia species (ironweeds)

Plants That Attract Hummingbirds

Aesculus pavia (red buckeye)
Aquilegia canadensis (columbine)
Bignonia capreolata (cross vine)
Campsis radicans (trumpet-creeper)
Clinopodium coccinea (scarlet calamint)
Impatiens capensis (jewelweed)
Lobelia cardinalis (cardinal flower)
Lonicera sempervirens (coral honeysuckle)
Monarda didyma (bee balm)
Salvia coccinea (scarlet sage)
Silene virginica (fire pink)
Spigelia marilandica (Indian-pink)

Plants with Colorful Fruits That Attract Birds

Note: (*) Asterisk indicates fruit that lasts into winter.

VINES

Parthenocissus quinquefolia (Virginia creeper)
Vitis species (grape)

SHRUBS

Amelanchier species (serviceberries)
Aronia species (chokeberries)
Callicarpa americana (beautyberry)
Chionanthus virginicus (fringe tree)
Cornus species (dogwoods)
Ilex species (hollies)
Lindera benzoin (spicebush)
Myrica cerifera (wax-myrtle)
Prunus species (cherries)
Rhus species (sumacs)
Rosa species (rose)
Rubus species (blackberries, raspberries, dewberries)
Sambucus canadensis (elderberry)
Symphoricarpos orbiculatus (coralberry)
Vaccinium species (blueberries)
Viburnum species (viburnums and haws)

TREES

Amelanchier species (serviceberries)
Aralia spinosa (devil's-walkingstick)
Crataegus species (hawthorns)
Juniperus virginiana (red-cedar)
Prunus species (cherries)
Sassafras albidum (sassafras)

Plants with Consistently Outstanding Fall Color

SHRUBS—PURE YELLOW

Aesculus parviflora (bottlebrush buckeye)
Calycanthus floridus (sweet shrub)
Clethra alnifolia (sweet-pepperbush)
Hamamelis virginiana (witchhazel)
Ilex verticillata (winterberry holly)
Lindera benzoin (spicebush)

SHRUBS—RED AND MIXTURES OF RED-YELLOW-ORANGE

Amelanchier species (shadbush)
Aronia species (chokecherry)
Croton alabamensis (Alabama croton)
Diervilla sessilifolia (bush honeysuckle)
Fothergilla gardenii (dwarf witch-alder)
Hydrangea quercifolia (oak-leaf hydrangea)
Itea virginica (Virginia-willow)
Prunus species (plum)
Rhododendron species (azalea)
Rhus species (sumac)
Vaccinium species (blueberry)
Xanthorhiza simplicissima (yellowroot)

TREES—YELLOW AND ORANGE-YELLOW

Acer pensylvanicum (striped maple)
Asimina triloba (pawpaw)
Betula lenta (sweet birch)
Carya species (hickories)
Cercis canadensis (redbud)
Chionanthus virginicus (fringe tree)
Cladrastis kentukea (yellowwood)

TREES—RED AND MIXTURES OF RED-PURPLE-YELLOW-ORANGE

Acer barbatum (southern sugar maple)
Acer leucoderme (chalk maple)
Acer rubrum (red maple)
Amelanchier species (shadbush)
Cornus florida (flowering dogwood)
Cotinus obovatus (smoke tree)
Diospyros virginiana (persimmon)
Franklinia alatamaha (Franklin tree)
Fraxinus americana (white ash)
Liquidambar styraciflua (sweetgum)
Nyssa sylvatica (black gum)
Oxydendrum arboretum (sourwood)
Prunus species (cherry)
Quercus coccinea (scarlet oak)
Quercus shumardii (Shumard oak)
Sassafras albidum (sassafras)

Plants with Distinctive Bark

SHRUBS

Clethra acuminata (cinnamon bark sweet-pepperbush, brown-peely)
Euonymus americanus (strawberry bush, green))
Philadelphus inodorus (mock-orange, peely)
Physocarpus opulifolius (ninebark, peely)
Prunus species (cherries, smooth, flakey)
Vaccinium species (blueberries, green)

TREES

Acer pensylvanicum (striped maple, striped)
Amelanchier species (shadbush, striped)
Aralia spinosa (devil's-walking-stick, spiny)
Betula nigra (river birch, very peely)
Carpinus caroliniana (ironwood, smooth)
Carya tomentosa (shagbark hickory, shedding in strips)
Celtis laevigata (hackberry, isolated warty & bumpy)
Cornus florida (flowering dogwood, blocky)
Cyrilla racemiflora (titi, winged ridges)
Diospyros virginiana (persimmon, blocky)
Fagus grandifolia (beech, smooth)
Fraxinus americana (white ash, ridged-reticulate)
Gymnocladus dioicus (Kentucky coffee tree, platy swirled)
Nyssa sylvatica (black gum, blocky)
Ostrya virginiana (hop hornbeam, shreddy)
Oxydendrum arboreum (sourwood, very deeply ridged)
Platanus occidentalis (sycamore, jigsaw puzzle)
Quercus rubra (red oak, broad grey-striped)
Zanthoxylum clava-herculis (toothache tree, very large prickles)

USDA Hardiness Zone Temperatures

USDA ZONES & CORRESPONDING TEMPERATURES

Temp °F			Zone	Temp °C		
-60	to	-55	1a	-51	to	-48
-55	to	-50	1b	-48	to	-46
-50	to	-45	2a	-46	to	-43
-45	to	-40	2b	-43	to	-40
-40	to	-35	3a	-40	to	-37
-35	to	-30	3b	-37	to	-34
-30	to	-25	4a	-34	to	-32
-25	to	-20	4b	-32	to	-29
-20	to	-15	5a	-29	to	-26
-15	to	-10	5b	-26	to	-23
-10	to	-5	6a	-23	to	-21
-5	to	0	6b	-21	to	-18
0	to	5	7a	-18	to	-15
5	to	10	7b	-15	to	-12
10	to	15	8a	-12	to	-9
15	to	20	8b	-9	to	-7
20	to	25	9a	-7	to	-4
25	to	30	9b	-4	to	-1
30	to	35	10a	-1	to	2
35	to	40	10b	2	to	4
40	to	45	11a	4	to	7
45	to	50	11b	7	to	10
50	to	55	12a	10	to	13
55	to	60	12b	13	to	16
60	to	65	13a	16	to	18
65	to	70	13b	18	to	21

Find hardiness maps on the Internet.

United States—*http://www.usna.usda.gov/Hardzone/ushzmap.html*
Canada—*http://www.planthardiness.gc.ca/*
or *http://atlas.nrcan.gc.ca/site/english/maps/environment/forest/forestcanada/planthardi*
Europe—*http://www.gardenweb.com/zones/europe/*
or *http://www.uk.gardenweb.com/forums/zones/hze.html*

Metric Conversions

INCHES TO CENTIMETERS

1 in.	2.5 cm
2 in.	5 cm
3 in.	7.5 cm
4 in.	10 cm
5 in.	13 cm
6 in.	15 cm
7 in.	17 cm
8 in.	20 cm
9 in.	23 cm
10 in.	25 cm
12 in.	30 cm
15 in.	38 cm
18 in.	45 cm
24 in.	60 cm
30 in.	75 cm
36 in.	90 cm
48 in.	120 cm

FEET TO METERS

1 ft.	0.3 m
2 ft.	0.6 m
3 ft.	0.9 m
4 ft.	1.2 m
5 ft.	1.5 m
6 ft.	1.8 m
7 ft.	2.1 m
8 ft.	2.4 m
9 ft.	2.7 m
10 ft.	3 m
15 ft.	5.4 m
20 ft.	6 m
25 ft.	7.5 m
30 ft.	9 m
35 ft.	10.5 m
40 ft.	12 m
45 ft.	13.5 m
50 ft.	15 m
55 ft.	17 m
60 ft.	19 m
65 ft.	20 m
70 ft.	21 m
75 ft.	22.5 m
80 ft.	24 m
85 ft.	25.5 m
90 ft.	27 m
95 ft.	28.5 m
100 ft.	30 m

FAHRENHEIT TO CELSIUS

40°F	4°C
65°F	18°C
70°F	21°C

Bibliography

Armitage, Allan M. 2006. *Armitage's Native Plants for North American Gardens*. Portland, Oregon: Timber Press.

Barry, John M. 1980. *Natural Vegetation of South Carolina*. Columbia, South Carolina: University of South Carolina Press.

Bell, C. Ritchie, and Anne H. Lindsey. 1990. *Fall Color and Woodland Harvests*. Chapel Hill, North Carolina: Laurel Hill Press.

Bir, Richard E. 1992. *Growing and Propagating Showy Native Woody Plants*. Chapel Hill, North Carolina: University of North Carolina Press.

Bridwell, Ferrell M. 2003. *Landscape Plants: Their Identification, Culture, and Use*, 2d ed. Albany, New York: Delmar.

Case, Frederick W., and Roberta B. Case. 1997. *Trilliums*. Portland, Oregon: Timber Press.

Cobb, Boughton. 1984. *Peterson Field Guide to the Ferns*. Boston: Houghton Mifflin Company.

Cullina, William. 2008. *Native Ferns, Mosses and Grasses*. New York: Houghton Mifflin.

Cullina, William. 2002. *Native Trees, Shrubs, and Vines*. New York: Houghton Mifflin.

Cullina, William. 2000. *Guide to Growing and Propagating Wildflowers of the United States and Canada*. New York: Houghton Mifflin.

D'Amato, Peter. 1998. *The Savage Garden*. Berkeley, California: Ten Speed Press.

Darke, Rick. 2002. *The American Woodland Garden*. Portland, Oregon: Timber Press.

Darke, Rick. 1999. *The Color Encyclopedia of Ornamental Grasses*. Portland, Oregon: Timber Press.

Deno, Norman C. 1993. *Seed Germination, Theory and Practice*. State College, Pennsylvania: Norman Deno.

Dirr, Michael A. 2009. *Manual of Woody Landscape Plants*. 6th ed. Champaign, Illinois: Stipes Publishing.

Dirr, Michael A., and Charles W. Heuser, Jr. 2006. *The Reference Manual of Woody Propagation*. 2d ed. Cary, North Carolina: Varsity Press.

Foote, Leonard E., and Samuel B. Jones. 1989. *Native Shrubs and Woody Vines of the Southeast: Landscape Uses and Identification*. Portland, Oregon: Timber Press.

Galle, Fred C. 1987. *Azaleas*. Portland, Oregon: Timber Press.

Godfrey, Robert K. 1988. *Trees, Shrubs and Woody Vines of Northern Florida and Adjacent Georgia and Alabama*. Athens, Georgia: University of Georgia Press.

Gupton, Oscar W., and Fred C. Swope. 1981. *Trees and Shrubs of Virginia*. Charlottesville, Virginia: University Press of Virginia.

Hemmerly, Thomas E. 2000. *Appalachian Wildflowers*. Athens, Georgia: University of Georgia Press.

Hériteau, Jacqueline. 1999. *Trees, Shrubs, and Hedges for the Home Landscape: Secrets for Selection and Care*. Upper Saddle River, New Jersey: Creative Homeowner Press.

Horn, Dennis, and Tavia Cathcart. 2005. *Wildflowers of Tennessee, the Ohio Valley and the Southern Appalachians*. Auburn, Washington: Lone Pine Publishing.

Justice, William R., C. Ritchie Bell, and Anne H. Lindsey. 2005. *Wildflowers of North Carolina*. Chapel Hill, North Carolina: University of North Carolina Press.

Kirkman, L. Katherine, Claud L. Brown, and Donald J. Leopold. 2007. *Native Trees of the Southeast*. Portland, Oregon: Timber Press.

Ladendorf, Sandra F. 1989. *Successful Southern Gardening*. Chapel Hill, North Carolina: University of North Carolina Press.

Lance, Ron. 2004. *Woody Plants of the Southeastern United States: A Winter Guide*. Athens, Georgia: University of Georgia Press.

Leopold, Donald J. 2005. *Native Plants of the Northeast: A Guide for Gardening and Conservation*. Portland, Oregon: Timber Press.

Loewer, Peter. 2003. *Ornamental Grasses for the Southeast*. Nashville, Tennessee: Cool Springs Press.

Mickel, John T. 2003. *Ferns for American Gardens*. Portland, Oregon: Timber Press.

Midgley, Jan. W. 1999. *Southeastern Wildflowers*. Birmingham, Alabama: Crane Hill Publishers.

Nelson, Gil. 2010. *Best Native Plants for Southern Gardens: A Handbook for Gardeners, Homeowners and Professionals*. Gainesville, Florida: University Press of Florida.

Nelson, Gil. 2006. *Atlantic Coastal Plain Wildflowers*. Guilford, Connecticut: Morris Book Publishing.

Newcomb, Lawrence. 1977. *Newcomb's Wildflower Guide*. Boston: Little, Brown and Company.

Olsen, Sue. 2007. *Encyclopedia of Garden Ferns*. Portland, Oregon: Timber Press.

Phillips, Harry R. 1985. *Growing and Propagating Wild Flowers*. Chapel Hill, North Carolina: University of North Carolina Press.

Porcher, Richard D., and Douglas A. Rayner. 2001. *A Guide to the Wildflowers of South Carolina*. Columbia, South Carolina: University of South Carolina Press.

Radford, Albert E., Harry E. Ahles, and C. Ritchie Bell. 1968. *Manual of the Vascular Flora of the Carolinas*. Chapel Hill, North Carolina: University of North Carolina Press.

Schnell, Donald E. 2002. *Carnivorous Plants of the United States and Canada*. 2d ed. Portland, Oregon: Timber Press.

Schrock, Denny, ed. 2004. *Complete Guide to Trees and Shrubs: Selecting, Planting, Growing, Pruning*. Des Moines, Iowa: Meredith Books.

Sorrie, Bruce A. 2011. *A Field Guide to the Wildflowers of the Sandhills Region: North Carolina, South Carolina, Georgia*. Chapel Hill, North Carolina: University of North Carolina Press.

Spira, Timothy P. 2011. *Wildflowers and Plant Communities of the Southern Appalachian Mountains and Piedmont*. Chapel Hill, North Carolina: University of North Carolina Press.

Taylor, Walter K. 2013. *Florida Wildflowers in the Natural Communities*. Gainesville, Florida: University Press of Florida.

Tallamy, Douglas W. 2007. *Bringing Nature Home*. Portland, Oregon: Timber Press.

Timme, Stephen L. 2007. *Wildflowers of Mississippi*. Jackson, Mississippi: University Press of Mississippi.

Towe, L. Clarence. 2004. *American Azaleas*. Portland, Oregon: Timber Press.

Wasowski, Sally. 1994. *Gardening with Native Plants of the South*. Dallas, Texas: Taylor Publishing.

Weakly, Alan S. 2011. Flora of the Southern and Mid-Atlantic States. Working draft of 15 May 2011. Chapel Hill, North Carolina: UNC Herbarium. www.herbarium.unc.edu/flora.htm

Wells, B. W. 1967. *The Natural Gardens of North Carolina*. Chapel Hill, North Carolina: University of North Carolina Press.

Wherry, Edgar T. 1961. *The Fern Guide*. Garden City, New York: Doubleday and Company.

Wofford, B. Eugene, and Edward W. Chester. 2002. *Guide to the Trees, Shrubs and Woody Vines of Tennessee*. Knoxville, Tennessee: University of Tennessee Press.

Acknowledgments

I would like to thank everyone with whom I have ever talked about native plants. I have always been inspired by sharing what I know and hearing what others have to say. Here is a finite list of an infinite assemblage of folks who have done a little more to further my knowledge and understanding, and my apologies to anyone whose name I have omitted—it was not intentional.

Ann Armstrong, Tony Avent (Plant Delights Nursery), Dick Bir, Steve Broyles, Hank Bruno, Wes Burlingame, Connie Byrne, the late Fred Case, Mike Creel, Bill and Jennifer Cure (Cure Nursery), Ed Davis, Sharon Day (Mellow Marsh Farm), Ron Determann (Atlanta Botanical Garden), Nancy Wilson Duncan, Bill Finch (Mobile Botanical Gardens), Frank Galloway, the late Rob Gardner, Peter and Jassmin Gentling, Barry Glick (Sunshine Gardens), Kim Hawks (Niche Gardens), Bill and Bob Head (Head-Lee Nursery), Jill Heaton, Meredith Hebden, Peter Heus (Enchanter's Garden), John Hummer, Jack Johnston, Ron Lance, Laurie Lawson (Niche Gardens), Rick Lewandowski, Dick Lighty, Peter Loewer, Bill Logan, Peter Loos, Robert and Julia Mackintosh (Woodlanders Nursery), John Manion (Birmingham Botanical Garden), Jim Matthews, Bob McCartney (Woodlanders Nursery), Gil Nelson, Tom Nunnenkamp, Tom Patrick, Dan Pittillo, Stefan Ploszak, Johnny Randall, Mark Rose, Susan Scholly (for undertaking to write a history of the UNC Charlotte Botanical Gardens), Bruce Sorrie, Fred Spicer, Ted Stephens (Nurseries Caroliniana), Michael and Michelle Styers, Bill Thomas (Sherwood Forest community), Nick Tropeano, Jan Truitt, John Turner (Southern Highlands Reserve), Carla Vitez, Lynda and George Waldrop, Andy Walker, Richard and Teresa Ware (Georgia Botanical Society), Alan Weakley (UNC Herbarium), Charlie Williams, Lindie Wilson, Jean Woods, Glenda Zahner, Jeff Zahner (Chattooga Gardens Nursery), and Margaret and Price Zimmermann. I also thank all those who have worked with native plants at botanical gardens and nature centers and lead field trips to see native plants.

Will and I want very much to thank several freelance photographers who generously shared their photos with us when we needed something. Alan Cressler is a natural history photographer from Decatur, Georgia. He works as a technician for the U.S. Geological Survey and has been using digital cameras since 2004. Alan has traveled over much of the United States photographing native plants. His images are publicly displayed at *http://www.flickr.com/photos/alan_cressler/sets/*.

Jim Fowler is a native orchid enthusiast, an early adapter of digital photography, and the author of *Wild Orchids of South Carolina: A Popular Natural History*. His photographs have appeared in the *North American Native Orchid Journal* and numerous magazines, newsletters, and websites. Jim can be reached via email at Jimstamp@aol.com.

Will and I also thank Connie Byrne, Mike Creel, Michael Drummond, Jim Ellis, Paula Gross, Michael Huft, John Hummer, Larry R. Lynch, Jim Petranka, Annkatrin Rose, and Brad Wilson, who also provided images for the book.

Thanks to Timber Press for wanting to produce this book in the first place, especially to our editor Juree Sondker for believing in us, and to Linda Willms and the company's diverse and professional staff who make all of the hard work coalesce into a beautiful product.

As usual in such an intense undertaking as writing a book, one's family and colleagues suffer from the workaholic requirements needed to bring the project to fruition. My wife, Audrey, has been longsuffering in this effort, and I thank her for her undying love and for keeping the home fires burning. Our daughter, Suzanne, has had to carry on with much less help from me. I love you both.

The hard-working staff at UNC Charlotte Botanical Gardens have had to operate and make decisions on their own, and they have done such a good job without me that I may just need to retire early.

Finally, this book would not have been possible without the able and tireless support of my assistant director at the gardens, Paula Gross. Not only did she read and comment on most of the text, but she also provided that all-important role of muse to help flesh out ideas and corral wild thinking before it got in the way. Paula has a way with words and ideas, and knows how to lead me to clearer writing. Her unwritten name is all over this book.

Photo Credits

All photos by **WILL STUART** unless otherwise noted here.

CONNIE BYRNE, page 272.
BRANT CASTEEL, page 323 left.
MIKE CREEL, pages 77 top right, 254 middle left.
ALAN CRESSLER, pages 47 top and bottom, 85 top left, 91, 95 right, 106 bottom, 154 right, 266 bottom.
DAVID CULP, page 267 bottom.
MICHAEL DRUMMOND, page 85 top right.
JIM ELLIS, page 112 right.
JIM FOWLER, pages 86 right, 87 left, 117 left.
PAULA GROSS, pages 139 middle right, 175 bottom.
MICHAEL HUFT, page 144 right.
JOHN HUMMER, pages 105 left, 109.
LARRY R. LYNCH, page 148 bottom left.
LARRY MELLICHAMP, pages 225 bottom right, 257 top right.
JIM PETRANKA, page 340 bottom.
ANNAKATRIN ROSE, page 35.
BRAD WILSON, page 156 bottom right.

Index

Bold-face indicates main entry pages.